FLIGHT of the HUIA

FLIGHT of the HUIA

Ecology and conservation of New Zealand's frogs, reptiles, birds and mammals

Kerry-Jayne Wilson

First published in 2004 by
CANTERBURY UNIVERSITY PRESS
University of Canterbury
Private Bag 4800
Christchurch
NEW ZEALAND

mail@cup.canterbury.ac.nz
www.cup.canterbury.ac.nz

ISBN 0-908812-52-3

Designed and typeset by Richard King
at Canterbury University Press

Printed by SRM Production Services Sdn. Bhd., Malaysia

A catalogue record for this book is available from
the National Library of New Zealand

Front cover: J. G. Keuleman's illustration of a pair of huia
from *A history of the birds of New Zealand*,
Sir Walter Buller, 1888

CONTENTS

PREFACE

This book tells the story of New Zealand's frogs, reptiles, birds and mammals from their origins in the Cretaceous era to the much-altered fauna of the present day. The first three chapters set the scene so the story can unfold. Chapter 1 traces the origins of the New Zealand land mass and its vertebrate animals (excluding fishes, the only vertebrate group beyond the scope of this book). Chapters 2 and 3 examine the biogeography of the country's frogs, reptiles, non-marine birds and bats, and the ecological factors that are important to understand the changes that have taken place since people first visited then settled New Zealand. Chapter 4 describes the vertebrate communities in forest and wetland habitats during the last two and a half million years – the Pleistocene and Holocene eras – up to the point when, about 2000 years ago, humans apparently visited and left their rats.

Next I look at the effects that humans and introduced animals have had on the communities described in Chapter 4. Chapter 5 reviews the extinctions that have occurred, and Chapter 6 examines the vertebrate animals introduced to New Zealand since European contact. The forest vertebrate communities present today are described in Chapter 7. I discuss the ecology and conservation of marine birds and mammals in Chapter 8. During the last 50 years there have been many often heroic attempts made to save those species now endangered. In Chapter 9 I review the efforts that have been made to conserve what remains of this once-distinctive, now-tattered biota and the ways that conservation management has changed over that period. Finally, Chapter 10 reviews the changes to the biota and puts the efforts made to conserve the indigenous biota into a global perspective.

The pace of change has not slowed and further species are even now declining towards extinction. Habitats continue to be lost or modified by activities such as forestry, mining, agriculture and urban sprawl. As the human population grows and the pressure on existing conservation lands intensifies, New Zealand's 'clean, green' image attracts a growing number of tourists who put ever-increasing demands on the natural environment they have come to enjoy. The risk of introducing rats and cats to islands where they do not yet occur, or introducing new species of animals, plants and micro-organisms to this country, intensifies as people and freight move freely between New Zealand and other parts of the world.

As we shall see, two facts of history are crucial to the story this book tells. First, New Zealand was isolated from the rest of the world for 80 million years; and second, it was the last land mass of any significant size to be discovered by humans. Today, New Zealand is part of the global village and contributes to, and is influenced by, problems such as overpopulation, economic growth, pollution and biodiversity loss.

In the face of all these pressures, New Zealanders have a responsibility to conserve as much as possible of this country's distinctive biota and unique ecological communities. Each generation views change in relation to what things were like in their childhood, so the baseline for measuring change is steadily eroded. It must be measured not by what we, our parents or even our grandparents knew, nor by 1840 when the Treaty of Waitangi was signed and Maori and Pakeha supposedly became united under the British Crown. The real baseline against which we should measure change is the fauna and ecological communities that were present at the time the first humans discovered this country and left their rats behind. I believe that effective conservation requires knowledge of the factors that shaped the ecological communities of that time and the changes that have occurred since.

We may regret the loss of moa and giant eagles and the other species now extinct, and the likely impending loss of endangered species such as kakapo, kiwi and black stilt. However, if we appreciate what makes our animals different and what makes them vulnerable, we will be better able to conserve those taonga that remain.

The title of this book was chosen for several reasons. The huia belongs to the endemic family Callaeatidae, whose affinities and origins have been lost in the mists of time. Like many New Zealand birds, the huia had reduced wings, so its flight was short, as was its time to extinction. It was revered by Maori, and huia tail feathers were worn by chiefs as a sign of rank. The bird was also of special interest to Pakeha. Male and female huia had very different bills and, according to early naturalists, different methods of feeding. Huia pairs were eagerly sought after by museums, and large numbers were collected, thus hastening, if not directly causing, their demise. Huia intrigue modern-day biologists. If the sexes did indeed have different methods of feeding, this would be unique among the world's birds. Their cream-coloured bills contrasted with their glossy black plumage, suggesting that these might also have been used to attract mates. Like so many of New Zealand's unique animals, there are aspects of this bird's biology we may never understand.

ACKNOWLEDGEMENTS

A book of this type cannot be written without help from many people. Over the years numerous individuals have influenced my ideas, and it is literally impossible to list them all. Some sent me reprints of their research, some presented ideas at conferences, others over drinks at cafés or pubs, but of particular importance were discussions with colleagues in the field. In preparing this book I have read the work of hundreds of naturalists, ecologists, biologists and other natural scientists. I sincerely hope that those workers whose research I have cited will consider justice was done to their labours. It was impossible to tell all the fascinating stories I wanted to include in this book, and it was with regret that so much of the good work I read about could not be covered. I sincerely acknowledge the work and contributions of all those people.

I am grateful to my friends and colleagues in the Ecology and Entomology Group at Lincoln University. All members of the academic and support staff have assisted me in some way or another. They have all been willing to discuss ideas, read sections of manuscript, lend me books or journals, ferret out references or help in numerous other ways, but Adrian Paterson and Graham Hickling deserve special mention, for I sought advice from them so often. Thanks also to Eric Scott and Bruce Chapman, who were heads of department during the time this book was written. They gave me time to work on this project and provided departmental support. Eric's set of the *New Zealand Journal of Zoology* saved me many a trip to the library. Most chapter drafts have been available to Lincoln University ecology students, and I appreciate the useful feedback I received from them. Certain students, some anonymously, drew my attention to references I had overlooked.

In order to write such a book as this, one needs access to a vast library or the support of exceptionally helpful library staff. Lincoln University has the latter. I thank various members of the university's library staff for processing hundreds of interloans, for locating some obscure and ancient references, for their interest and encouragement throughout this project and for forgiving the occasional misdemeanour when borrowed material was returned late.

Chapter drafts were reviewed by Warren Chinn, Dr Richard Duncan, Dr Janet Greive, Dr Graham Hickling, Euan Kennedy, Dr Kim King, Dr Mike Imber, Dr Elaine Murphy, Dr Colin O'Donnell, Dr Shaun Ogilvie,

Dr Adrian Paterson, Dr Ralph Powlesland, Dr Murray Williams, Tony Whitaker and Trevor Worthy. I appreciate the time these busy people devoted to reviews of the manuscript and thank them for their helpful comments. Euan Kennedy, Adrian Paterson and Trevor Worthy each reviewed two or more complete chapters. I think Euan deserves my thanks for goading me into writing certain sections of Chapter 10.

Parts of this book were written during periods of study leave. I thank Lincoln University for granting me leave and the following institutions and people for hosting me: Edward Grey Institute for Field Ornithology; Oxford University (Professor Christopher Perrins and Dr Andrew Gosler); School of Environmental and Information Sciences, Charles Sturt University, Albury, Australia (Professor Nicholas Klomp); and The Key Centre for Tropical Wildlife Management, Northern Territory University, Darwin, Australia (Dr Peter Whitehead, Don Franklin and Dr Richard Noske). This book could not have been written had I not enjoyed the uninterrupted time that today's academics enjoy only during precious periods of study leave.

Special thanks go to Pauline Morse for the four beautiful habitat paintings that appear in this book. I was unable to offer the financial rewards her time and skill deserved. Thanks also to Craig McNeill of the Small Design Company for the text illustrations.

Permission to reproduce figures and quotations was kindly given by the respective authors (acknowledged in the text) and the following publishers and organisations: Department of Conservation, Elsevier Science, Ornithological Society of the New Zealand, Royal Society of New Zealand, John McIndoe, Ministry for the Environment and A. H. & A. W. Reed, and the editors of *Condor*, *Herpetologica* and the *New Zealand Journal of Ecology*.

Finally, I thank Mike Bradstock, Richard King and Kaye Godfrey at Canterbury University Press. Mike was managing editor during most of the years this book was in preparation, and his encouragement, advice, editorial skills and friendship are greatly valued. Richard ably took the reins when Mike departed and gave me invaluable advice and firm but friendly editorial direction as *Flight of the Huia* neared completion.

Kerry-Jayne Wilson

Church Bay
Banks Peninsula

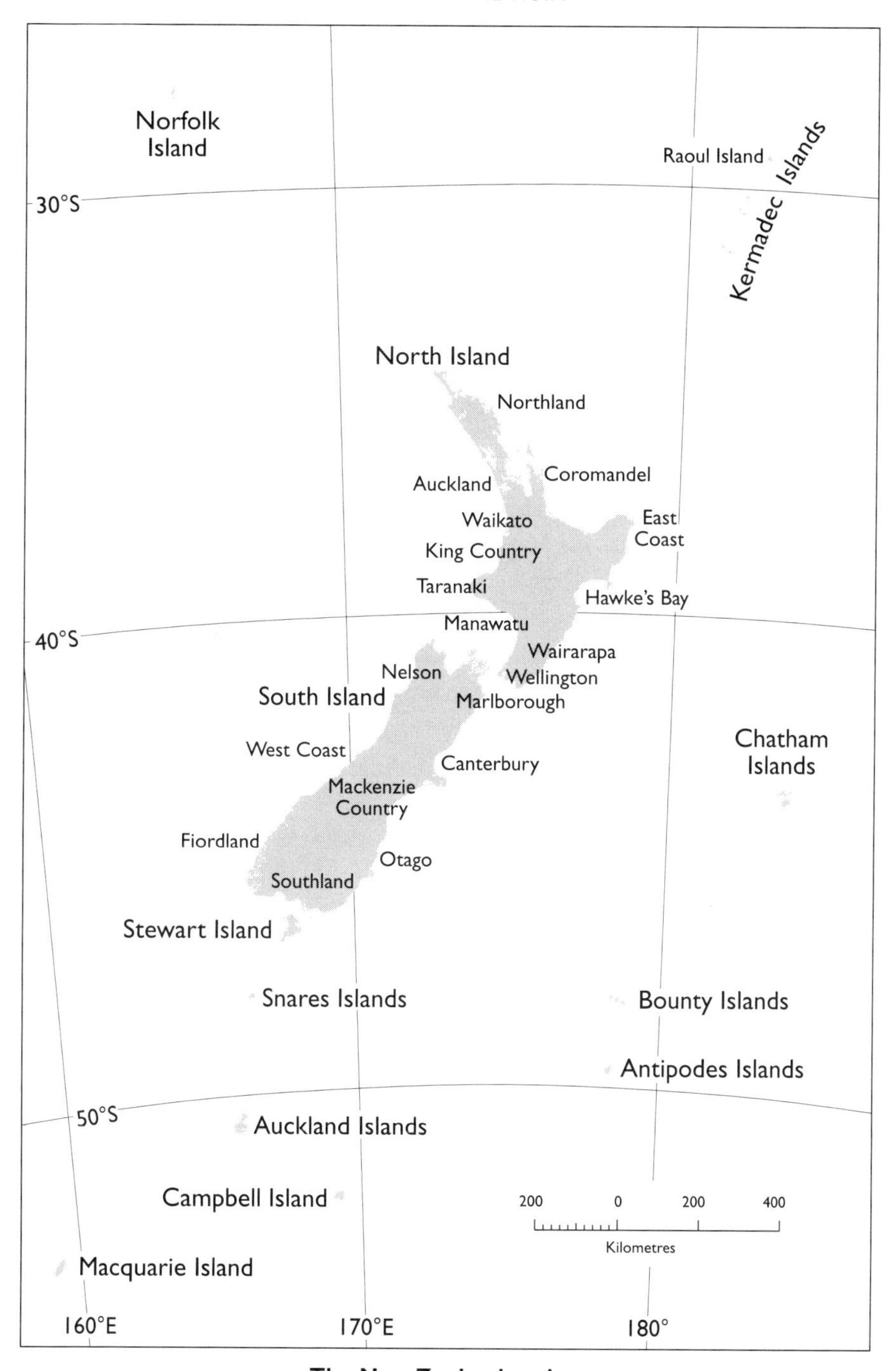

The New Zealand region

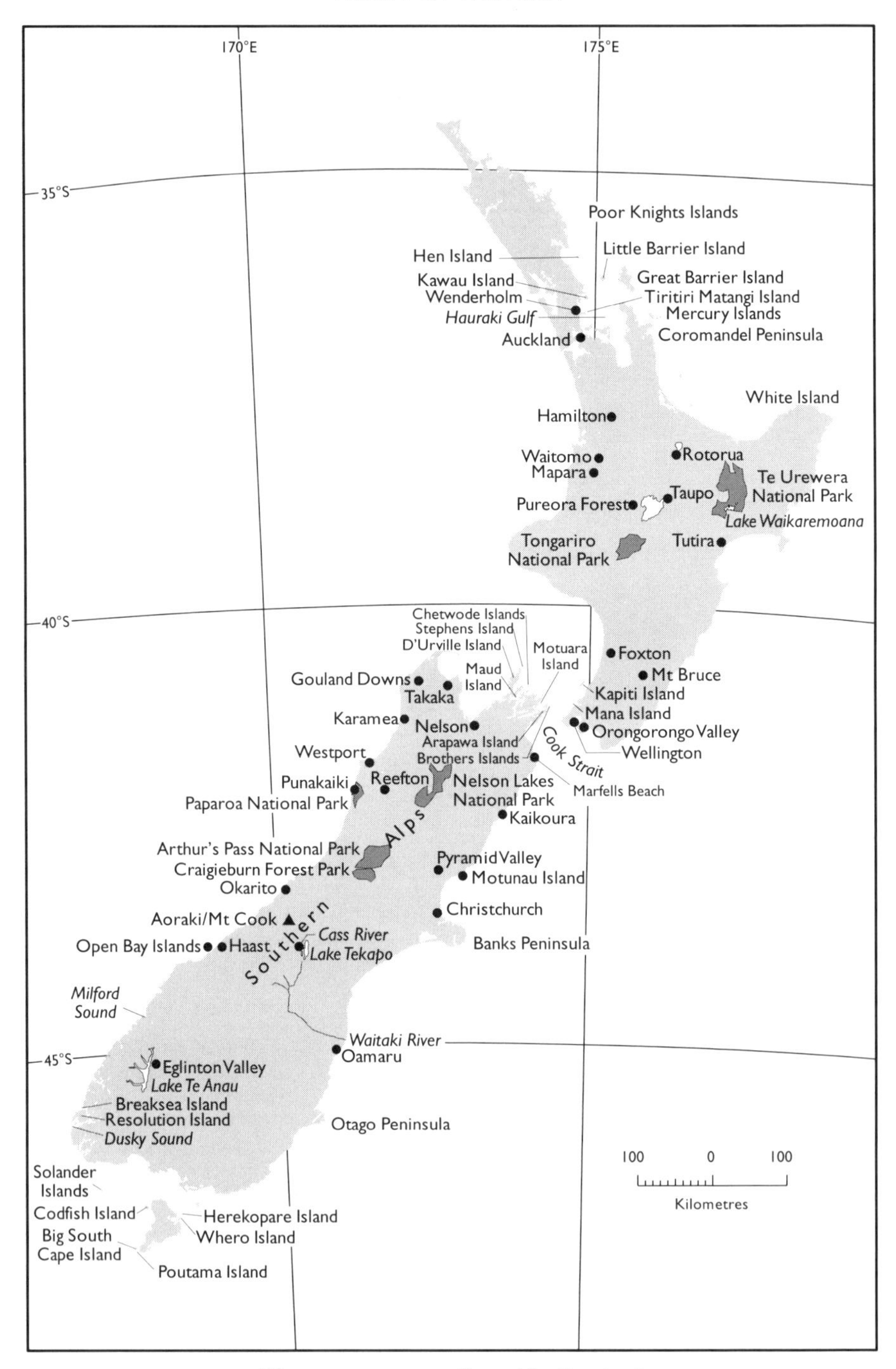

Place names mentioned in the text

CHAPTER 1

New Zealand: archipelago and mini-continent

> Taking into consideration the peculiarities of the flora and fauna of these islands, and the entire absence of fossil remains indicating a former connection with other continents, we are justified in concluding that, during the whole Tertiary period at least, if not for much longer New Zealand has maintained its isolation from all other extensive tracts of land.
>
> – Alfred R. Wallace (1883)

In the bottom left-hand corner of the world's largest ocean lies a land once home to some most unusual animals and plants. These species were so unlike those elsewhere that Jared Diamond, the well-known American biologist and author of the book *Guns, Germs and Steel,* suggested that studying life in New Zealand was the closest he could come to researching on another planet.[1] Imagine how our understanding of what is biologically possible would be enhanced by an experiment in evolution independent of life as we know it. We are unlikely to get any such opportunity, and certainly not in our lifetimes, so the best experiments in the independent evolution of life accessible to today's biologists are to be found on isolated islands such as New Zealand.

Most oceanic islands are geologically young or insufficiently isolated to prevent plants and animals colonising from larger land masses. Diamond identified four island groups that have been isolated for long enough to evolve dramatically different life forms, and are large enough not to be plagued with high natural extinction rates. These are Hawaii, New Caledonia, Madagascar and New Zealand. Of these, he suggests New Zealand is the most interesting. Hawaii is the smallest and youngest; New Caledonia is ancient but small; and Madagascar, while both ancient and large, is too close to Africa to prevent the influx of that continent's mammals.

Once part of the enormous southern continent known as Gondwana, New Zealand is large, remote, has been isolated from other land masses for 80 million years and lacks the mammals that are dominant elsewhere. At least 80 per cent of the species belonging to most non-marine animal and plant groups are endemic. Some of these species belong to families or orders that occur nowhere else, which illustrates how long they have been separated and how distant their relationships have become. Only other remote islands like Hawaii have a similar proportion of endemic species (Table 1.1).

New Zealand consists of two large islands surrounded by loosely clustered small to moderately large islands. Effectively it is both an archipelago and a very small continent. As an archipelago New Zealand is of particular interest because it includes smaller islands of two distinct types. There are numerous land-bridge islands lying close inshore that were connected to the main islands during glacial advances (such as Stewart and Kapiti Islands). Further offshore there are truly oceanic islands that have never had a mainland connection. They range from the subtropical Kermadec Islands to subantarctic Campbell Island, and have been colonised by a subset of species found on the mainland.

Since humans discovered New Zealand about 2000 years ago, and permanently colonised this Gondwanan liferaft almost 1200 years later,

Table 1.1 Numbers of native and endemic non-marine bird species breeding on some oceanic and land-bridge islands

	Total species	*Endemic species*
Fragments of ancient continents		
New Zealand	104	91 (87%)
New Caledonia	72	21 (29%)
Remote, oceanic islands		
Chatham Islands	49	18 (37%)
Hawaiian Islands	54	49 (90%)
Land-bridge islands*		
Tasmania	136	13 (10%)
Borneo	358	37 (10%)

* Land-bridge islands have a smaller proportion of endemic species, owing to colonisation from the neighbouring continent.

Sources: Pratt et al. 1987, Doughty et al. 1999, Simpson & Day 1993, MacKinnon & Phillipps 1993. New Zealand and Chatham Island figures are from Appendix 1 and include mainland species present before 1800.

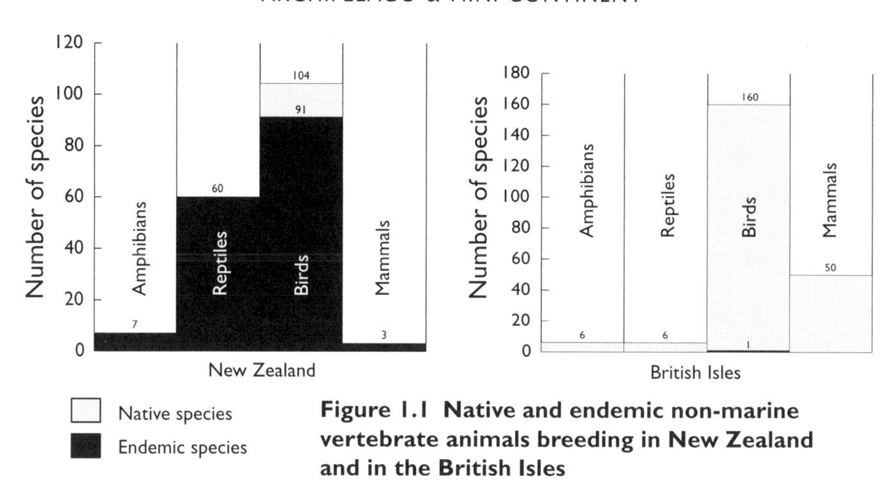

Figure 1.1 Native and endemic non-marine vertebrate animals breeding in New Zealand and in the British Isles

the natural environment has been greatly altered. Almost a half of the native non-marine birds are extinct, and about half of those that remain are threatened or endangered. Of the endangered or threatened birds listed by Birdlife International in 2000, 5.3 per cent are endemic to New Zealand.[2] Considering the country does not have a long list to start with, this is a disproportionately high figure: there are 62 New Zealand species on the list, placing New Zealand eighth equal when counties are ranked this way. All the other countries in the top 25 are larger, and most are tropical nations with far greater species diversity. As well as losing so many species, New Zealand's ecosystems have been much altered in other ways, with many foreign species introduced and habitats fragmented. Other catastrophic changes are less obvious; for example, the loss of many pollinators and seed dispersers. The integrity of our unique ecosystems is under severe threat.

We can appreciate how different New Zealand is from the intensely studied northern hemisphere ecosystems by comparing our numbers of vertebrate animals with those of the British Isles, a similar-sized temperate archipelago (Figure 1.1). Not only are many New Zealand species endemic, but many belong to families or orders that are restricted to this country (see Chapter 3).

The foundations of ecology were developed in the European and North American temperate zones, and even today most of the world's ecologists and the headquarters of most environmental organisations are based in that part of the world. Conversely, most endemic birds, most endangered species and the ecosystems under most immediate threat are in the southern hemisphere or the tropics.[3]

A foreign land

The first Polynesians who came to New Zealand found landscapes, plants and animals very different from those they had previously known. They must have been thrilled to discover a land mass much larger and with a far greater variety of animals than the small, scattered islands whence they came. Not only were there more kinds of animals, but a special bonus was the large, easily hunted, meaty birds. European settlers also found New Zealand to be a foreign land, but for different reasons. They probably expected something resembling their temperate island homeland, and indeed there were superficial similarities, but the ancient evergreen Antipodean forests and their animals were fundamentally different from the open, deciduous forests of Europe.

Initially, neither Polynesian nor European settlers had time to contemplate the strange land, plants and animals, because they were faced with the pressing needs of food and shelter. Just like colonists elsewhere in the world, Maori and Pakeha alike tried to adapt the land to their previous lifestyles, rather than they themselves adapting to the new land. The European settlers were especially assiduous, and it would take several generations for either group to evolve a conservation ethic reflecting the needs of the native animals of this land.

It was long believed that the Maori settlers were the first people to visit these islands. However, during the 1990s it was discovered that kiore (Polynesian rat) had been in New Zealand for about 2000 years, more than a thousand years longer than the earliest likely date of Maori settlement.* They could not have arrived without human assistance, and appear in the subfossil record at about the same time on both main islands. We will probably never know just who those people were that brought the rodents here, as at that time kiore were present on most other South Pacific islands. In a land bereft of mammals, the small ground-dwelling birds and reptiles were defenceless against even this small rat, and we can picture a plague of kiore sweeping across the land with no pied piper to come to the rescue. By the time the Polynesians settled New Zealand around AD1200 (see Chapter 5), many of the small ground-dwelling birds, reptiles, frogs and invertebrates were already extinct.

When Europeans encountered New Zealand, the bird fauna, in particular the large meaty species such as the famous moa, had become

* This early date for the arrival of kiore in New Zealand remains contoversial and the evidence for it is discussed in Chapter 5.

further depleted by Maori. By the time of Captain Cook's brief visit in 1769, at least 40 species of bird had become extinct, along with three species of native frog. Many other birds and reptiles had also disappeared from large parts of their original range.[4] Almost a third of the forests, mostly in the drier eastern parts of the main islands, had been destroyed (Figure 1.2). Many New Zealand birds had evolved to become giants of their kind, with reduced powers of flight, long lifespans and small clutches. Such an existence had served them well for millions of years but left them ill-prepared for the threats posed by rats and humans.

The first European visitors did not venture far from their ships but, like those unknown first people, they brought their own, larger rat that

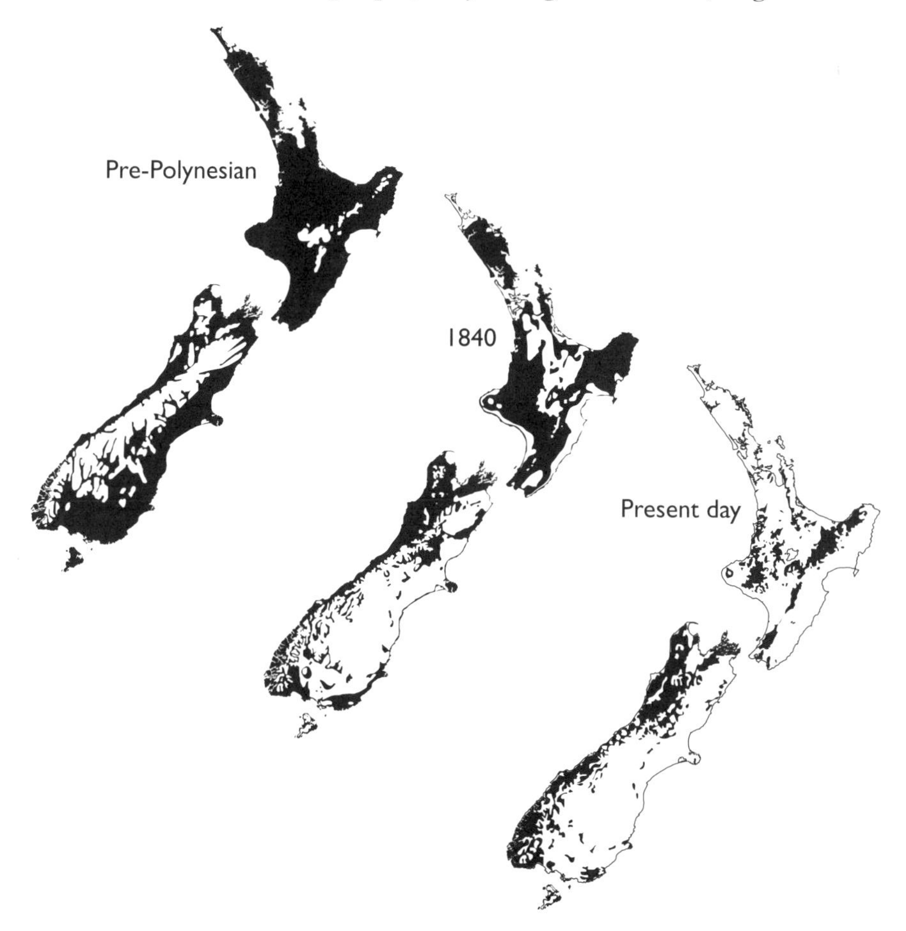

Figure 1.2 New Zealand vegetation cover

Sources: Ministry for the Environment 1997, McGlone and Wilmshurst 1999

posed additional threats to native life forms. By 1800, explorers, sealers, whalers, missionaries and traders were regular visitors. Settlers began arriving early in the nineteenth century, and in 1839 the first of the planned British settlements was founded at what is today Wellington. Other settlers quickly followed. While the Polynesians and first Europeans had been largely hunters, the new colonists brought foreign seeds, animals and technology – and the will to carve out new British settlements

Era	Period	Epoch	Time
CENOZOIC	QUATERNARY	HOLOCENE	10,000 years ago
CENOZOIC	QUATERNARY	PLEISTOCENE	2 million years ago
CENOZOIC		PLIOCENE	5 million years ago
CENOZOIC		MIOCENE	24 million years ago
CENOZOIC		OLIGOCENE	37 million years ago
CENOZOIC		EOCENE	53 million years ago
CENOZOIC		PALEOCENE	65 million years ago
MESOZOIC		CRETACEOUS	135 million years ago
MESOZOIC		JURASSIC	190 million years ago
MESOZOIC		TRIASSIC	235 million years ago

Figure 1.3 Geological time scale spanning that part of the world's history relevant to this book

in the Antipodes. The European arrival heralded an era of extra-ordinarily rapid and extensive habitat change. Between 1800 and 1950, 90 per cent of all wetlands and more than half the remaining forests were lost, and about 16 more bird species become extinct. Indigenous habitats were fragmented into small patches that subsequently became highly modified by introduced mammals. Numerous species of animals and some plants now survive only on predator-free offshore islands.

Why were so many of New Zealand's species so vulnerable to hunting and habitat change? To answer this question we need to understand their ancient origins, and that means going back into the depths of time, before New Zealand existed as such.

Break-up of Gondwana

During most of the Mesozoic the lands that make up the bulk of the southern hemisphere land masses, plus additional lands now located in the northern hemisphere, were united in the supercontinent Gondwana. On the eastern fringe of Gondwana, flanked by west Antarctica on one side and Australia on the other, lay the ancient proto-New Zealand land mass.[5] During the Cretaceous, Gondwana gradually separated into smaller continents, eventually leaving what is today Antarctica centred on the South Pole, while offshoots became Africa, India, Madagascar, South America, Australia and New Zealand. The processes of plate tectonics and drifting continents are not well understood by many people, and an excellent explanation of this seemingly improbable process, with particular reference to New Zealand, is presented by Graeme Stevens in his 1985 and 1988 books *Lands in Collision* and *Prehistoric New Zealand*.

There has been land in the New Zealand region since mid-Jurassic times, about 160 million years ago, and at its peak size, 135 million years ago, the ancient land mass was almost half the size of present-day Australia and completely different in shape. It extended north to New Caledonia, west to the Lord Howe Rise, east to the Chatham Islands and south to the edge of the Campbell Plateau.[6] At that time warm, temperate conditions prevailed over much of Gondwana, and the New Zealand region was virtually contiguous with what was to later become Australia and Antarctica.

Late in the Jurassic, Gondwana began to rift into the sectors that would eventually become today's continents. The initial split separated a larger western part (South America, Africa, Arabia, Madagascar and

India) from a smaller eastern part, which included Antarctica, Australia, New Guinea, New Caledonia and New Zealand. The fragmentation of West Gondwana has little bearing on our story and can be summarised briefly. A rift opened between India/Madagascar and Africa, then in the late Cretaceous India separated from Antarctica and began to move northward, eventually to collide with what is now Asia. Next, Africa and South America pulled apart, then Africa separated from the remaining Gondwanan lands.

The history of East Gondwana is of much greater relevance to New Zealand. It remained a single continent long after the western part had split up. New Zealand and South America were still connected via Antarctica until about 85 million years ago, which explains the close affinities of some New Zealand and South American plants and animals; for example, southern beech trees and some freshwater insects and fishes. The presence of marsupials in both Australia and South America suggests that land connections between these two continents and Antarctica were not broken until the Eocene period, around 50 million years ago.

A rift between Australia and the greater New Zealand region formed much earlier, about 120 million years ago in the early Cretaceous, so the Tasman Sea began to form while both lands were still attached to Antarctica. During the mid- to late Cretaceous, the climate over much of Gondwana (including the New Zealand region) cooled but remained temperate as the Australasian part of Gondwana rotated southwards.[7] The land links between the greater New Zealand region (including New Caledonia) and Gondwana eroded, and the land connections with Antarctica were finally severed 80–85 million years ago.[8] At the end of the Cretaceous, the land mass destined to become New Zealand still lay close to the eastern edge of Australia, and it appears that the ancestors of the kiwi colonised New Zealand about this time or shortly thereafter.[9] The Tasman Sea continued to widen until it reached its present size about 60 million years ago.

New Zealand has drifted northwards throughout the Tertiary period. By the end of the Cretaceous, the greater New Zealand region, which still incorporated New Caledonia and the Chatham Islands, was smaller than it had been earlier in the Cretaceous, low-lying and partly covered by ocean. The last major land link to be severed was between eastern Antarctica and Australia, about 55 million years ago. Since this was one of the key events in shaping the New Zealand fauna, some knowledge of the flora and fauna of the Cretaceous age is necessary to understand subsequent events in New Zealand's ecological history.

100 MILLION YEARS AGO

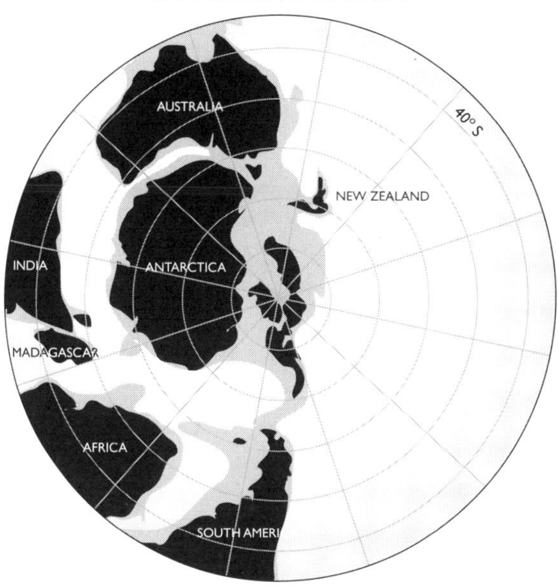

40 MILLION YEARS AGO

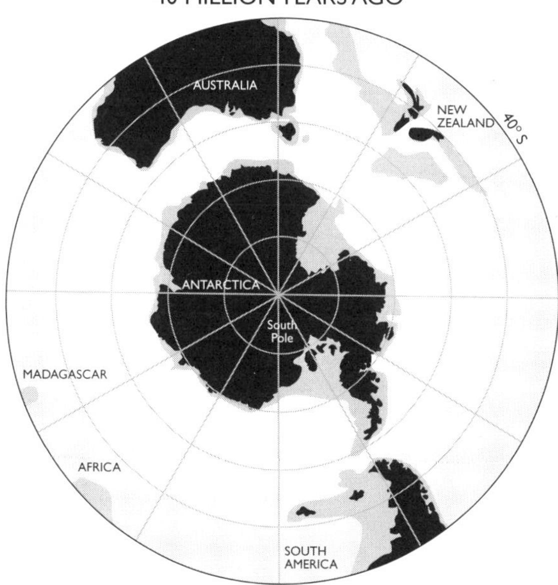

Figure 1.4 The break-up of Gondwana, showing the position of New Zealand in the mid-Cretaceous and late Eocene

Based on *New Zealand Historical Atlas*, plate 4

New Zealand's Cretaceous flora and fauna

Reptiles were the dominant terrestrial vertebrates at the time New Zealand severed its land connections with Gondwana. The Cretaceous reptiles were a diverse and magnificent lot, including terrestrial dinosaurs, flying pterosaurs, marine ichthyosaurs and plesiosaurs. The Cretaceous was an especially interesting period for New Zealand to begin this experiment in evolutionary isolation. Soon the reptiles were to suffer mass extinction. Mammals had appeared on the scene a hundred million years before the demise of the dinosaurs, but despite many of the lineages having evolved, they were of little consequence and their major radiation took place after the demise of the ruling reptiles.

Unfortunately, our knowledge of New Zealand's Cretaceous vertebrates is fragmentary and few land animals of that time appear in the fossil record. Since that period New Zealand has undergone tectonic and volcanic upheavals that have destroyed or distorted most rocks that could contain Cretaceous fossils of terrestrial animals.

When New Zealand separated from Gondwana, it took along a selection of the animals and plants that lived during the reign of the dinosaurs. No New Zealand dinosaurs, pterosaurs or marine reptiles appear to have survived the Cretaceous/Tertiary mass extinction,[10] but other species did and their descendants comprise an important element in the present-day biota. Modern plant and animal groups derived from ancestors that lived on the proto-New Zealand land mass during the Cretaceous include moa, New Zealand wrens, tuatara and native frogs; probably geckos, kauri, podocarps, weta, peripatus, the giant land snails and, possibly, skinks.[11] Some of these species, including tuatara and native frogs, belong to primitive groups that have since become extinct in other parts of the world but soldiered on in New Zealand, isolated from other evolutionary pressures. New Zealand's beech, podocarp and kauri forests date back to the Cretaceous when these forest types covered much of Gondwana. Today, southern beeches (*Nothofagus*) occur in New Zealand, the wetter, cooler parts of southern South America and eastern Australia, the montane forests of New Guinea and much of New Caledonia – all lands formerly part of Gondwana.[12] *Nothofagus* fossils have also been found in Antarctica.

One peculiarity of the New Zealand biota is the absence of terrestrial mammals apart from three species of bats. During the Cretaceous, mammals were present over much of Gondwana, and it has always been a puzzle that there were not more in New Zealand. However, in 2002 as-yet-unidentified fossils were discovered in Otago of mammals that became extinct sometime during the last 20 million years.[13]

Africa and South America collided with northern continents in later epochs, so their Gondwanan heritage is mixed with species from the north; however, the ostriches of Africa and rheas in South America remind us of these continents' Gondwanan origins. Australia and South America separated after the extinction of the dinosaurs and after the monotremes and marsupials had dispersed across parts of Gondwana.

New Zealand during the Tertiary age

At the beginning of the Tertiary, the sea flooded large areas of land, breaking the New Zealand region into a series of islands.[14] The Tasman Sea had almost attained its full width and the oceans south of New Zealand had formed, creating a barrier to the dispersal of animals from Antarctica, which still had a temperate climate. New Zealand and New Caledonia, for so long parts of the same land mass, now became separated as the ocean flooded the intervening land. Islands along the Norfolk Ridge, which still runs under the sea between New Zealand and New Caledonia, appeared and disappeared over the course of time, facilitating the dispersal of certain animals and plants between the two archipelagos. The ancestors of the short-tailed bats,[15] piopio and wattlebirds may have arrived early in the Tertiary before the ocean crossing became too wide. By the late Eocene, Antarctica had cooled, glaciers were forming on it and, 30 million years ago, ice sheets began to spread out and blanket the continent. Gradually the beech and podocarp forests of Antarctica and the species inhabiting them were extinguished.[16]

The New Zealand land mass continued to erode until, by the Oligocene (35 million years ago), all that remained was a chain of small islands surmounting a drowned plateau. The area of dry land was possibly just a fifth that of present-day New Zealand,[17] with the largest island about the size of Canterbury. The land was not only small in area but also low-lying, so it offered a reduced range of habitats. Although fossil evidence is lacking, we can imagine the extinction of many species that could not adapt to changing habitats. Conversely, snails, insects and perhaps small vertebrates such as geckos may have evolved into many more species during this time as isolated island populations diverged genetically from one another. The populations of kiwi, moa and New Zealand wrens were presumably reduced to low numbers and much of the previously present genetic variability may have been lost.[18] It is interesting that recently discovered reptile and mammal fossils found in Otago date from the Miocene (15–22 million years ago), showing that some animals survived the Oligocene drowning, only to become extinct at a later date.

Volcanoes and earthquakes

A period of volcanism and mountain building began about 20 million years ago, just in time to save this Gondwanan liferaft from sinking beneath the waves. This geological activity was caused by the collision of the Pacific and Indian-Australian tectonic plates, the boundary of which New Zealand straddles. It has increased in tempo and continues today. During the last 20 million years, new habitats have been created that have greatly influenced the biogeography of invertebrates and plants but have had little influence on the vertebrates. The environment and conditions during the Pleistocene (the last two to three million years) have been vastly different from the previous 60 million years, with a cooler and less even climate, and a more mountainous landscape than at any previous time during the Tertiary.

It may seem surprising that such dramatic events had so little impact upon the vertebrate species, but it is not so when you consider the volcanic activity of the last century, during which White Island, Ngauruhoe and Ruapehu have erupted from time to time. The 1889 eruption of Tarawera ejected heavy falls of scoria for 16 kilometres and ash for 24 kilometres. While some forests close to the mountain were totally destroyed, just a few kilometres further away there was, within a few years after the eruption, little lingering evidence of damage. Bigger still was the Taupo eruption of 1800 years ago, the impact of which was recorded as far away as China. The pumice, ash and rock ejected almost totally destroyed vegetation over an area of 20,000 square kilometres, yet just 300 years later, tall forest once again covered the devastated area.[19] The ash and mud from volcanic eruptions actually enrich the soil, so the damage is short term and soon obliterated by regeneration. Volcanoes can build up quickly; for example, Ngauruhoe is just 2500 years old and was probably just a small hill when those Pacific explorers visited these shores and left their rats. Rangitoto emerged from the Hauraki Gulf after Maori settled the Auckland isthmus.

Similarly, earthquakes are seldom catastrophic to whole species or populations. New Zealanders can expect to experience at least one major quake during their lifetime. Some of the most significant have been at Wellington (1855), Murchison (1929), Napier (1931) and Inangahua (1968).

Earthquakes and volcanic eruptions are memorable events, but actual mortality is low compared with, say, the arrival of new predators.

The mountains that are so much a feature of New Zealand are only a few million years old. Most uplift of the major mountain ranges has occurred during the last five million years, and the present rate of uplift is probably as rapid as it ever was in the past. The main ranges continue to grow in height at an average rate of about a centimetre a year, and New Zealand continues to move north about six centimetres a year.

The West Wind Drift and Australian immigrants

When Australia moved away from Antarctica it completed the circumpolar ring of water that is the great Southern Ocean, thus during the Eocene a westerly current circling Antarctica began to develop.[20] Since the Miocene, prevailing westerly winds have encircled the globe between 40 and 60 degrees south. This has been, and still is, the main factor influencing the biogeography of marine birds and mammals throughout southern latitudes, helping to transport animals and plants across the seas and enabling circumpolar movement of marine life[21] – the importance of which will later be seen. Dispersal across the Tasman has been the main source of land vertebrates for the past 20 million years, and most birds with Australian affinities have probably colonised New Zealand some time during that period. Several species have even colonised more than once, the first arrivals evolving into new species before the second influx took place. For example, the takahe, weka and black stilt evolved from earlier colonisations of New Zealand by the pukeko, banded rail and pied stilt, all of which now co-occur in New Zealand with their daughter species.

Compared with many oceanic islands, New Zealand is not especially isolated, yet it has surprisingly few terrestrial and freshwater vertebrates. New Zealand is only 1800 kilometres downwind from the biologically rich Australian continent, and must have been receiving a steady flow of wind-blown animals since the Oligocene. The paucity of Australian-derived species in the New Zealand fauna has never been satisfactorily explained. Getting here was one thing, but finding suitable habitat and a mate were also necessary in order to leave descendants. In prehuman times the habitats that greeted the Australian immigrants were mostly dense forests with very different plants and habitats to those in Australia. Even though Australia and New Zealand have some closely related trees, such as their southern beeches (*Nothofagus*), these mostly occur in quite

Table 1.2 Relative abundance of native birds, self-introduced silvereyes and introduced birds in unmodified indigenous forests on Little Barrier Island and in mainland indigenous forests modified by browsing mammals and other factors

	Little Barrier Island	North Island	South Island
Native birds	98.8%	70%	71.4%
Silvereye	0.1%	12%	14.6%
Introduced birds	0.1%	18%	14%

Adapted from Diamond & Veitch 1981

different associations in the two countries. Even the open habitats available in New Zealand were very different to those the animals were adapted for in Australia. Just how effectively this excluded Australian immigrants is illustrated by the fact that 14 bird species have successfully colonised New Zealand since human settlement (most during the last 150 years) – far more than in previous millennia.[22]

Human-induced changes have enabled natural colonisation by these species as habitats became modified so they more closely resembled those in Australia. In 1886, the silvereye was first recorded in New Zealand and since then they have become one of our most common birds. On Little Barrier, one of the few large islands where forests have not been modified by browsing mammals or logging, silvereyes and introduced birds are very rare in the forests, yet they are numerous in the modified habitats around the ranger's house. These birds are common in indigenous mainland forests that have been altered by browsing mammals and other factors (see Table 1.2). Most foreign species arriving in New Zealand before humans came on the scene were unable to adapt to the ancient Gondwanan forests, so the unique biota persisted until humans modified those habitats.

The Pleistocene ice ages

> Our conclusion is, therefore, that New Zealand is the remains of one of the most ancient . . . of the islands of the globe; that it has undergone many fluctuations in area; that the two islands have been quite recently united, and that at some remote epoch it was many times more extensive than it is now . . .
>
> – Alfred R. Wallace (1883)

About two million years ago, the Earth's climate cooled and glaciers covered much of the temperate lands in both hemispheres. We can picture the following period as one of fluctuating temperatures, causing the glaciers to vary in size. When they advanced, sea levels fell by up to 150 metres, so that the North, South, Stewart and most land-bridge islands were all joined in a single land mass. The most recent major advance reached its fullest extent 20,000 years ago and finished as the climate once again warmed some 14,000 years ago.[23] The ice ages may not be over: the climate could again cool and glaciers advance once more.

With each glacial advance, vegetation changed as forests gave way to montane grasslands and herbfields, which were then re-invaded by forest when the temperatures rose again. On the continents, habitats and species would have shifted north or south as the glaciers advanced or retreated, but in New Zealand northward movement was limited and during glacial advances forest animals and plants were restricted to ice-free refugia. During major glacial advances, podocarp forest south of the Waikato would have been restricted to small, isolated fragments.[24] Beech forest would have been more extensive, but large parts of the country would have supported only grasslands or shrublands. As the glaciers retreated and the climate warmed after the last glaciation, forest again covered most of the country long before humans arrived. Although the vegetation changes were dramatic, no bird species is known to have become extinct during the Pleistocene. The alpine habitats that first appeared during the Pliocene and the Pleistocene allowed the alpine kea and rock wren to diverge from their forest-dwelling ancestors, the kaka and bush wren.

New Zealand is thus a place where ancient animals and plants, survivors of past geological ages, live in a youthful landscape created during the last few million years by tectonic uplift and volcanic eruptions, and subsequently shaped by glaciers and erosion.

Rats, humans and other aliens

The arrival of rats, humans and other introduced species was the most momentous event in the history of New Zealand since the break with Gondwana, 80 million years earlier. The rate of ecological change during the last 2000 years is orders of magnitude greater than during any earlier era. Passengers on the Gondwanan liferaft were completely unprepared for the massive disruptions, and New Zealand's ecological communities have been greatly altered by the loss of many species, including the major browsers, predators, insectivores and frugivores.

Even our national parks, reserves and offshore island reserves contain only a subset of the endemic species once present. Today, many of the species most familiar to New Zealanders are those that occur naturally in Europe or Australia. Only a few of the most widespread and hardy native birds still survive in the urban or rural habitats where most of us live, which bear little resemblance to those of 150 years ago. The most common city-dwelling native birds, such as silvereyes and welcome swallows, actually established here after the Europeans arrived.

The change from a mammal-free sanctuary to the heavily modified landscape of today took place over an amazingly short time, and the intensity and pace of change is unrecognised by most people. New Zealand was the last habitable land to be colonised by people. When the Polynesians discovered these islands and later settled here, they had already discovered almost every other island scattered across the world's largest ocean. Australia had been inhabited by aboriginal people for at least 40,000 years (probably much longer), and the Americas for over 10,000 years. In Europe, the Greek and Roman civilisations had flourished then fallen; while in the Americas the great Mayan civilisation was in decay and the Inca civilisation was on the ascendant. Literate societies had existed in Egypt, Mesopotamia, India and China for at least 3000 years. At the time Polynesians settled New Zealand, Hadrian's Wall and the Great Wall of China were already at least a thousand years old and castles were under construction in Britain.

CHAPTER 2

New Zealand's frogs and reptiles

> I believe we have no reason to be especially proud of our knowledge about New Zealand's herpetofauna. Except perhaps for the tuatara, this group of vertebrates has remained very much the neglected child of the wildlife biologists in this country. We know little or nothing about the living habits, habitat requirements or distribution of many species, the taxonomic status of some is still unresolved and the discovery of new forms, even new species is quite possible.
>
> – Christoph Imboden (1982)

New Zealand has none of the brightly coloured frogs or large, spectacular reptiles so familiar in other parts of the world. Our frogs and reptiles are small and seldom seen, but among these cryptic animals are some of our greatest zoological treasures. The tuatara and native frogs are primitive animals, long isolated from the mainstream of evolutionary change. The tuatara belongs to a group of reptiles that had its heyday before the rise of the dinosaurs, and the native frogs are the world's most primitive. These animals are well known, but until recently the lizards have attracted little attention. Between 1980 and 1994, the number of lizard species recognised in New Zealand grew by more than a half, to over 60 species;[1] and since then more new species have been found but not yet named, while still others almost certainly await discovery. New Zealand once had about as many lizard species as terrestrial bird species, but since so many of the latter have become extinct, there are now more species of endemic reptiles. Although the lizards belong to only two families and four genera, three of those genera are endemic.

Almost half of the reptiles are threatened or endangered, yet until 1981 no lizards had legal protection, and only in 1996 were the last few species given protected status. It is now illegal to hold in captivity or to handle any native reptile or amphibian without a permit. However, conservation action to protect these endemic species still lags far behind measures taken to protect native birds.

In this chapter I introduce each of the amphibian and reptile groups found in New Zealand and discuss their ecology, their place in our ecosystems and their current status.

Frogs, reptiles and Gondwana

Both tuatara and the native frogs belong to groups that, prior to the break-up of Gondwana, occurred in other parts of that supercontinent. They survived in isolation in New Zealand while elsewhere they were replaced by more advanced animals. The tuatara belong to the order Sphenodontida, a group of reptiles that originated in the early Triassic about 225 million years ago.[2] Fossil sphenodontids have most commonly been found in Triassic and Jurassic rocks in South America, North America, Africa, Europe and Asia,[3] and evidently declined during the Cretaceous, so that by the end of that era they apparently survived only in New Zealand. The fossil most similar to tuatara was *Homeosaurus*, from Jurassic deposits 140 million years ago in Europe. At the time the first sphenodontids evolved, Gondwana and most northern hemisphere lands were united in another supercontinent called Pangaea.

A lot of misinformation has been published about tuatara. They are not dinosaurs, as many popular accounts suggest, but are even more ancient. They evolved before the dinosaurs and their decline during the Cretaceous suggests they were displaced by these more advanced animals. Tuatara belong to the same major branch of the reptilian dynasty – the Diapsids – as the snakes, lizards, crocodiles, dinosaurs and birds (Figure 2.1). Tuatara are of particular interest because they are the least specialised of the living reptiles, so they have some features in common with early, primitive reptiles. However, tuatara have not persisted unchanged for over 200 million years as some books have suggested: in fact they have specialised as well as primitive features (Table 2.1). Nor are they members of the most ancient surviving lineage of reptiles: that distinction belongs to the turtles and tortoises.

Less is known about the prehistoric distribution of New Zealand's native leiopelmatid frogs. The bones of small frogs seldom fossilise and the only known fossil leiopelmatids are in Jurassic rocks from Argentina. Presumably these frogs were once found in other Gondwanan lands.

Table 2.1 Some primitive and specialised features of tuatara

Primitive characteristics

- Primitive skull
- Concave articulations on both ends of each vertebra
- Teeth are bony protuberances on the jaw – not implanted in sockets as they are in lizards
- The spleen is the only lymph organ
- No palate between mouth and nasal passages
- Low metabolic rate – 55 per cent of a same-sized lizard
- No penis
- So-called 'third eye' – the pineal organ, with a vestigial lens and retina; becomes covered by opaque scales when the animal is about six months old
- Ability to shed the tail when pursued by predators is less well developed than in lizards

Specialised characteristics

- Possesses a tapetum (light-reflecting surface on the retina, which serves to intensify the available light); similar structures occur in other nocturnal animals
- Heavy, strong skull with fewer bones, but with primitive features
- Cold-adapted, able to maintain activity at temperatures as low as 7°C

Adapted from Crook 1975, Dawbin 1982

Another primitive frog, *Ascaphus truei* from western North America, was formerly included in this family but is no longer thought to be closely related.[4]

Skinks and geckos have an almost worldwide distribution, occurring even on some of the most remote oceanic islands. Thus, until recently, New Zealand lizards were considered island waifs, relatively recent additions to the biota. It had been thought that the geckos colonised New Zealand during the Miocene, and the skinks in the Pliocene or Pleistocene. However, recent research indicates they have been here far longer, as the geckos belong to an ancient, possibly Gondwanan group.[5] It is now thought that skinks arrived in New Zealand at least 25 million years ago. The origins and taxonomy of the New Zealand lizards are discussed on page 32.

New Zealand's dinosaurs and other Cretaceous reptiles

If Jurassic-style frogs and tuatara survive in New Zealand, this raises the question whether there were ever dinosaurs here. For many years it appeared not, but in 1975 Joan Wiffen, a keen amateur palaeontologist, discovered a tailbone later identified to be from a bipedal carnivorous dinosaur. Subsequently, bones from three or possibly four other species of dinosaurs, a pterosaur and one freshwater turtle have been found, all by Joan Wiffen or her co-workers and all from the same late Cretaceous (65 million years ago) deposits in Hawke's Bay.[6] Each dinosaur discovered so far belongs to a different family, suggesting that a variety of Mesozoic reptiles may have inhabited this corner of Gondwana. Marine reptiles are better represented, with several species of ichthyosaurs, plesiosaurs, mosasaurs and turtles found in Mesozoic rocks in several parts of New Zealand.

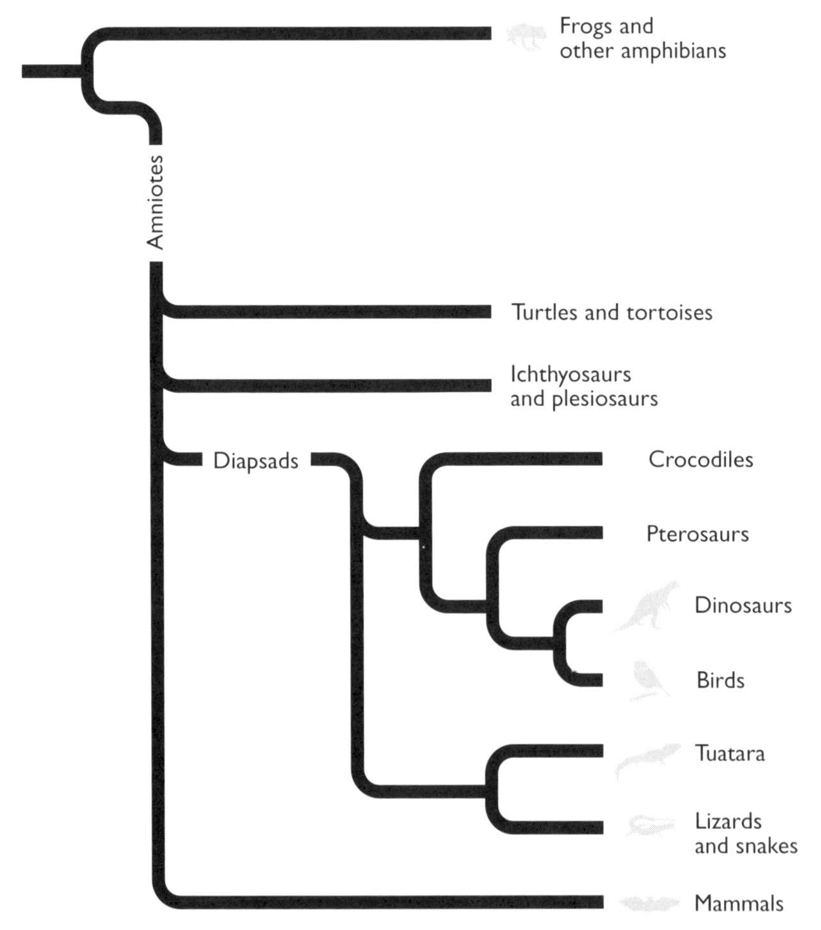

Figure 2.1 Evolutionary relationships of reptiles, birds and mammals

Note that only groups still living and some of the most familiar groups of extinct reptiles are shown: there were also many other groups.

Adapted from Archer 1984

Native frogs

The native frogs are small, quiet, dull-coloured, nocturnal and secretive, yet their inconspicuous appearance belies their tremendous zoological interest. Being primitive is not their only claim to fame. Three of the living species are terrestrial, and the fourth lives beside, not in, water. All show peculiar breeding strategies: they pass the tadpole stage inside the egg and hatch directly into tailed froglets. The primitive features of these frogs include lack of tympanic membranes (eardrums), eustachian tubes (connecting the mouth and the ear) and vocal sacs. Consequently, they have little power of hearing and the only sounds they can make are shrill chirping calls when disturbed and sometimes when mating.[7] Although tailless like all other frogs, the *Leiopelma* frogs retain two small tail-wagging muscles that have been lost in the more advanced species.

Today, the native frogs are restricted to a few scattered locations in the northern half of the North Island and two islands in the Marlborough Sounds.[8] Hochstetter's frog is more widespread than the other species, being most common on the Coromandel Peninsula, in the Auckland region and near East Cape, but a few occur at some other northern localities (Figure 2.2). Archey's frog was thought to be restricted to the Coromandel Ranges until an isolated population was discovered at Whareorino in the King Country in 1991.[9] The Maud Island frog occurs only on that island, in Pelorus Sound, while Hamilton's frog, perhaps the rarest of all frogs, is restricted to a small part of Stephens Island, at the entrance to Pelorus Sound.

These species were formerly much more widely distributed and three now-extinct species were also present. It is difficult to build up a clear picture of their pre-human distribution because their bones do not generally preserve well. They are most often found in limestone caves or the middens built up over many generations at laughing owl nest sites, but the absence of bones in other places does not necessarily mean the frogs were not present there as well. Suffice to say that native frogs were apparently once widespread and common through most of the country. Although now extinct in the South Island, they were probably common, at least in Nelson, Westland and Fiordland. Three species are known to have co-existed on Takaka Hill, near Nelson. They were probably absent from seasonally dry lowland Canterbury.[10] The declines have apparently occurred within the last 2000 years and are attributed to habitat loss and predation, initially by kiore and subsequently by other introduced predators.[11] Direct evidence of this was obtained when five Archey's frogs were found to have been killed by rats and the remains of another was recovered from the stomach of an introduced frog.[12] Why they surivied on the North Island but became extinct in the South Island remains a mystery. Perhaps they would have been less vulnerable to introduced predators if they had an aquatic lifestyle.

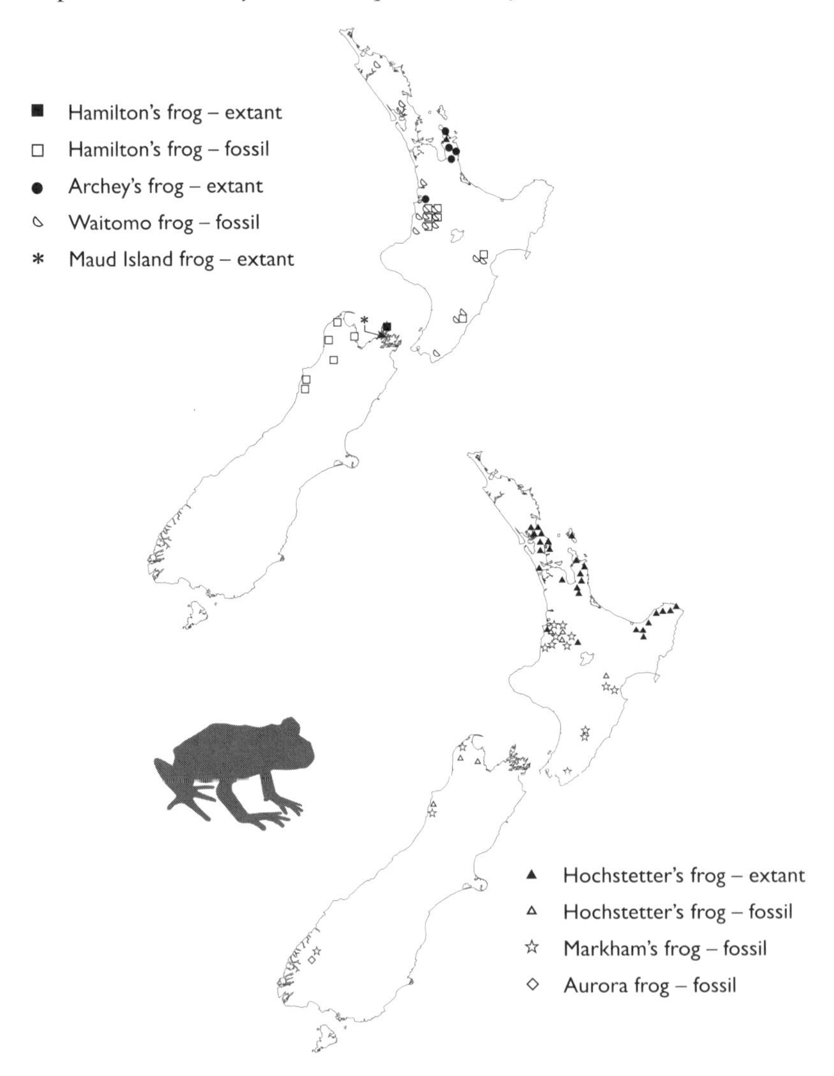

Figure 2.2 The present and past distributions of all seven New Zealand frog species

Adapted from Bell et al. 1998b

Relationships among the native frogs

Until 1987 the taxonomy of the leiopelmid frogs was simple: there were three species, all with restricted distributions. That year Trevor Worthy described three species of fossil frogs, all of which were larger than the living ones. Subsequently, biochemical studies showed that the frogs on Maud and Stephens Islands, both previously considered to be Hamilton's frogs, were separate species.[13]

Hochstetter's frog is more robust than the other living species and differs from them in appearance, ecology and mode of reproduction.[14] Worthy grouped two of the extinct species – Markham's frog and the Aurora frog – with Hochstetter's, implying that these were similar to Hochstetter's in breeding biology and ecology. He suggested that the extinct Waitomo frog was similar to Archey's, Hamilton's and the Maud Island frogs. These two species groups are sufficiently different from one another that it may be appropriate to place them in separate genera.[15]

The relationship between Archey's, Hamilton's and Maud Island frogs is complex. Archey's and Hamilton's are genetically closer to one another than either is to the Maud Island frog, even though the only surviving populations of Maud and Hamilton's are on islands 40 kilometres apart and the two populations were previously regarded as a single species.[16]

Not everyone agrees that the Maud and Stephens Islands frogs are sufficiently different to be separate species.[17] Recent fossils of Archey's-like frogs have been found at many sites in the North Island and northern South Island, but their identification now has to be reassessed. Bones discovered in caves near Waitomo were identified as Hamilton's frog by Worthy, but Ben Bell considered these bones belonged to the recently discovered King Country population of Archey's frog.[18] Bell suggested that Archey's and Hamilton's frogs are sister species separated in the Pliocene by the strait that at the time flooded the lower North Island. However, now that two species are recognised from the Marlborough region, the South Island fossils could be Hamilton's frog, Maud Island frog or yet another species.[19] Variation also occurs between populations of Hochstetter's frogs, not in the biochemical make-up but in their chromosomes and sex-determination system,[20] but to date these have not been reclassified as separate species.

Watersiders and mist-dwellers

> This frog was discovered by my brother, Richie Smith in 1916. We as children were sheltering from the wind – in the crater-like boulder-

> strewn hollow on the summit of Stephens Island. My brother was shifting some rocks to make a seat, when he found this wee frog. We took it home and our father . . . sent it to Mr Hamilton who . . . identified it as a previously unknown species and thanked us for forwarding it to him.
>
> – Emily Scott (daughter of a Stephens Island lighthouse keeper)

The native frogs are not denizens of ponds and swamps like most frogs, but instead live a terrestrial or semi-terrestrial life.[21] Among the living species, Hochstetter's frog comes closest to an aquatic lifestyle but lives under rocks and logs or in vegetation alongside cool, shaded, rocky forest streams. Archey's frogs are restricted to forests 400–800 metres above sea level on rainy, mist-dampened mountains, sheltering by day beneath rocks and logs, and at night climbing about in the trees to seek their insect prey. At Whareorino, where logs and stones are rare, they shelter in dense vegetation or fern crowns.[22] Their distribution is patchy, reflecting the frog's need for damp habitats.

The Maud Island frogs occur in coastal forests. Hamilton's frogs are even more restricted in the habitats they occupy. Presumably, they too were once forest-dwellers, but since the forest on Stephens Island was cleared they have become restricted to a rock pile almost on the summit of this dry and windswept island. They live in spaces between the boulders, retreating into deep crevices when conditions are dry. Archey's, Maud Island and Hamilton's frogs all rely on dew, rain or mist to prevent dehydration and are most active on wet nights.

Hochstetter's frog is more sturdily build than the others. It has partially webbed hind feet, while the more terrestrial species have lost all traces of this webbing.

None of the native frogs has a free-swimming tadpole: the tadpole stage occurs before the egg hatches. Hochstetter's frog lays its eggs in wet, secluded sites beneath logs, rocks or vegetation, and these hatch into froglets with a tadpole-type tail but well-developed hind legs (Figure 2.3).

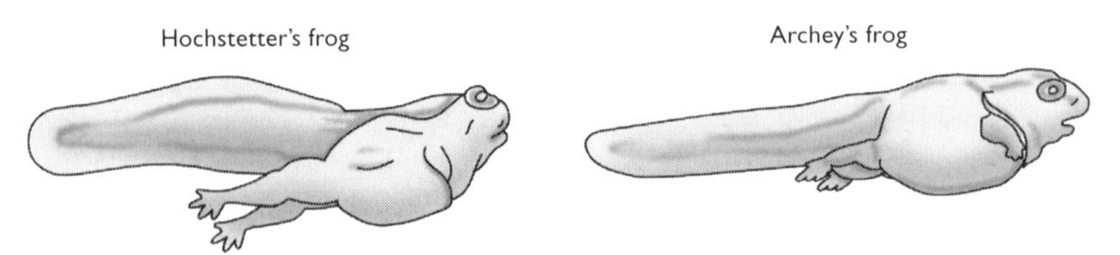

Figure 2.3 Newly hatched froglets

Adapted from Bell 1978

When the larvae hatch, their forelegs lie beneath the membrane that covers the front part of the body, but over the next few weeks these forelimbs develop fully and the tail is absorbed. Initially, the newly hatched froglet remains in water seepages or similar semi-aquatic habitats, but by 11 weeks after hatching it is a fully developed, though still tiny, frog.[23] The froglets of the more fully terrestrial species hatch with smaller tails and better-developed forelimbs than Hochstetter's, and are never found near water.

The eggs of Archey's and Maud Island frogs are laid in shallow, damp depressions beneath logs or stones; but at Whareorino, where such sites are rare, they are laid among leaf litter accumulated in fern crowns.[24] The males provide some care for the eggs and newly hatched froglets. This paternal care is more extensive in the terrestrial species than in Hochstetter's frog because eggs and young of the former are more susceptible to dehydration.

Native frogs probably do not breed until they are three years old. They are long-lived and each breeding season lay fewer than 20 large, yolk-rich eggs. Other frogs typically lay many more eggs than this. Archey's frogs have been known to live more than 17 years, and Maud Island frogs over 29 years.[25]

Tuatara

> Besides these there is a lizard-like animal *Hatteria punctata*, of so peculiar a structure as to form a distinct order of reptiles called Rhynchocephalina, intermediate between lizards and crocodiles.
>
> – Alfred Wallace (1883)

Tuatara were once widespread throughout New Zealand: their bones have been found in dunes, caves and middens at many places, almost literally from North Cape to the Bluff (Figure 2.4, page 26). They probably preferred relatively dry ground and were apparently rare in western rain forests.[26] When kiore arrived in New Zealand, tuatara were widespread, but by the time Europeans arrived they were absent or at best very rare on the North and South Islands, and gone from many offshore islands.[27] Tuatara continue to decline and are now extinct on 10 of the 40 islands they occupied a century ago.[28] They are susceptible to all introduced predatory mammals and are significantly more common on islands free of kiore. Furthermore, on most kiore-infested islands no juvenile tuatara (less than 180 millimetres long) are found.[29] Tuatara cannot maintain their populations in the presence of kiore, although local extinction will take decades or longer to occur. Don Newman and Ian McFadden[30] suggested that tuatara survive longer in the presence of kiore on forested islands, where rat densities are generally low, than in cleared habitats, which kiore favour.

Two species of tuatara are recognised. In 1990, Gunther's tuatara was restricted to about 1.7 hectares of scrub on tiny North Brother Island in Cook Strait.[31] New populations have subsequently been established on two other islands. The other, more common, species remains on about 30 islands off the northeastern coast of the North Island and in Cook Strait. Vulnerable they may be, but this species is still relatively numerous. The population may exceed 100,000 individuals, with an estimated 30,000–50,000 on Stephens Island alone. This high number

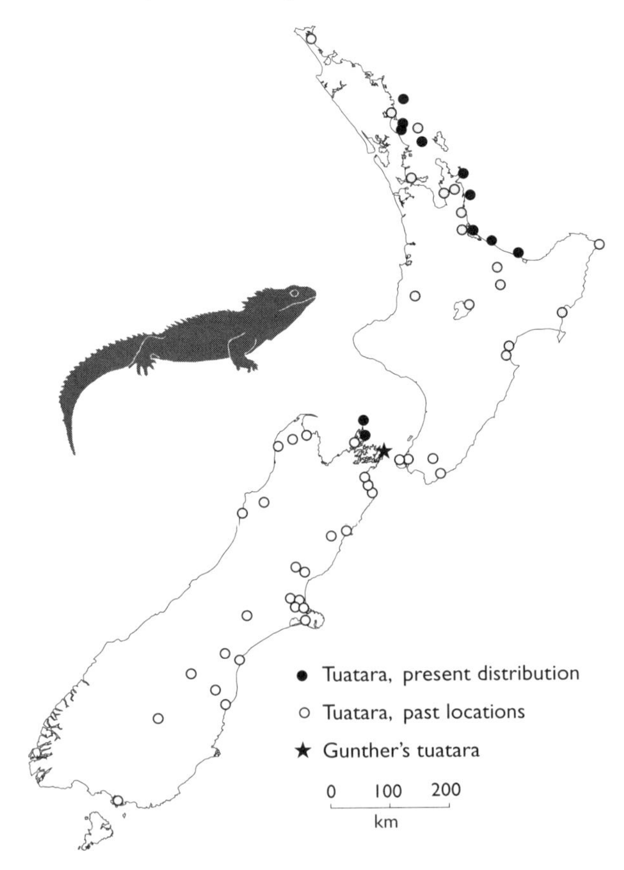

Figure 2.4 Past and present distribution of tuatara

Source: Newman 1987a, Gaze 2001

gives a misleading impression of their conservation status. On most islands there are very much fewer (half of the islands where tuatara survive are smaller than 10 hectares), and kiore are present on most of the larger islands. Tuatara remain on only six islands that are both rat-free and larger than 10 hectares.[32]

The common species includes two subspecies: the northern tuatara and the Cook Strait tuatara. The former is the most widely distributed, being present on 25 islands, but most of these are small or have kiore present.

Rats have subsequently been eradicated from five tuatara islands.[33] The Little Barrier population was once considered to be a third subspecies and was feared extinct in 1990, but a few individuals have since been found. All known Little Barrier tuatara are currently held (and are successfully breeding) in a predator-free enclosure on that island awaiting the eradication of kiore. The Cook Strait tuatara is the most abundant and, while it occurs on only four islands, is the most secure subspecies.

After a century when tuatara were considered to comprise a single species spread over many islands, and thus safe from extinction, conservation efforts are at last under way to secure existing populations by eradicating kiore, and to establish new tuatara populations on rat-free islands, including some with unrestricted public access. The recovery plan for tuatara[34] identifies the conservation needs of each species or subspecies as being quite different. Conservation management of tuatara is discussed on page 302.

Tuatara can tolerate some degree of habitat change. It has been suggested that tuatara numbers may be artificially high on some islands with both forest and grassland habitats, in particular on islands with lighthouses and associated pasture land.[35] Adult tuatara may live in forests but lay eggs in open habitats where warmer temperatures speed up their development. On Stephens Island, tuatara are found in forest remnants, on pasture grass and even around the lighthouse buildings. Numbers can be as high as 2000 per hectare in some forest remnants, but little more than a third of this on grassland.

Tuatara and petrels

Many of the islands on which tuatara survive are also home to large numbers of burrow-nesting petrels such as fairy prions and diving petrels. In the past it was suggested that tuatara may be dependent on the petrels, but the evidence is against this view. Petrel colonies were certainly once much more widespread, but tuatara occurred in places where petrels were not present; and at Punakaiki, where burrow-breeding petrels used to be numerous, tuatara were rare. Tuatara no doubt benefit from the presence of seabirds, but their co-existence is not due to the dependence of one on the other: it is because neither can withstand introduced predators so they simply share the same sanctuaries.[36]

Tuatara benefit from the birds in several ways. They may use the same burrows, which were in most cases excavated by petrels. The seabirds import nutrients to the island, and the guano-enriched soils encourage a high biomass of invertebrate food that can support large numbers of tuatara. The open nature of the ground, a result of seabird disturbance, may create better conditions for foraging tuatara. They also occasionally eat petrels, their eggs or chicks. Usually these comprise a small proportion of their diet, but in one study the reptiles ate more than a quarter of the eggs and chicks produced.[37] Fairy prions and other petrels attack any intruder that enters their burrow and tuatara have been seen retreating from burrows with angry prions clinging to their tails.

There appears to be an optimum prion density above which tuatara density declines. Tuatara do not use burrows while prions are nesting in them, and only one tuatara will inhabit a burrow at a time, so Don Newman suggested that tuatara numbers best relate to the number of burrows left vacant by seabirds.

Reproduction and growth

Reproduction in tuatara is a long-drawn-out affair. Female tuatara usually produce eggs only once every four to five years, though some lay every second year.[38] Egg development in Stephens Island tuatara begins within a year after the previous clutch was laid, but the development process can take several years, a situation unique among reptiles.[39] Once the eggs are fully developed, tuatara mate in late summer (January–March), but the eggs are not laid until early the following summer (October–December). The eggs are fertilised soon after mating, but shell production takes a further six to eight months. The eggs are buried in chambers or tunnels and take 12–15 months to hatch. This means the interval between mating and hatching is about two years. Usually, 7–11 eggs are laid, but up to 18 have been recorded. Gunther's tuatara lay clutches of only four to eight eggs at about four-yearly intervals.

Growth after hatching is slow, and this relatively small reptile takes between 9 and 13 years to reach sexual maturity (9–11 years on Lady Alice Island in the north; 11–13 years on Stephens Island, which lies five degrees further south). They continue to grow for a further 20 or more years.[40] Males grow to 1 kilogram, and females about 0.5. The smaller Gunther's tuatara may grow even more slowly.[41] This is a phenomenally slow rate of growth for a reptile – even slower than for giant turtles, which are by no means speedsters in the reptilian world and whose eggs also take less time to hatch. Most large turtles first reproduce at a younger age than tuatara. Tuatara can live at least 60 years and possibly more than a hundred.

The unusually long period between laying and hatching, and the very slow growth, make tuatara vulnerable even to small predators like kiore. First, the eggs may be eaten and are exposed to this risk for a long time. Second, for several years tuatara remain small enough to be vulnerable to predation by kiore, which may also compete with them for food.

Foods of tuatara

Tuatara are carnivorous and fairly unselective hunters, taking most appropriately sized animals that remain close enough for long enough to be captured. Tuatara move about on the ground searching for prey or sit motionless, often in the entrance to their burrows, waiting for prey to pass by. They mostly eat invertebrates but when opportunity arises they will prey on small vertebrates including native frogs, lizards, smaller tuatara and seabirds[42] (Figure 2.5, page 30). The invertebrates they most commonly eat are more than 10 millimetres long, slow-moving and ground-dwelling. On rodent-free Stephens Island, where the range of food animals is correspondingly diverse, more than a hundred food species have been recorded, about half of them large beetles. Small beetles, weta, other insects, spiders, millipedes and earthworms were also commonly taken. Lizards, young tuatara, and seabird eggs or chicks altogether made up less than 10 per cent of the food items eaten. Stephens Island tuatara living in the bush preyed mainly on tree weta and darkling beetles, whereas those living in paddocks took isopods, craneflies and small beetles.[43]

The diet of tuatara on kiore-infested Lady Alice Island was similar, which suggests that competition for food between the two species is of little significance.[44] These tuatara only ate invertebrates, and 85 per cent of their food items were arthropods, including beetles, moth larvae and weta.

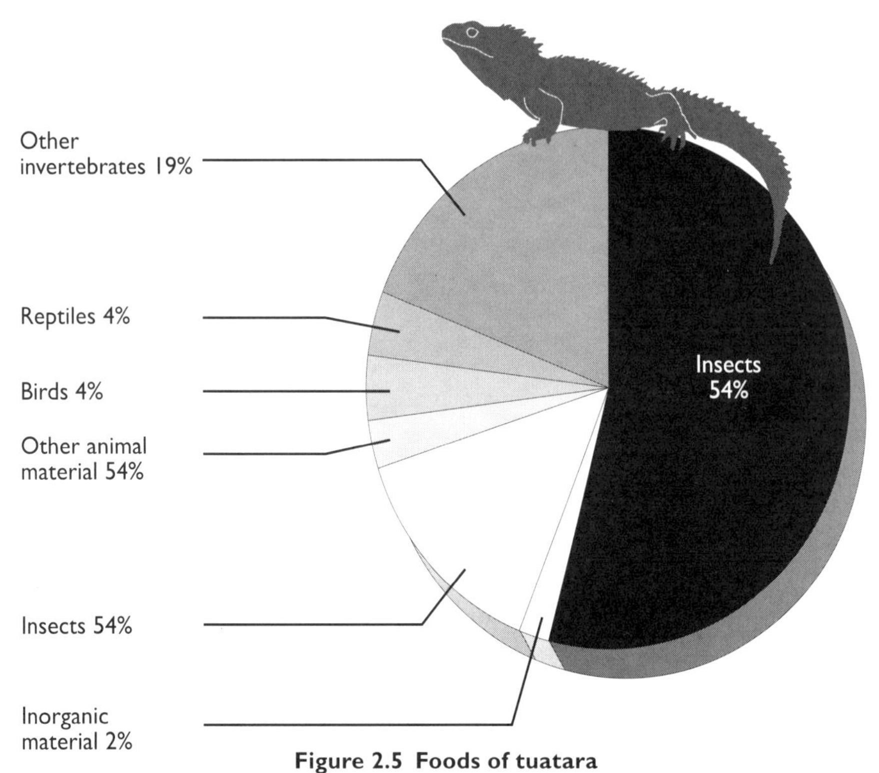

Figure 2.5 Foods of tuatara

Adapted from Newman 1987a

Each tuatara normally has one or several own home burrows, which, if multi-chambered, may be shared with petrels. Most tuatara will have several burrows they regularly visit. Only one tuatara will be found in a burrow at once, although several may make use of any particular burrow.[45] Tuatara emerge from their burrows on most summer evenings, but during winter they are active only when temperatures are warmest. Even in winter they do most foraging at night. The metabolic rate of tuatara is only about 55 per cent that of a similar-sized lizard, and their body temperatures are lower than for lizards living alongside them.[46] Tuatara can tolerate low temperatures, and a few will forage during evenings when the temperature is as low as 4°C, though they are most active at 17–20°C. The ability to forage in cold conditions is an adaptation for a nocturnal reptile living in cool temperate climes and may partly explain the high numbers of tuatara found on some islands.

New Zealand lizards

> The ecologically and taxonomically diverse lizard fauna is probably one of New Zealand's best kept biological secrets.
>
> – Daugherty et al. (1994)

New Zealand has lizards on the main islands, the Three Kings and Chatham Islands, but not on the oceanic Kermadec Islands or the cold subantarctic islands. Lizards live in virtually all habitats from intertidal rocks to alpine bluffs; from semi-arid Central Otago to Westland rain-forests. New Zealand has more species of lizards than do California or Texas, and, per area of land, more species than Australia[47] (Figure 2.6). Middle Island, a 10-hectare islet in the Mercury Islands, has 10 species of lizards plus tuatara, and this is probably the highest diversity of reptiles on a land-area basis anywhere in the world.

New Zealand's lizards all feed primarily on insects, but some also eat fruit, nectar, seabird regurgitations and smaller lizards. They must

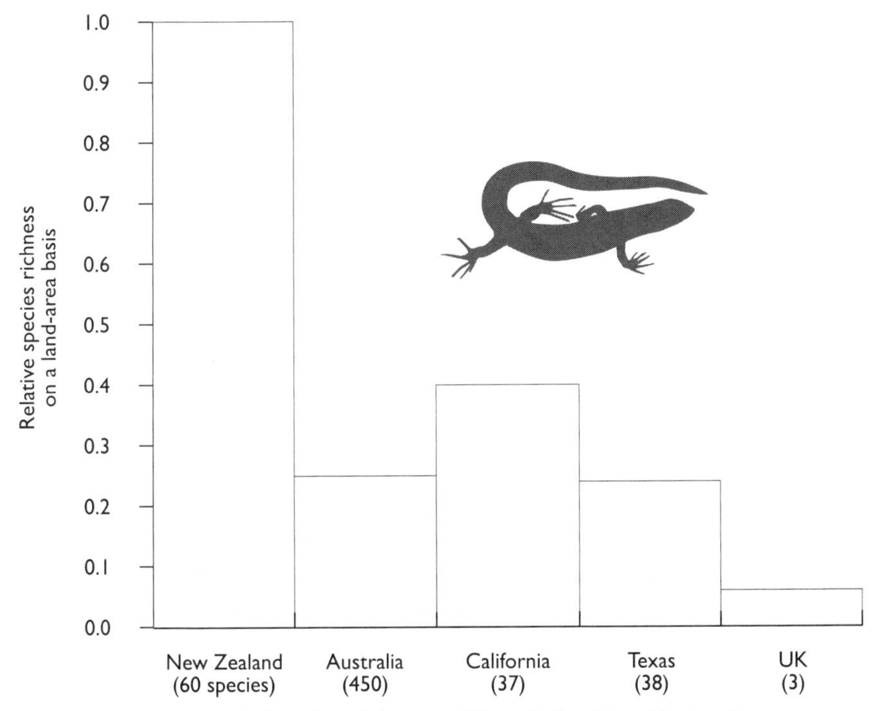

Figure 2.6 Species richness of lizards for New Zealand and other temperate regions

Adapted from Daugherty et al. 1990 and recalculated on the basis that New Zealand has 60 species

have been ubiquitous in pre-human New Zealand and their role in these ecological communities has been under-rated. Lizards pollinate and disperse the seeds of some native plants, which appear to have evolved fruits specifically to attract them (see page 199).

The most recent review of the New Zealand lizard fauna listed 59 living species,[48] but since 1994 more have been discovered. In the 1950s, only 28 species were recognised and, until 1996, field guides to the lizards included little more than half of the species now recognised, so it was virtually impossible for the non-specialist to identify them. Even the most recent guide[49] does not include all the then-known species. It is likely that further species await discovery.

In the past, New Zealand lizards, which belong to two cosmopolitan groups – the skinks and geckos – have been overlooked in favour of the tuatara and native frogs. As the taxonomy and ecology of these animals has been unravelled, however, a diverse and unique assemblage of ecologically fascinating animals has been discovered.

At least three species of lizards are believed to be extinct, 24 are rare and many are now island refugees.[50] The distribution of many lizard species is now patchy, and comparatively young fossils have been discovered far outside the species' present range. These indicate that the North and South Island lizard faunas were once much more diverse and lizards more abundant than today.

New Zealand lizards: island waifs or Gondwana relics? Are they older than the hills?

It used to be thought that the lizards were relatively recent arrivals – that the geckos had arrived by over-ocean dispersal in the Miocene, and the skinks in the Pliocene – but recent research[51] has elevated their status to longer-term residents. The New Zealand geckos belong to a particular subfamily, the Diplodactylinae, a primitive group found only in Australia, New Caledonia and New Zealand.[52] Geckos from other, more modern groups, and some skinks, have reached even the most remote Pacific islands aboard flotsam. However, the diplodactylids are less able to survive long ocean voyages than some modern groups. Molecular differences between the New Zealand and Australian diplodactylids suggests that the geckos, like the tuatara, are Gondwanan survivors.[53] The New Zealand geckos are more closely related to the New Caledonian species than to the Australian.

The skinks are more recent arrivals than the geckos, but they still must have arrived long before the Pliocene because there are much greater genetic differences between the species than would be likely had they arrived so recently. Earlier studies suggested there had been a series of colonisations from both Australia and New Caledonia, and described northern, central and southern groupings of skinks within New Zealand.[54] More recent studies of the distribution and genetic variance of skinks are incompatible with these traditional ideas.

Until 1996, most New Zealand skinks were placed in the genus *Leiolopisma*, along with superficially similar lizards from Australia and New Caledonia. They are now considered to be sufficiently different that the New Zealand species are placed in the endemic genus *Oligosoma*.[55] The other skink genus, *Cyclodina*, occurs only in New Zealand, Lord Howe Island and Norfolk Island. It had been thought that the skinks colonised New Zealand before the hypothetical land bridge with New Caledonia broke up, perhaps during the late Eocene. More recent studies suggest that the skinks may be even older, possibly also having Gondwanan origins.[56] Despite similarities of fauna and flora, there is not strong evidence for a land bridge between New Zealand and New Caledonia. However, up until 40 million years ago islands lying between the two land masses may have facilitated the interchange of animals.[57]

Early in the Oligocene, sea levels rose, reducing the then-low-lying land to a series of small islands totalling less than a fifth of the present land area. The evolutionary radiation of lizards, and perhaps other animals, may have begun as the land area increased late in the Oligocene, before mountain-building tectonic and volcanic activity began in the Miocene.[58] While molecular differences between New Zealand, New Caledonian and Australian geckos indicate Gondwanan origins, the divergence between the New Zealand species suggests that speciation within this country followed the Oligocene drowning.[59] If this scenario is correct, New Zealand's gecko lineage is far older than the hills they inhabit, but the species only a little older than those hills.

What is a species?

Taxonomy used to be much simpler in the days when species were defined by simple characteristics such as appearance, choice of habitat, and behaviour. Now the picture has become complicated, with biochemical tools revealing that some populations that seem to be identical are sufficiently different genetically to comprise separate species. These are called cryptic species, and many of the newly recognised New Zealand lizard species are of this kind. Cryptic species often turn out, on closer examination, to have slight differences in form and appearance, but these are so subtle, and often there is enough overlap in the range of variation between a pair of species, that such differences are unreliable for distinguishing them.

Gunther's tuatara was recognised as distinct from the other tuatara, and the Stephens and Maud Island populations of Hamilton's frog were found to be separate species, after biochemical analyses showed genetic divergence between populations previously thought to be single species.[60] The tuatara, frogs and nine new species of gecko previously lumped together as *Hoplodactylus maculatus* and three species previously called *H. pacificus*[61] are all geographically separated forms, evolved in each case from a common ancestor. These are equivalent to the geographically isolated North, South and Chatham Island robins. Other cryptic species co-exist but have ecological differences that maintain their reproductive isolation.

Other cryptic species in New Zealand include mudfish, parakeets and brown kiwi.[62] Poor taxonomy can compromise good conservation, and in all of these cases the changed taxonomic status of the cryptic species has had important implications for conservation.

While the small grey geckos comprise a number of cryptic species, the green geckos show quite the reverse: they exhibit differences in coloration and pattern that are not reflected in their genes.[63] For instance, Banks Peninsula populations of jewelled geckos have brown males and green females, while on Otago Peninsula both the sexes are green; yet these populations are genetically indistinguishable. As the results of new molecular studies become available, the story gets ever more intriguing. It now appears that the evolutionary radiation of green geckos occurred within the last five million years. Despite their differences from the grey geckos, the green geckos are more closely related to some species of grey geckos than those grey geckos are to some others that look just like them. This means that, in order to accurately represent the genetic variation within the present genus *Hoplodactylus*, two new genera need to be erected.[64]

But what is the real significance of cryptic species if only the biochemists can tell them apart? The generally accepted test of whether two populations belong to the same species is whether they can interbreed and produce fertile offspring. However, this is only one of several species concepts, and in ecology or conservation it is not always the most useful. For example, in the much-altered habitats of today, pied stilts sometimes mate with black stilts and produce fertile offspring, and the same is true of red-crowned and Forbes' parakeets; but all these species prefer mates of their own kind when these are available.[65]

Probably the most useful species concept with regard to the issues considered in this book is the phylogenetic species concept, and this was the concept adopted when compiling Appendix 1. A biological species is a group of interbreeding individuals in a natural population that is reproductively isolated from other such groups. Small differences in genetic make-up can result in huge differences in ecology, behaviour and appearance; for example, humans and chimpanzees share more than 98 per cent of their genes, yet look very different, do not interbreed and are placed in separate genera.

A New Zealand example is illustrated by the skinks that used to be known as the subspecies *Oligosoma nigriplantare maccanni*. These lizards have now been found to exhibit such great genetic variation that they have been divided into four cryptic species, up to three of which may co-exist at the same location.[66] The genetic differences between these cryptic species are as great as the differences between any one of them and the spotted skink or green skink, both of which are larger and look quite unlike the cryptic species[67] (Table 2.2) On the other hand, the geographically isolated and differently marked spotted and green skinks appear, on the evidence so far, to be genetically identical.

Ecological studies of the three cryptic species shown in the table below provide further evidence that they are indeed separate species. In Central Otago, all three co-exist and, despite overlapping in the habitats they use and foods they eat, there are ecological differences between them.[68] The common skink, the most numerous species, is most often found in grasslands, while the other two species strongly prefer herb or

Table 2.2 The genetic distance between three coexisting cryptic species of skinks formerly all included within the taxon *Oligosoma nigriplantare maccanni* and the much larger and differently marked spotted and green skinks

	Common skink	Cryptic skink	McCann's skink	Spotted skink	Green skink
Common skink	–	.82	.73	.51	.51
Cryptic skink	.44	–	.17	.89	.89
McCann's skink	.48	.84	–	.89	.89
Spotted skink	.60	.41	.41	–	0
Green skink	.60	.41	.41	1.0	–

Adapted from Towns et al. 1985

scrub habitats, although both also occur on rocky substrates. The cryptic skink prefers rocks lying on top of the soil, and McCann's skink prefers rock outcrops. Cryptic skinks are more often found in gullies, and McCann's in more open habitats. The common skink is displaced from the habitats preferred by the other species. The three species also differ in the foods they eat. This reflects differences in the invertebrate prey available in the habitats each species occupies, as well as their choice from the foods available.

Alastair Freeman[69] studied two cryptic species of skink co-existing on apparently homogenous terrain at Kaitorete Spit in Canterbury. These species show clear differences in the habitats they use. The common skink is found only in scrubby habitats on the older, more stable sand dunes, while McCann's skink predominantly lives on foredunes covered in marram grass and pingao (a grass-like sedge).

Field identification of these co-existing skinks poses problems. In Central Otago, common skinks live in grassy habitats and are strongly striped in pattern. There the stripes on McCann's skinks are broken and often indistinct, which is an appropriate pattern for camouflage in the scrub and rocks they inhabit. On Kaitorete Spit, common skinks live mostly in scrub and were speckled, while McCann's skinks occur most often in grasslands and are striped. The choice of vegetation possibly reflects a preference of McCann's for drier habitats. In Central Otago, the dry habitats are on rock outcrops; on Kaitorete Spit they are on the foredunes. The pattern is variable but provides camouflage to match whatever vegetation the lizard inhabits.

The kawekaweau

Maori legends refer to a giant forest lizard called the kawekaweau, but scientific evidence for the existence of this species is scanty and it remains one of the world's most mysterious animals. Only one specimen is known: an unlabelled stuffed animal in the Musée d'Histoire Naturelle in Marseille, France. It is 620 millimetres long, slightly larger than a fully grown male tuatara and larger than any other gecko, living or extinct. The animal belongs to the genus *Hoplodactylus* and it has been assumed that the specimen was collected here, as all known species in this genus are endemic to New Zealand.[70] However, where it was obtained and how it ended up in a French museum is not known. Two bones collected in South Canterbury last century, said to be from a large lizard and once used as evidence for the existence of kawekaweau, are no longer attributed to this species. South Island bone deposits have now been so well studied that if the species existed and was present in that island, its bones ought to have been found, like those of tuatara and other now-extinct lizards.[71] In 1870 an Urewera chief was reported to have killed a large lizard similar in size, colour and pattern to the Marseille specimen.

The ecology of lizards

Of all the animal groups covered in this book it is the lizards whose ecology is least well known. Tony Whitaker[72] divided the lizards into four broad ecological groups:

- Diurnal tree-dwelling green geckos in the genus *Naultinus*: seven species found throughout New Zealand.
- Nocturnal ground- or tree-dwelling geckos in the genus *Hoplodactylus*: up to 25 species in virtually all habitats throughout the country and on the Three Kings Islands.
- Diurnal skinks in the genus *Oligosoma* that live mostly in open habitats: 21 species throughout New Zealand and on the Chatham and Three Kings Islands.
- Nocturnal forest skinks in the genus *Cyclodina*: six species, mostly in damp, thickly vegetated lowland North Island habitats and on the Three Kings Islands.

At least one species of small diurnal skink and a small nocturnal gecko can be found throughout New Zealand in most habitats from the coast to subalpine grasslands. Throughout the country in primeval times these lizards would have been part of a diverse reptile community. Some of the widespread lizards have quite general habitat and dietary requirements. Other species are highly specialised as to habitat or restricted in distribution; for example, Fiordland skinks are restricted to foreshore rocks and boulder beaches on the exposed coast, but absent from the sheltered coast inside the fiords where forest grows almost down to the water's edge.[73] The egg-laying skink is found near rock pools and in the intertidal zone of the northern North Island.

In contrast to these two maritime species, some others are restricted to montane habitats. The scree skink is found only on unstable screes in mountains of the eastern South Island, while the black-eyed gecko lives only on alpine bluffs at 1100–2200 metres above sea level.[74] The Harlequin gecko, arguably the most beautiful of all New Zealand reptiles, is restricted to windswept scrub in southern parts of Stewart Island.

Lizards are easily overlooked animals but can be surprisingly common. In suitable microhabitats there may be one per square metre, which equates to a biomass of 50 kilograms per hectare. On the larger scale, 1000–2000 lizards per hectare is not uncommon, although in places they are much scarcer.[75] At Pukerua Bay, near Wellington, there are about 4900 common skinks per hectare, plus lesser numbers of several other species.[76] Today, lizards are generally more abundant on islands than on the mainland.

Most New Zealand lizards have very small home ranges, although some, in particular Otago and grand skinks, may range over several hundred metres. At Turakirae Head, near Wellington, 70 per cent of marked common geckos were recaptured in the same trap in which they were first caught. Only seven per cent of them had travelled five metres or more.[77] Decades after it was originally marked, a Duvaucel's gecko was recaptured on North Brothers Island was recaptured, possibly on the same rock it had inhabited years before.[78]

The diet of lizards

Most New Zealand lizards are insectivorous but also many eat other foods. Some eat fruit or nectar, and their role in pollination and seed dispersal is discussed in Chapter 7. On seabird islands, lizards have been seen eating partly digested food spilt by birds when feeding their chicks.[79] The large scree skink includes smaller lizards in its omnivorous diet. Young tuatara and larger lizards may overlap in the foods they take.

Co-existing lizard species apparently eat many of the same foods. For instance, in Auckland City the diets of the ornate skink and the smaller copper skink were essentially similar except that only the larger lizards ate small land snails.[80] In the Manawatu sand country, brown and common skinks ate many of the same species of invertebrates. However, brown skinks preferentially selected arachnids (spiders and harvestmen), while common skinks were more opportunistic and fed more on insects.[81] In Central Otago, differences in the foods taken by cryptic skinks mostly reflected variation between the insects available in the different microhabitats the three species occupied. Diversity in foraging strategies and competition between the species probably also influenced their diets.[82]

A little-known alpine gecko

Black-eyed geckos are known only from a few localities in the Kaikoura mountains and the Arthur Range, near Nelson. They are found only on bluffs and rocky outcrops 1300–2200 metres above sea level, where it can be extremely cold, heavy snowfalls are frequent and snow lies for several months each year. Even in summer, temperatures can fall to 5°C. Geoff Harrow discovered the species in 1970 while studying Hutton's shearwaters on Mt Tarahaka. Despite subsequent searches they were not seen again until 1983, when Tony Whitaker found several at the Kahutara Saddle. All of the 70 or so black-eyed geckos found have been on rock bluffs or outcrops, some distance from the tussock and herbfield vegetation of these altitudes. The geckos may inhabit these bluffs because the rock heats up during the day and retains warmth into the night.

The black-eyed is the largest nocturnal gecko remaining on the mainland, and it is likely that its extreme habitat enabled it to survive while predators eliminated similar-sized nocturnal species from more equitable climes. In the Kaikoura mountains, five other species of lizards occur above 800 metres, but the black-eyed gecko lives up to 400 metres higher than these.[83] It remains one of New Zealand's least-known lizards, and Whitaker's 1984 paper contains virtually all that we know about this enigmatic species.

Breeding biology of New Zealand lizards

New Zealand lizards, apart from one native skink, are unusual in being viviparous – they give birth to live young. Elsewhere in the world some cool-climate skinks are viviparous, but only one other species of gecko is. Most of New Zealand's lizards are ovoviviparous – they retain the fully formed eggs in the oviduct, where they hatch before the young are born. In common skinks, a 'placenta' forms and the developing embryo receives nourishment from its mother.[84] Vivipary presumably evolved as an adaptation to the cooler Pleistocene climates, as most modern viviparous lizards live in cool climates.

On average, New Zealand lizards lay fewer eggs than do most overseas skinks and geckos.[85] The geckos normally have two young each time they breed. Some populations of common and Duvaucel's geckos do not breed every year, but it is assumed that other species reproduce annually. The Macraes Flat (Central Otago) form of the common gecko breeds at two-yearly intervals, and one litter in four produces only a single offspring, so these lizards produce on average fewer than one young per year. This is an unusually low rate of reproduction, lower even than for the tuatara. The common gecko is a complex of several cryptic species, and the Macraes species is larger and less fecund than the common geckos studied in warmer parts of the country, which reproduce annually. Most New Zealand geckos mate in spring or summer and give birth in autumn or winter. Gestation appears to last at least seven months in Duvaucel's gecko, and the Macraes Flat common geckos give birth about 14 months after mating.

Skinks usually have between two and five young (though some species can have larger litters) and generally breed each year, but some *Cyclodina* skinks may not breed annually.[86] Skinks mate in spring and give birth in summer or autumn.

All New Zealand geckos are limited to a single litter each year, whereas most overseas species produce two or more clutches. Vivipary may be the factor that constrains New Zealand geckos to a single litter each year. Egg-laying species overseas are done with their eggs quickly, leaving time to produce a second clutch. The reproductive outputs of New Zealand skinks are at the low end of the range for skinks but are similar to other small-bodied viviparous skinks. Possibly the lizard lineages that colonised New Zealand already had small clutches and their reproductive rates became further constrained by the viviparous mode of reproduction and the cool climate, which slows development, thus permitting only one 'pregnancy' each year.[87]

The egg-laying skink lives only on stony beaches and rocky shores in northern New Zealand, where it nests in small chambers under stones or buried in damp substrates just above high-tide mark. Usually a nest contains about five eggs, but because each female probably lays only three eggs each year, larger clutches must be the result of communal nesting.[88]

New Zealand lizards appear to live unusually long lives. Duvaucel's geckos first breed when about seven years old, and one wild male is known to have lived for at least 36 years, and a female at least 19 years.[89] The smaller common gecko has a similar lifespan. Of 133 common geckos marked on Motunau Island, Canterbury, between 1967 and 1973, 16 were still alive during the summer of 1996–97. This suggests that lifespans of more than two decades are not uncommon.[90] Judging from their size at first capture, 10 of these lizards were probably at least 36 years old. Skinks tend to have shorter lives than geckos; for example, the large Otago and grand skinks live about 10 years.[91]

Reptile communities on New Zealand's northern islands

In the north of New Zealand, up to 12 species of lizards plus tuatara co-exist on some rat-free islands. Fewer species remain on islands where kiore occur, and the range of habitats occupied by most remaining lizards is also reduced.[92] On islands with kiore, the egg-laying skink, moko skink and shore skink are confined to boulder beaches and coastal cliffs, where kiore are least numerous. On kiore-free islands, these three skinks show less overlap in the habitats they use. Egg-laying skinks are the most coastal species and are restricted to the islands' rocky shores and adjacent coastal zone, largely feeding at night under rotting seaweed. On the Poor Knights Islands, they have been observed swimming in tide pools, presumably hunting invertebrates. Of these three small skinks, the moko occurs furthest from the seashore, mostly utilising the forest fringe as its habitat. Three nocturnal geckos, plus the country's smallest lizard, the copper skink, frequent most available habitats in the absence of kiore, but on islands where kiore are present these lizards are common only in coastal habitats.

Three large nocturnal *Cyclodina* skinks occur on rat-free islands but are absent on all small islands where kiore occur, although a few individuals remain on Little Barrier, Great Barrier and Mana Islands and at Pukerua Bay – all localities where kiore or other rodents are present. The marbled skink is the least specialised of the three and occurs in most habitats. Whitaker's skink is the most specialised, requiring warm, moist habitats such as seabird burrows or deep boulder banks,[93] and it forages only on warm, humid nights. The robust skink is most common in densely forested areas, where it hunts in deep leaf litter.

The interactions between the lizard species present are not clear. Where there are no nocturnal skinks, the nocturnal geckos forage both on the ground and in vegetation.[94] Where nocturnal skinks are present, the skinks forage only on the ground, and the geckos in vegetation and up trees.

The diversity of lizards on these offshore islands may seem remarkable, but not so when you consider how diverse they once were on the adjacent North Island mainland. Before the arrival of predatory mammals, the coastal broadleaf forests of Northland probably supported four species of native frog, one tuatara species, six geckos and 11 skinks.[95] Of these 22, only eight lizard species and two frogs survive on the mainland today.

Kiore and lizards on the Mercury Islands

It has long been suggested that lizard abundance and diversity are reduced on islands with kiore,[96] yet until recently the evidence was circumstantial and not everyone was convinced that kiore predation alone caused local extinctions. David Towns[97] investigated the relationship between lizards and kiore on the Mercury Islands, and found the highest diversity and numbers of lizards on two smaller mammal-free islands. Ten species of lizards were present on Middle Island (10 hectares) and at least seven species on tiny Green Island (3 ha). However, just five species of lizards were present on Korapuki (18 ha) and Stanley (100 ha) Islands, which were infested by kiore and rabbits. The species that were absent from kiore-infested islands were large and/or nocturnal, whereas most of the species that co-existed with kiore were small or active by day.

Kiore were exterminated from Korapuki in 1986 and, in response, most lizard species both increased in abundance and expanded their range to occupy more habitats. Even shore skinks, which had been relatively common before rats were removed, increased hugely in abundance in the nine years that followed.[98]

The presence of kiore was not the only factor, as there were rabbits on Korapuki and Stanley Islands and some degree of habitat change had occurred on all the islands. However, after rabbits were exterminated on both islands, the lizard numbers only increased on Korapuki, from which kiore had also been removed. The same five lizard species survived on both rat-infested islands, despite marked differences in the size of the islands and the habitats available. This suggests it was the ability to co-exist with rats (rather than rabbits) that determined which lizard species were present. After kiore had been exterminated, four of the absent species were successfully reintroduced to Korapuki. Subsequently, kiore have been exterminated from Stanley Island.

Lizards of South Island tussocklands

Lizards are generally animals of warm climates, and by and large the New Zealand species conform to this rule, with more species in the north than in the south. However, the South Island's montane and subalpine tussocklands support a surprisingly diverse lizard fauna. At least five species are restricted to these regions. The alpine black-eyed gecko (see page 39) occurs only in truly alpine habitats, and four large, colourful skinks live only in montane and subalpine rocky habitats, screes and tussocklands.[99]

These regions experience a wide climatic range: snow may fall even in summer, and in winter the ground may be snow-covered for weeks. Summer temperatures can exceed 30°C by day, then drop below freezing at night. In winter, skinks can be seen basking on wet rocks while snow lies around them.[100] How these comparatively small reptiles can cope with such extreme conditions is not known. Geoff Patterson noted that all of the skinks specialised for high-country habitats lived on rocky substrates, and he suggested that the rocks may be important to lizards either by retaining heat or by providing crevices that offer protection from predators.

The grand skink and Otago skink occur only in Otago. The scree skink is more widespread, also occurring in the semi-arid tussocklands of inland Marlborough and Canterbury. Scree skinks live on open rocky areas such as screes, dry streambeds and rock outcrops. All these skinks are now less common and their distributions more restricted than before.

In some Otago schistlands, there may be up to eight species of lizards: the local form of the nocturnal common gecko, the jewelled gecko, and six species of diurnal skinks.[101] The three large skinks may be found at the same location but not in the same habitats: the Otago skink prefers bluffs; the grand skink prefers rock outcrops among tussocks; and the slightly smaller green skink lives in those tussocks. Three smaller, superficially similar skinks may also be present. These were once thought to belong to a single species and differences in their habitats and foods are described earlier (page 35).

Conservation of frogs and reptiles

Three species of frogs and three lizards are known to have become extinct since kiore and people arrived in New Zealand. All four living frogs, both tuatara and 40 per cent of the lizard species are now rare, threatened or endangered.[102] Two frogs, both tuatara* and eight species of lizards are now island refugees (Figure 2.7, page 44), while on the mainland the ranges of the frogs and most of the surviving lizards have been greatly reduced. It is impossible to determine the full impact predators have had on lizards and frogs, because the tiny bones of these prey are seldom preserved and only a few of them can be reliably identified.

Duvaucel's gecko illustrates the impact predators and habitat change

* I have assumed that Gunther's tuatara is not naturally confined to North Brothers Island and once had a wider distribution that included parts of the South Island mainland.

	Frogs	Tuatara	Geckos	Skinks
Island endemics*	–	–	4	1
Now restricted to offshore islands	2	2	2	6
Remnant mainland distribution	2	–	4	7

*Naturally restricted to Stewart Island or other land-bridge islands

Figure 2.7 The conservation status of New Zealand's frogs and reptiles
Adapted from Towns & Daugherty 1994

have had on reptile distribution. This gecko is now restricted to islands along the northeastern coast of the North Island and in Cook Strait. Such patchy distribution suggests that the species once occurred over much of the North Island and in northern parts of the South Island. Fortunately, this is one species whose bones can be identified, and they have been found as far south as Otago, suggesting that the lizards lived across most of the country until kiore arrived.[103] The decline in range of Hochstetter's frog may have occurred before humans arrived,[104] but for all other species, and maybe also for Hochstetter's frog, the declines have been caused by people and introduced predators.

The main threat to frogs and reptiles has been predation. Kiore had almost 2000 years to take their toll before the larger, more ferocious predators arrived. European rats, mustelids and cats must have triggered a wave of local extinctions, but their impact on reptiles and frogs is poorly documented. Two examples indicate how vulnerable reptiles were to these recent arrivals. Tuatara disappeared from Whenuakura Island within two years after Norway rats arrived; and the Fiordland skink, found only on rat-free islands, recolonised Breaksea Island soon after Norway rats were exterminated.[105]

The most vulnerable species were large and nocturnal.[106] All the extinct frogs were larger than the species that survived. The only large lizards that survived on the mainland were those that are active by day; similar-sized nocturnal species are now confined to rodent-free islands. All the large, diurnally active skinks that remain on the mainland occur in fragmented populations and occupy only part of their original range. Some, such as the spotted skink, still have widespread (though patchy) distributions, but others, such as the Otago and grand skinks, are now endangered.

The decline of some mainland lizards continues. In the 1960s, spotted skinks were common on Kaitorete Spit, but by the 1990s they were rare. Even the common gecko is affected by the presence of predators. Alastair Freeman found fewer and smaller geckos on Kaitorete Spit than on nearby Banks Peninsula or predator-free Motunau Island.[107] He suggested that on the peninsula geckos could occupy crevices too small for introduced predators, but on Kaitorete there were few, if any, secure hideaways. Skinks and geckos were collected from the spit for the pet trade, research and education, and Freeman suggested that this also contributed to their decline.

Habitat change presumably contributes to the decline of some reptiles and frogs, but by itself does not appear to have caused the extinction of any species. Hochstetter's frog is threatened by localised habitat changes. Mining and forest clearance have caused streams to silt up, and introduced mammals change the frogs' favoured microhabitats by trampling and killing vegetation alongside steams.[108] People searching for frogs by rolling over the rocks they live under may inadvertently squash animals or, unless they replace the rocks carefully, change the spaces beneath the rock and make them unsuitable for the frogs.

The Otago and grand skinks are two of New Zealand's most threatened lizards. They are restricted to montane tussocklands so they have no safe island retreats. They are now restricted to only a tenth of their original range, and even there they do not occur in all suitable habitats, though the reasons for this are not clear.[109] Predators are certainly one threat, especially cats in winter when they are hungry and the lizards are slow-moving. Fire is another threat. Tony Whitaker found that on 'improved' pasture the population density of grand skinks was half that of adjacent less modified tussockland. The reason for this was not clear, but the loss of shrubby vegetation that provided cover and food was probably a factor.[110] Whitaker suggested that each small group of lizards living on isolated rock outcrops was at risk of local extinction, but an outcrop would subsequently be repopulated by animals moving in from rocks nearby. It appears that grand skinks will readily move through tall tussock but not across closely grazed pasture. Whatever the reason, skinks do not survive long after the surrounding native vegetation has been removed.

In comparison with the heroic efforts made to save some bird species, until recently few attempts have been made to preserve rare or endangered frogs and reptiles. The tuatara is one of New Zealand's most distinctive animals, yet it was only in 1990 that the second species was recognised and found to be critically endangered. Frogs and lizards have received even less attention. However, since the early 1990s there has been increasing management effort to benefit threatened reptiles, including the eradication of rodents from some islands, recovery plans for various reptiles and management of some of their key mainland sites.[111] Conservation management of frogs and reptiles is discussed in Chapter 9.

CHAPTER 3

Ecology of birds and bats

> The huia never leaves the shade of the forest. It moves along the ground, or from tree to tree, with surprising celerity by a series of bounds or jumps. In its flight it never rises, like other birds, above the tree-tops, except in the depth of the woods when it happens to fly from one high tree to another.
>
> – Sir Walter Buller (1888)

Ornithological time travellers visiting New Zealand 2000 years ago would have been amazed by the abundance and variety of albatrosses, shearwaters and petrels they saw as they approached New Zealand. Once ashore, even greater surprises would have awaited them. In the forests, not a single species would have been familiar. Some very strange flightless birds wandered there, and even the smaller birds, some of which appeared at first glance superficially similar to those of other lands, were unique. The time-travelling ecologists would have found that some birds behaved more like mammals than birds and differed in many other ways from birds found elsewhere.

Ornithologists who visit New Zealand today are still struck by the diversity of seabirds, but on land much has changed. Species from other countries have been introduced and many native birds are now extinct, but some distinctive birds with unusual ecological adaptations remain.

Chapter 1 identified the features that make the New Zealand fauna unique and described the historical factors that shaped the biota. In this chapter I will describe the composition of our bird fauna, discuss its origins and investigate some aspects of the ecology of New Zealand's birds and bats. The emphasis will be on ecological features that differ from the accepted norms derived from the intensively studied northern hemisphere species.

Bats are the only non-marine mammals native to New Zealand. The three species present are fewer than might be expected for a relatively large temperate archipelago. Two of those species – one of which is, alas, now extinct – are among the most unusual of all bats.

Composition of the New Zealand bird fauna

Compared with bird faunas in most parts of the world, the New Zealand fauna is very rich in marine and wetland birds. More than a third of New Zealand's birds are seabirds, 10 times the global figure. Birds of wetland habitats, which include estuaries, lakes, swamps and rivers, make up almost a third of New Zealand's species, but less than seven per cent of the world's bird species are found primarily in those habitats. Most of the world's birds are terrestrial, yet in New Zealand only one-third of the native species were found in such habitats. Over half of the world's bird species are passerines (songbirds), but 2000 years ago New Zealand had only 30 species of passerines, making up just 12 per cent of its native birds.

Why such a diversity of seabirds? The short answer is that it results from the diverse and productive marine environment that surrounds New Zealand, and this is explored further in Chapter 8. Most of the seabirds belong to groups that are widespread throughout the southern hemisphere. The relative diversity of wetland birds and paucity of terrestrial species presumably reflects the greater ease with which wetland

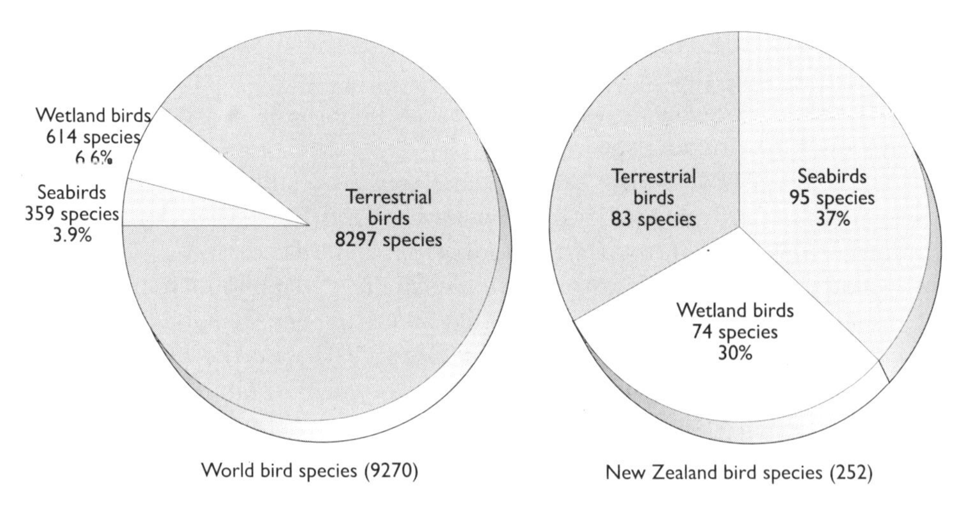

Figure 3.1 Composition of global and New Zealand bird faunas

species could colonise New Zealand. Wetland habitats are much the same everywhere, so such birds arriving here had a higher probability of finding suitable habitats than was the case with terrestrial birds. Furthermore, in the past the area of wetlands was greater. Those parts of New Zealand that are likely landfalls for wind-blown birds (particularly in the west) were dominated by closed-canopy rainforest, very different from habitats in Australia and other source countries. Thus terrestrial birds arriving in this country would have had little chance of finding a place where they could survive.

Although there were few terrestrial birds, they were the most distinctive species, reflecting New Zealand's long period of isolation. Most were endemic and some belonged to endemic families or orders. Some terrestrial birds have, or had, ecological roles that differ from their counterparts elsewhere, and these birds will be discussed in Chapter 4.

Biogeography

> The birds of which there is a tolerable stock, are almost entirely peculiar to the place.
>
> – Captain James Cook

The New Zealand bird fauna consists of a Gondwanan base on to which have been added successive waves of immigrants, mostly from Australia.[1] The Gondwanan base comprises those species whose ancestors were stranded on these islands when the greater New Zealand region (which then included New Caledonia and the now-submerged Lord Howe Rise[2]) separated from Gondwana during the Cretaceous. Just which species may claim membership to this exclusive club is open to debate. Traditionally, moa and kiwi were uncontested members, but recent evidence[3] suggests that they may not be as closely related as was previously thought and that the kiwi's ancestors may have been flighted birds that arrived in New Zealand shortly after the separation from Gondwana. Moa, on the other hand, were descended from ancestors stranded here during the Cretaceous.

The New Zealand wrens apparently also have Gondwanan origins, having diverged from the mainstream of passerine evolution during the late Cretaceous, soon after the passerine birds first evolved.[4] The ancestors of kea and kaka may also have been in New Zealand at that time.[5] Seventy-five million years ago the gap between the greater New Zealand region and Australia was relatively narrow, but this increased until the Tasman Sea attained its present width 60 million years ago. The ancestral wattlebird and short-tailed bat, and presumably the kiwi, probably colonised New Zealand during this era of sea-floor spreading.[6] The ancestral short-tailed bat may have arrived then or later, during the Oligocene.

The prevailing westerly winds that now influence the dispersal of animals to New Zealand began when Australia and South America separated from Antarctica. Since the Miocene, wind-assisted trans-Tasman dispersal has been the most important agent in New Zealand biogeography.

The ancestors of the piopio and whitehead were presumably early arrivals. The piopio are the sole members of a family closely allied to the bowerbirds of Australia and New Guinea. It is thought that they diverged from the bowerbirds about 27 million years ago, and presumably the ancestral piopio flew the Tasman at about this time.[7] The whitehead and its kin are related to the Australian whistlers but are different enough to be placed in a subfamily of their own. The degree of divergence from their Australian kin suggests that the whitehead lineage arrived in New Zealand later than the piopio lineage.

Other forest birds were presumably later arrivals, as they are more closely related to Australian birds. The tui and bellbird are members of the Australian honeyeater family, but each is sufficiently distinctive to be placed in its own genus. The hihi is currently also placed in the honeyeater family, but it is so different from other members that it may not truly belong in this group. If it is not a honeyeater, its real origins are unknown. New Zealand's robins and tomtits belong to the genus *Petroica*, species of which also occur in Australia, New Guinea and Melanesia. It has been suggested that the robin's ancestor arrived during the Pliocene, and the tomtit's ancestor during the Pleistocene. The fantail is so similar to its Australian counterparts that fantails from both countries belong to the same species, indicating an even more recent arrival in New Zealand.

The enigmatic aptornis, an extinct giant rail-like bird, may be distantly related to the equally enigmatic kagu of New Caledonia. About 40 million years ago there was an island chain between New Zealand and New Caledonia, and during the early Tertiary there may even have been a land bridge between them. The lizards of New Zealand are most similar to those of New Caledonia (see Chapter 2), and some invertebrates share ancestory with those of New Caledonia,[8] but aptornis and the New Zealand parakeets (see later) are the only birds with New Caledonian connections.

Trans-Tasman dispersal continues today but at an accelerated rate. Since European settlement, 10 terrestrial or freshwater bird species have successfully colonised New Zealand by trans-Tasman dispersal, and six other species are probably recent but pre-European colonists.[9] This colonisation rate is orders of magnitude greater than that in previous millennia and has presumably been made possible by humans in creating open habitats that more closely resemble those of Australia.

The crested grebe, coot, pukeko and marsh crake are species whose range extends from Europe across Asia and Australia to New Zealand, with varying numbers of subspecies spanning the range. The South Island pied oystercatcher, New Zealand scaup and Auckland Island merganser are closely related to northern hemisphere birds, and it has been suggested that they may have colonised New Zealand directly from there – not by way of Australia.[10] In my view, colonisation via Australia seems a more likely scenario for the oystercatcher and scaup. There have been two useful introductions to New Zealand biogeography,[11] although much of the detail in these has been superseded by later research.

Colonisation of outlying islands

Several birds that presumably evolved on mainland New Zealand have colonised outlying islands, and three or four species that presumably evolved here have spread further afield. Pycroft's petrel, now restricted to New Zealand, formerly bred on Norfolk and Lord Howe Islands. Norfolk Island once had a pigeon and a parrot closely related to the New Zealand kereru and kaka.[12]

Parakeets with red crowns occurred on New Caledonia, Norfolk, Lord Howe and Macquarie Islands, as well as the Kermadec, Chatham and most New Zealand subantarctic islands. It used to be assumed that these birds colonised the islands from the New Zealand mainland,[13] but recent molecular studies have changed our understanding of their biogeography. It is now considered that the birds on Norfolk Island, New Caledonia, Antipodes and Macquarie Islands are sufficiently different from the New Zealand birds that they should be treated as separate species.[14] It appears that the New Caledonian species is the most basal of the various forms (Figure 3.2, page 52), indicating that the red-crowned parakeet complex originated in New Caledonia and dispersed to New Zealand via Norfolk Island, perhaps 500,000 years ago. The Antipodes and Chatham Islands were apparently colonised twice by *Cyanoramphus* parakeets. On the Antipodes, there is the larger, all-green Antipodes Island parakeet, plus a red-crowned bird that is probably most

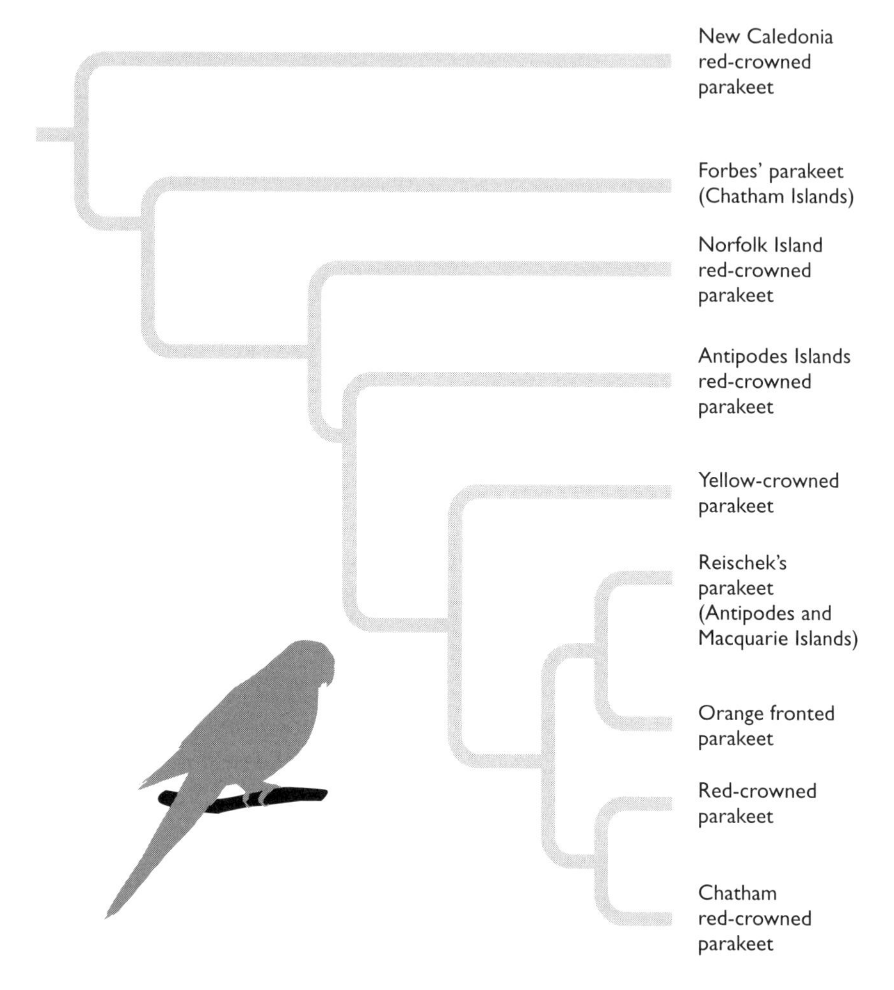

Figure 3.2 Phylogenetic tree for the *Cyanoramphus* parakeets

Adapted from Boon et al. 2001a

closely allied to the Macquarie parakeet. On the Chatham Islands, Forbes' parakeet has head markings similar to those of mainland yellow-crowned parakeets, and for years it was treated as a subspecies of that form.[15] However, molecular studies show that the species is distinct from both red- and yellow-crowned parakeets and apparently colonised the Chathams soon after the ancestral parakeet reached New Zealand, and long before red-crowned parakeets reached those islands.[16] The only other place *Cyanoramphus* parakeets were found was French Polynesia. The Lord Howe, Macquarie Island and French Polynesian parakeets, plus the Norfolk Island pigeon and kaka, are now extinct and the Norfolk Island parakeet is endangered.

Various species including the kereru, pipit, tomtit, tui, bellbird and teal colonised some outlying island groups (see Appendix 1). The Chatham forms of the kereru, robin, fernbird, bellbird, warbler and some extinct birds differ sufficiently from their mainland counterparts to be treated as distinct species, whereas most populations on the subantarctic and Kermadec Islands show, at best, differences that warrant their recognition as subspecies. This suggests that the Chatham populations have been isolated for longer than have those on other island groups. The explanation for this is that the Chatham Islands were not ice-covered during the Pleistocene, but the subantarctic islands were.[17] Molecular studies of these species may change our understanding of avian biogeography within the greater New Zealand region. For example, the New Zealand pipits were treated as several subspecies of a species that also occurs in Africa, Eurasia and Australia. However, molecular studies suggest that New Zealand mainland pipits are better treated as a distinct species, with at least subspecific recognition of the Chatham, Auckland, Antipodes and Campbell Island populations.[18]

Endemism

There are 133 species of birds that breed only in the New Zealand region. Eighty-seven per cent of the terrestrial birds and 44 per cent of all breeding seabirds are endemic. The long-tailed cuckoo and some seabirds only breed in New Zealand (and thus are considered endemic) but migrate elsewhere after the breeding season. Other petrels and albatrosses breed here but, even during the breeding season, may forage far beyond the 200-mile Exclusive Economic Zone.

Many species are endemic at higher taxonomic levels. There are two orders, one infra-order, three families, three subfamilies and 38 genera of birds that occur only in New Zealand (Table 3.1, page 54). In addition, species native to New Zealand that also occur elsewhere are represented by 36 subspecies endemic to this country. This number would be even greater if Macquarie and Norfolk Islands, which in a biogeographical sense are more properly part of New Zealand, were included. Of the world's 218 endemic bird areas (zones with lots of endemic birds identified by Birdlife International), four – North Island, South Island, Chatham Islands and the subantarctic islands – are in New Zealand.[19]

Table 3.1 Higher vertebrate taxa endemic to New Zealand

Amphibians
Family LEIOPELMATIDAE, *Leiopelma* (native frogs), 4 living, 3 extinct species

Reptiles
Order SPHENODONTIDA, *Sphenodon* (tuatara), 2 species

Birds
Order DINORNITHIFORMES, moas, 11 species
Order APTERYGIFORMES, *Apteryx* (kiwi), 5 species
Order GRUIFORMES (rails), about 7 families, 1 endemic
  Family APTORNITHIDAE, *Aptornis* (adzebills), 2 extinct species
Order PSITTACIFORMES, parrots, 2 families, 2 endemic subfamilies
  Subfamily STRIGOPINAE, *Strigops* (kakapo)
  Subfamily NESTORINAE, *Nestor* (kea and kaka)
Order PASSERIFORMES, about 80 families, 1 infra-order, 2 families, 1 subfamily endemic
  Infra-order ACANTHISITTIDAE, New Zealand wrens, rifleman, rock wren and 4 extinct species.
  Family CALLAEATIDAE, huia (extinct), kokako, saddleback
  Family, TURNAGRIDAE, *Turnagra* (piopio), extinct
  Subfamily MOHOUINAE, whitehead, yellowhead, brown creeper

Mammals
Order CHIROPTERA, bats, many families, 1 endemic
  Family MYSTACINIDAE, short-tailed bats, 2 species

Not only birds are endemic, but so are all the native frogs, reptiles and bats, with endemic families or orders in each of these groups. More than 90 per cent of New Zealand's native insects and 81 per cent of the plants are endemic.

Recent taxonomic reviews

Taxonomy – the assigning of scientific names to species – and systematics – the study of phylogenetic (evolutionary) relationships among species – are considered by many people to be dry and unfashionable pursuits. Yet in this book I stress the importance of names and relationships. Why is it so important, what light does it shed on the ecology of New Zealand birds, and how may it influence conservation priorities?

In studying ecology and evolution, we want to know if animals with similar adaptations to similar ways of life have these because they shared a common ancestor or if these adaptations have evolved independently in response to a similar environment. For instance, is New Zealand's brown creeper related to the brown creeper of North America and the various creepers of Central and South America? Or are they unrelated birds that simply evolved to look similar because their needs were the same? All these birds forage by creeping over trunks or through foliage, picking insects from the tree surface or probing crevices in the bark. In fact, their similarities are superficial: they belong to quite different families that have independently evolved similar methods of exploiting tree-dwelling insects.

Understanding phylogeny helps set conservation priorities. It is important to conserve tuatara and kiwi because they are very distantly related to any other living animals. If they become extinct, nothing like them will remain. Conservation of crested grebes is given lower priority. While New Zealand has fewer crested grebes than kiwi or tuatara, the same subspecies occurs in Australia and there are other subspecies on other continents. To determine conservation priorities, we need to know what it is we are conserving and how distinctive that species is. Systematics provides us with much of this information.

Some New Zealand birds, such as kiwi, moa and kakapo, have long been recognised as very different to those in other parts of the world. Others, notably the passerines, are superficially similar to species elsewhere and were merged into family groups that obscured their true relationships. They were known to be endemic species but were thought to belong to widespread orders or families. We now recognise that many more species belong to groups unique to this country. This distinctiveness is an important guide in assigning conservation priorities.

Taxonomy was formalised in Europe in the eighteenth century, and when birds from the colonies began arriving in museums there, many animals that looked similar were 'shoehorned' into the same families as their European counterparts. Because northern hemisphere fossils were more abundant and better studied, it was assumed that birds first evolved then radiated in that part of the world some time after the Cretaceous–Tertiary mass-extinction event. Recent studies have changed this view. Molecular research suggests that birds evolved much earlier and that their initial radiation occurred in Gondwana.[20] This evidence indicates that ratites, penguins, galliformes, waterfowl, pigeons, parrots and passerines first evolved in the southern hemisphere during the Cretaceous. Subsequent studies have provided even stronger evidence that the passerine birds originated in Gondwana.[21] Ironically, the Australasian region had been considered a place where 'waifs and strays' found refuge from the more advanced biota of the northern hemisphere. Instead, it is probably the birthplace of the passerine birds.

DNA-DNA hybridisation, a taxonomic revolution

When attempting to determine the relationships of species, taxonomists must determine whether similarities are due to common descent or convergent evolution from separate origins. Their decisions are partly subjective and they have long tried to devise methods of answering these questions more objectively. In the 1980s, US scientists C. G. Sibley and J. E. Ahlquist devised a method to determine relationships in an objective manner by comparing the genetic material itself, rather than appearance or anatomy.[22] Basically, they separated the paired strands of DNA from each of two species then allowed single strands from each to combine as best they could. These hybrid pairs were then heated and the temperature at which they separated indicated how similar their DNA sequences were and thus, by inference, the relatedness of the species.

Over a decade, Sibley and Ahlquist extracted DNA from about 1700 bird species representing most families and genera, then built up a family tree. Assuming that the DNA of birds with similar generation times changes at a constant rate (an assumption challenged by many biologists), they then estimated when each evolutionary line diverged from its ancestral lineage. There are problems with this rough-and-ready method of determining relationships, but it has provided some important insights into the evolution and biogeography of the birds. Bird taxonomy is far from resolved and subsequent studies have continued to shed light on their evolution, but have yet to produce a definitive account of bird relationships.[23] Richard Holdaway[24] has listed the New Zealand avifauna as it would appear if all the Sibley and Alquist changes were adopted. I have adopted the family- and order-level taxonomy used for the Australian avifauna by Les Christidis and Walter Boles,[25] as this appears to be a sensible compromise between the molecular data presented by Sibley and Alquist and the relationships suggested by other lines of evidence. In the following sections I review some of their findings that help us understand the biogeography of New Zealand birds.

In the last decade, different molecular techniques have been used to determine taxonomic relationships within some genera, and the results have enhanced our understanding of biogeography and systematics (for example, the *Cyanoramphus* parakeets and geckos), and conservation (for example, kiwi, tuatara and parakeets). These results are discussed elsewhere in this book.

The ratites

The ratites have long been considered the classic Gondwanan group, represented on most of the former Gondwanan continents. However, ratite relationships had never been satisfactorily resolved. DNA-DNA hybridisation provided a method to test the relationships among this disparate group to see if they conformed to the relationships expected of long-isolated animals of common ancestry. Sibley and Alquist[26] placed all living ratites in a single order, as opposed to the more widely accepted practice (adopted here) of placing each group in an order of its own.

They suggested that the ostrich (Africa) and rhea (South America) diverged 75–80 million years ago, but that the time when kiwi diverged from the Australian ratites was long after New Zealand separated from Gondwana. Subsequent studies[27] present further evidence suggesting that kiwi are more closely related to emus and cassowaries than would be the case if their ancestors had been separated with the Gondwanan break-up. In Sibley and Alquist's scenario, the rhea became separated from all other ratites about 89 million years ago, and the moa, ostrich and emu/cassowary/kiwi group all diverged about 75 million years ago with the fragmentation of Gondwana. The current view suggests that the ancestor of the ostrich and Madagascan elephant birds dispersed from the Antarctic/Australian zone across what is now the submerged Kerguelen Plateau, to the Indo-Madagascar region.[28] The ancestral ostrich later was able to spread overland to Africa, which by then had long been isolated from other Gondwanan lands, after India collided with Asia.

Kiwi are now believed to have diverged from the Australian ratites about 68 million years ago.[29] At this time the greater New Zealand region, which included New Caledonia, was close to Australia, possibly even connected to it. The moa genera *Emeus* and *Dinornis* apparently diverged from one another about 13 million years ago.

New Zealand wrens

The New Zealand wrens have long been an enigma whose relationship to the other passerines has been subject to prolonged debate. DNA-DNA hybridisation studies supported the commonly held view that they are suboscine (primitive) passerines but failed to resolve their relationships within the suboscines.[30] The Old World suboscines include the pittas and broadbills, both of which are virtually confined to Southeast

Asia, and the asities of Madagascar.[31] The New World suboscines are a diverse assemblage of mostly South and Central American birds, including the tyrant flycatchers, tityras, cotingas, manakins, antbirds, woodcreepers and their kin.

If New Zealand wrens were most closely allied to the Old World suboscines, that would suggest they arrived from Australia. However, the only Australian suboscines are three species of pittas, a group of birds whose southernmost limit of distribution is northeastern Australia. If the New Zealand wrens were allied to the New World suboscines, it would suggest they dispersed early, from South America across Antarctica to New Zealand.

Sibley's study suggested that the New Zealand wrens were equally distantly related to both Old and New World suboscines. He placed the New Zealand wrens, the Old World and the New World suboscines in three different infraorders, but noted that the New Zealand wrens were sufficiently distantly related to any other passerines that they could be assigned to a suborder of their own.

The story did not end there. Subsequent anatomical studies indicated that the New Zealand wrens may be more closely allied to the oscine passerines than the suboscines.[32] This would suggest that the New Zealand wrens are a sister group to the oscines and that they, plus the oscine passerines, comprise a sister group to the suboscines. The mystery deepened even further when molecular studies suggested that the passerines developed very early in the evolution of the birds, and that the oscine and suboscine passerines are no more closely related to one another than either are to many non-passerine orders.[33] Where that would leave the New Zealand wrens was not discussed.

The latest view is even more intriguing: it indicates that the New Zealand wrens are the sole survivors of the sister group to all other passerines.[34] Whichever view is correct, the New Zealand wrens apparently diverged from other passerines soon after the passerines first evolved (perhaps about 80 million years ago). This makes them New Zealand's most ancient living birds, having separated from their ancestral line before the kiwi.

Other passerine groups

Until Sibley and Alquist's study, it was thought that the initial radiation of the passerines occurred in the northern hemisphere and that the Australasian passerine fauna was the result of waves of colonising species invading from the north a miscellany of waifs and strays, some of whose lineage later crossed the Tasman. Sibley and Alquist's research suggests instead that the Australasian fauna represents one of the two great radiations of oscine passerines (the parvorder Corvida). They suggest that the other major radiation (parvorder Passerida) probably originated in Asia or Africa, and species from both parvorders subsequently dispersed to other parts of the world. New Zealand and Australia have species belonging to both parvorders, although the Australo-Papuan Corvida predominate. Recent research suggests that the Passerida evolved from Corvida stock.[35] This, plus the fact that New Zealand wrens are endemic, provides strong evidence that the passerines originated in the Australasian region of Gondwana.

Unrelated passerines that exploit the same kinds of foods and habitats can be misleadingly similar. The fantail, tomtit, robin, fernbird, whitehead, yellowhead, brown creeper and grey warbler used to be placed in the family Muscicapidae, along with the introduced song thrush and blackbird.[36] At the time, this family included warblers, flycatchers, thrushes and similar-looking birds from Europe, Asia, Africa and Australia. The Muscicapidae were thought to have originated in the northern hemisphere and some of the Australasian members were considered aberrant. However, Sibley and Alquist showed that the Muscicapidae had lumped together species from both parvorders (see Appendix 1). The fernbird belongs to the same family as the northern hemisphere warblers and is one of the few native passerines that belongs to the parvorder Passerida. The fantail, robins, tomtits, grey warbler, whitehead, yellowhead and brown creeper are distributed among four different families within the parvorder Corvida.

The origins of New Zealand's bats

The two short-tailed bat species belong to the endemic family Mystacinidae, whose affinities had been so obscure that at one time or another they have been placed in seven of the 18 living families of insectivorous bats. They are now placed in a superfamily whose only other living members occur in the American tropics. Molecular studies suggest they split from this lineage at least 45 million years ago, but after New Zealand had separated from Gondwana.[37] Because large parts of Antarctica were still forested in the early Oligocene when the ocean gaps were not yet impossibly wide, dispersal from South America via Antarctica to New Zealand would have been feasible. Recently, Miocene fossil mystacinids

Table 3.2 Phylogeny of the New Zealand passerines before and after Sibley & Alquist 1990

On the left is the taxonomic system adopted in this book which follows Sibley & Alquist (1990) as modified by Christidis & Boles (1994). On the right is the taxonomic order used in the 1970 OSNZ checklist that predates any of the important molecular studies that have greatly changed our understanding of bird evolution. Introduced species are in *italics*.

The taxonomy adopted in this book	The taxonomy used in the 1970 OSNZ checklist
Order Passeriformes **Infraorder Acanthisittides** **Family Acanthisittidae, New Zealand wrens** Rifleman, bush wren and others.	**Order Passeriformes** **Suborder Tyranni** **Family Acanthisittidae, New Zealand wrens** Rifleman, bush wren and others.
Parvorder Corvida	**Suborder Passeres**
Family Pardalotidae, Australasian warblers Grey warbler and Chatham Island warbler	**Family Alaudidae, larks** *Skylark*
Family Meliphagidae, honeyeaters Hihi, bellbird and tui	**Family Hirundinidae, swallows** Welcome swallow
Family Petroicidae, Australasian robins Tomtits and robins	**Family Motacillidae, wagtails and pipits** Pipits
Family Pachycephalidae, whistlers and allies Subfamily Mohouinae, whitehead and allies Whitehead, mohua and brown creeper	**Family Prunellidae, accentors** *Dunnock*
Family Dicruridae, Australasian flycatchers Fantail	**Family Muscicapidae, warblers, flycatchers and thrushes** Subfamily Sylviinae, warblers Fernbirds
Family Artamidae, Australian magpies Australian magpie	Subfamily Malurinae, Australian warblers Whitehead, mohua, brown creeper, grey warbler and Chatham Island warbler Subfamily Muscicapinae, flycatchers Fantail, tomtits and robins
Family Corvidae, crows and ravens *Rook*, New Zealand raven and Chatham Island raven	Subfamily Turdinae, thrushes *Blackbird* and *song thrush*

Family Turnagridae, piopio
Piopio
Family Callaeatidae, New Zealand wattlebirds
Kokako, saddleback and huia

Parvorder Passerida

Family Alaudidae, larks
Skylark

Family Motacillidae, wagtails and pipits
Pipits

Family Passeridae, sparrows and grass finches
House sparrow

Family Fringillidae, finches
Chaffinch, greenfinch, goldfinch and *redpoll*

Family Emberizidae, buntings
Yellowhammer and *cirl bunting*

Family Hirundinidae, swallows
Welcome swallow

Family Sylviidae, Old World warblers
Fernbirds

Family Zosteropidae, white-eyes
Silvereye

Family Prunellidae, Accentors

Dunnock
Family Muscicapidae, thrushes and their kin
Blackbird and *song thrush*

Family Sturnidae, starlings
Starling and *myna*

Family Zosteropidae, white-eyes
Silvereye
Family Meliphagidae, honeyeaters
Hihi, bellbird and tui

Family Emberizidae, buntings
Yellowhammer and *cirl bunting*

Family Fringillidae, Finches
Chaffinch, greenfinch, goldfinch and *redpoll*

Family Ploceidae, sparrows and weavers
House sparrow

Family Sturnidae, starlings
Starling and *myna*

Family Callaeatidae, New Zealand wattlebirds
Kokako, saddleback and huia

Family Cracticidae, bell magpies
Australian magpie

Family Turnagridae, Piopio
Piopio

Family Corvidae, crows and ravens
Rook, New Zealand raven and Chatham raven

have been found in northern Australia and their morphology suggests that they were ancestral to the New Zealand bats.[38] A species of tick normally associated with Australian bats also lives on New Zealand's short-tailed bats.[39] This evidence indicates that mystacinid bats colonised New Zealand from Australia, further suggesting that during the early Tertiary mystacinid bats were found in several Gondwanan lands.[40] The ancestral short-tailed bat may have colonised New Zealand some time after the Oligocene when the land area once again increased.

The long-tailed bat belongs to a nearly cosmopolitan family and its genus includes five other species that occur in Australia, New Guinea or New Caledonia.[41]

Ecology of New Zealand birds

> Most remarkable of all the birds . . . is the kiwi or Apteryx, of which there are three or four species in the two larger islands. These are totally wingless and tailless birds, with feathers resembling hairs, and altogether unlike our usual idea of a bird.
>
> – Alfred R. Wallace (1883)

As noted earlier, the foundations of ecology were laid in Europe and North America, and most ecological theory has been developed from studies in the northern hemisphere. Because the Australasian faunas have had very different evolutionary histories to their northern counterparts, it is not surprising that their ecology and life histories also differ. Northern temperate climates are intensely seasonal and most birds in those regions are strongly migratory. The less seasonal southern climes enable birds to remain in their home ranges year-round, so southern species show a more diverse range of social systems and breeding strategies.[42] In *Ecology of birds: An Australian perspective*, Hugh Ford highlights some of the important differences between Australian and northern hemisphere birds. In the rest of this chapter I will discuss some of the more distinctive ecological features shown by New Zealand birds.

Islands have fewer species of terrestrial animals than do continents, and lack certain influential groups of animals, so island species can exhibit ecological features quite different from their continental counterparts. However, the absence of influential continental players can result in ecological radiation of those animal groups present. Waterfowl living on small islands tend to be smaller, have a more carnivorous diet, make greater use of terrestrial habitats, lay fewer eggs and be more sedentary than their continent-dwelling ancestral forms. The Australasian teal illustrate these characteristics well.[43] The Auckland and Campbell Island teals are smaller than the New Zealand brown teal or the Australian chestnut teal, and have reduced wings (both the subantarctic species are flightless.) All three New Zealand species make greater use of terrestrial habitats than does their Australian counterpart. The Auckland Island teal lays fewer but larger eggs than the brown teal, which in turn lays fewer but larger eggs than the Australian teal. The island-dwelling teals live in smaller groups and have less conspicuous courtship displays than the brown teal. Australian chestnut teals form larger flocks, are more social and show more pronounced courtship displays than any New Zealand teal.[44] New Zealand-breeding great crested grebes are less likely to migrate between inland summer breeding grounds and coastal wintering areas and have simpler courtship displays than their European counterparts.[45]

Flightlessness

> I have heretofore asked the question concerning Mauritius henns and dodos, thatt seeing these could neither fly nor swymme, beeing cloven footed and withoutt wings on an island far from any other land – how they shold come thither.
>
> – Peter Mundy (seventeenth century)

More than half of New Zealand's endemic terrestrial and freshwater birds were flightless, had reduced powers of flight or flew reluctantly. Moa and probably kiwi evolved from flightless ancestors, but all other species lost the power of flight after their ancestors colonised New Zealand. Not only have birds have lost the ability to fly, 10 orders of insects have flightless species in New Zealand, and in some of those orders flight has been lost in more than one evolutionary lineage.[46] The factors that make flightlessness advantageous have seldom been clearly identified. One popular explanation – that flighted birds have a much higher risk of being blown out to sea – may be relevant to very small islands but not to those the size of New Zealand's islands. Most flightless birds occurred on oceanic islands where predatory mammals were absent. Many are now extinct or endangered, and the introduction of predators has been the main reason for their demise. The absence of mammals does not cause birds to lose the power of flight, but can allow flightlessness to evolve if other factors make it advantageous. In this section I set out to identify those factors.

Flight requires the expenditure of a lot of energy, and the muscles and skeletal structure that make flight possible comprise a fifth to a quarter of the bird's weight.[47] Flight allows birds to travel long distances quickly, to locate dispersed food sources, to escape predators and to exploit three-dimensional habitats such as forests more readily than non-flighted animals can. However, birds that exploit evenly dispersed food sources that occur at ground level (or in water), in a place where flight is not an appropriate method to escape predators, would find flight of little advantage. In New Zealand, such birds probably included snipe, the ground-dwelling wrens, kiwi and those rails that ate soil or leaf-litter invertebrates. A reduction in the large, metabolically demanding flight muscles would represent huge energy savings.[48] Jared Diamond likens a winged rail on a predator-free island to a backpacker placing 15 kilograms of bricks in her pack then tramping on half-rations.

Flightless species are in most cases larger than their flighted ancestral species. A large body hinders flight but is advantageous in cool climates because it takes proportionally less energy to keep warm; and for some species large size is a consequence of diet. A leafy diet and the power of flight are almost mutually exclusive because browsers and grazers need to consume large quantities of bulky, low-energy food. Leaves are slow to digest, so leaf-eating birds need a much longer, thus heavier, digestive tract and tend to be larger than species that eat more nutritious foods. Moa, takahe, kakapo and New Zealand geese were browsers or grazers, and were all flightless and giants among their kind. Kokako are arboreal herbivores that have limited powers of flight, while the more strongly flighted kereru show a preference for the more nutritious parts of plants.[49] The kakapo is the world's largest parrot, the only flightless parrot and has a higher proportion of low-calorie leaves in its diet than do flighted parrots. The body weight of kakapo may vary by 50 per cent from season to season because fat is laid down in times of plenty and used when food is scarce. Flighted birds can carry much smaller fat stores, and the ability to lay down large fat deposits may provide kakapo with a margin of safety against food shortages.[50] Having a diet dominated by leaves is probably the reason for the reduction in flight ability in these species. The absence of predatory mammals enabled flightlessness to evolve, and predation by mammals (including humans) is the major reason why virtually all flightless birds are now extinct or endangered.

Rails have colonised numerous remote oceanic islands, and many of these birds became flightless. The New Zealand region alone had nine species of flightless rails. Rails show several morphological, behavioural and ecological characteristics that make them successful colonisers, which, once established on an island, have a high propensity to become flightless.[51] They mostly forage on the ground, and even the species that can fly do so mainly to escape predators rather than as their main means of locomotion. Their flight is heavy and laboured, and if caught by a storm they are unable to fly against the wind back to the land, even though they can remain aloft for long periods. If they do find an island, their ability to exploit a wide range of foods and habitats means they are likely to be successful colonisers. Flight ability was likely to become reduced on islands where there was little predation pressure and the foods available made large size or development of big legs for walking advantageous. Mere disuse of wings alone will not cause flightlessness to evolve.

Two flighted rail species have dispersed to various Pacific Islands, including New Zealand, and given rise to flightless daughter species on the various islands they have colonised. The pukeko (or a closely related ancestor) independently colonised both the North and South Islands, giving rise to the North and South Island takahe and to other takahe-like species on New Caledonia and Lord Howe Island.[52] All four of these flightless island forms were larger than the ancestral pukeko, and only the South Island takahe is not now extinct. The North Island takahe was taller but less stout than its South Island counterpart, and probably diverged from its flighted ancestor more recently than the southern species.[53] Two flightless Chatham Island rails – Dieffenbach's and the Chatham Island rail – are thought to have evolved from the same flighted ancestor, probably the banded rail. They are but two of many daughter species of the banded rail that occurred on many islands in the western Pacific. All the island species were larger, flightless counterparts of the banded rail, the only known flighted member of the genus.[54] Most of the island species were probably morphologically and ecologically similar to one another, perhaps resembling the weka, which itself is one of these daughter species.

Like weka, these species probably exploited a wider range of habitats and foods than the ancestral banded rail. On the Chatham Islands – the only island group where two *Gallirallus* species co-existed – the larger Dieffenbach's rail was a generalist weka-like bird, while the small Chatham Island rail had a long bill and specialised ground-probing habits.

DNA analysis suggests that the extinct New Zealand and Chatham Island coots were sister species, both descended from the flighted

Australian coot. The small differences in DNA between these two flightless coots suggests that flight was lost over a relatively few generations, and not over a period of millennia as previously assumed.[55]

The Auckland and Campbell Island teals are the only flightless waterfowl that are smaller than their ancestral species, and it has been suggested[56] that the former species lost the power of flight while living in a 'regime of stable population densities and relatively constant (but harsh) environmental conditions'. Auckland Island teals live in small home ranges that are occupied year-round. Because food is available throughout the year, there is little need for flight.[57]

Numerous species of flightless rails and waterfowl have evolved on islands in many parts of the world, but only four species of flightless passerines are known, and three of these were New Zealand wrens. Two of them were ground-feeding insectivores; the third species apparently clambered about on tree trunks, probing into crevices for insect prey.[58] The only other flightless passerine was the long-legged bunting from the Canary Islands.[59] All four flightless passerines became extinct soon after predatory mammals were introduced. Other passerines endemic to New Zealand have reduced powers of flight.

Giants and dwarfs

Island species frequently evolve to become larger (or, less often, smaller) than their continent-dwelling ancestors. In New Zealand, we have a number of species that fit this pattern. For example, we had several species of moa that were larger than any other ratite; the kakapo is a heavyweight among parrots; the takahe is larger than its ancestral *Porphyrio* species; and the New Zealand and black robins are larger than any other species in the genus *Petroica*. The kereru is one of the world's largest pigeons, and the weka is one of the largest living species in its genus. Changes in body size have also occurred when mainland New Zealand species have colonised outlying islands: Chatham Islands warblers and pigeons are larger and the Chatham Islands robin is smaller than their mainland equivalents. Auckland and Campbell Island teals are dwarfs when compared with the ancestral Australasian teal. Kiwi are 'dwarfed giants': while the ratites in general are giants of the bird world, kiwi are dwarves among the ratites.

Table 3.3 Body weights of selected New Zealand giant and dwarf species with the weights of closely related or ecologically similar species

Giant species	**Ancestrial or equivelent species**
Takahe 2–3 kg	Pukeko 0.8–1.1 kg
Weka 0.7–1.25 kg	Banded rail 170 g
Kereru 550–850 g	
Parea 750 g	Topknot pigeon (Australia's heaviest pigeon) 475–600 g
Kakapo M 2.5 kg, F 1–2 kg	Sulphur-crested cockatoo (Australia's heaviest parrot) 815–975 g
South Island robin 35 g	
Black robin M 25 g, F 22 g	Scarlet robin (Australia's heaviest *Petroica*) 13 g
Chatham warbler M 10 g, F 8.5 g	Grey warbler 6.5 g
Dwarf species	
Campbell Island teal 315–500 g	Brown teal 500–665 g
Auckland Island teal 375–480 g	Chestnut teal 600–700 g

Source: *HANZAB*; Heather & Robertson 1996

K-selection

New Zealand birds tend to lay clutches with fewer eggs, and to lay fewer clutches each season, with the result that they fledge fewer young per pair (Table 3.4, page 68) than similar species in Europe or North America. Many New Zealand birds are also larger, longer-lived and delay breeding until they are several years of age, but compensate for their comparatively low numbers of progeny by investing high levels of parental care in them. This suite of characteristics is termed 'K-selection'. The converse strategy of producing many young during a short life span, with little parental care of those young, is termed 'r-selection'.[60] K-selected species generally live in stable populations, the size of which changes little from season to season, year to year. They can be thought of as species that live at or close to the carrying capacity of their environment. Species with r-selected breeding strategies live in environments that change markedly season to season, year to year. They are capable of rapid reproduction, so they can quickly increase in numbers to take advantage of times when conditions are favourable. K-selection and r-selection are relative terms and useful only when comparing similar species (Figure 3.3, page 68). Even the most K-selected birds produce

Table 3.4 The numbers of young fledged each year by some forest-dwelling insectivorous birds

Young fledged per pair per year	Location
Grey warbler 4	Kaikoura
Fantail 2.6 – 2.7	Tiritiri Matangi and Cuvier Islands
South Island robin 2.1–3	Kaikoura
Rifleman 3	Dunedin
Brown creeper 3.2	Kaikoura
Whitehead 1.1	Little Barrier Island
Mohua 2.4	Eglinton Valley, Fiordland

Adapted from Elliott 1996a

more young than do elephants, and r-selected birds lay fewer eggs than does virtually any insect species. The kakapo is the archetypal K-selected species. This giant parrot can live for several decades, first breeds when several years old, and produces a single clutch of two to four eggs only once every three or four years.[61]

Pelagic seabirds lay one small clutch each year and have evolved this K-type life history to meet the specific challenges posed by feeding at sea and breeding ashore – seabirds the world over have life histories of this kind. Their breeding sites may be tens, hundreds or even thousands of kilometres from where they feed, thus restricting the number of chicks they can raise. Albatrosses and petrels are among the most pelagic of birds, and they lay only one large egg at each breeding attempt.

	Takahe 	Pukeko 
Body weight	2–3 kg	0.8–1.1 kg
Flightedness	Flightless	Flighted
Age at first breeding	Usually 2 years	1 year
Clutch size	2 eggs	3–6 eggs
Clutches per year	1	2–3

Figure 3.3 Body weights and breeding parameters of the K-selected takahe and the r-selected pukeko

Data from *HANZAB*

Incubation takes 40 days in small petrels but almost 80 days in the albatrosses, and it takes a further 45–275 days to raise their chicks. This development takes longest in the largest species and in those that feed farthest offshore,[62] and prevents more than one egg being laid each year. Some albatrosses lay only every second year. Individuals of species that feed offshore will incubate for days or even weeks on end while their partner is at sea feeding. For example, a mottled petrel or grey-faced petrel may incubate for 15 days before its partner returns to take a similar spell ashore.[63] Once the eggs hatch, the parents return to feed their chicks at similarly infrequent intervals.

Inshore-feeding shags, on the other hand, lay more eggs, have shorter incubation spells and feed their chicks several times a day. Spotted shags feed up to 16 kilometres offshore, most commonly lay three eggs and feed their young up to four times a day.[64] The time from egg laying to fledging is about 90 days for the spotted shag, over five months for the sooty shearwater, and a year for the royal albatross. Most seabirds are long-lived and delay breeding until they are several years old. Yellow-eyed penguins first breed when two to four years of age,[65] sooty shearwaters five to seven years, and wandering albatrosses 11 to 12 years.

In New Zealand, most land birds are K-selected when compared with northern hemisphere equivalents, although the clutch size differs little from that of Australian species. No New Zealand passerine can match the four- to 12-egg clutches laid by the great tit of Europe, which sometimes raises two clutches, even during the short British summer.[66] The New Zealand tomtit, unrelated but like the great tit a small forest insectivore, typically lays one or two clutches of four eggs a year.[67] Its close relative the South Island robin produces only two- or three-egg clutches, while the Chatham Island black robin usually lays only two eggs.[68] New Zealand robins and tomtits can lay more eggs, a trait exploited during the black robin recovery programme when the robin's first and sometimes second clutches were given to tomtit foster parents and the robins responded by re-laying (see Chapter 9).

Many birds endemic to other mammal-free oceanic islands are also K-selected. For example, the Cook Islands kakerori, another small insectivorous forest bird, is long-lived (three are known to have lived for 17 years) and usually produces only one clutch of two eggs a year.[69] The relatively mild maritime climate of oceanic islands probably also influences K-type breeding strategies.

Species belonging to lineages that have been in New Zealand longest – for example. mohua and bellbird – tend to be longer-lived and lay fewer eggs than species such as fantail and grey warbler that colonised New Zealand more recently. Takahe and pukeko are closely related species, but the takahe has been in New Zealand much longer than the pukeko. Takahe are larger and flightless, lay only one clutch of two eggs a year and first breed when three years old; pukeko lay multiple clutches of five or six eggs and first breed when one year old (Figure 3.3).[70]

Before predatory mammals were introduced, New Zealand birds encountered fewer species of predators than did birds on continents and islands close to continents. The absence of predatory mammals and the paucity of predatory birds and reptiles have often been suggested as reasons why a smaller clutch was advantageous. However, related species in Australia have much the same clutch sizes despite having evolved alongside a suite of predatory mammals, birds and reptiles at least as diverse as those on northern continents. Most species of introduced passerines have smaller clutch sizes in New Zealand than in their native Europe,[71] and this change has occurred since predatory mammals became common. For example, the mean clutch of New Zealand house sparrows is 3.81 eggs, but in England it is 4.1. These introduced species apparently live longer in New Zealand than in Europe and, because the breeding season here is longer, some species are able to lay an extra clutch. Small clutches may compensate for lowered adult mortality and a longer breeding season.

K-selected traits generally prevail in seldom-disturbed, stable habitats or environments where seasonal differences in food availability are small. Conversely, r-selected species are adapted to colonising habitats after disturbance, or to highly seasonal environments where animals suffer food shortages at certain times of year while enjoying an abundance during other seasons.

New Zealand's past 20 million years of tectonic uplift, volcanism and glaciation, together with its often stormy climate, may not initially seem to favour K-type life histories. However, the time between geological disturbance events is far longer than the lifespan of even long-lived animals, and the effect of geologically induced disturbances tends to be localised rather than widespread. Climate is a more likely factor to influence breeding strategies.

The New Zealand land mass has experienced a maritime, temperate climate for many millions of years, and even during the Pleistocene glaciations the climate was less harsh than in northern temperate regions.[72] Winters here are mild, at least compared with the intensively studied temperate northern hemisphere continental locations so often used as a yardstick in ecological studies. In New Zealand, some food is available year-round, ensuring animal populations do not undergo the large annual fluctuations characteristic of northern temperate zones. Our evergreen forests may also serve to dampen the differences between winter and summer. Most terrestrial birds in northern hemisphere temperate zones are migratory, whereas New Zealand species, as well as those on oceanic islands and most tropical birds, are sedentary. A sedentary life allows, perhaps even favours, K-selection, while the higher mortalities suffered by migratory species favour higher reproductive rates. The explanation may be even simpler: most Australian birds lay few eggs each year and, as most New Zealand bird lineages were derived from Australia, this could just be an inherited trait.

Because Auckland Island teals lay fewer and larger eggs, their ducklings are proportionately larger than those of other teals when they hatch, and are probably able to regulate their body temperature at an earlier age.[73] Murray Williams suggested that the advantages of larger size at hatching meant that a large egg size was favoured and, for Auckland Island teals, the reduced clutch size was a consequence of this.

Clutch size and the age of first breeding usually fall within a clear range but may vary depending on environmental constraints in different parts of the species' range. For example, at Kowhai Bush, near Kaikoura, most South Island robins bred when a year old, whereas on the predator-free Chetwode Islands they lived at higher density and first bred when two years of age.[74] At Kowhai Bush, adult mortality was higher and birds laid more clutches each season than at the Chetwode Islands.

In New Zealand, K-selected life histories are not confined to birds. As we saw in Chapter 2, New Zealand frogs, geckos and tuatara lay small clutches and some species go years between breeding seasons. The giant land snails, many large insects and some other native animals also show K-type life-history strategies. K-selected species have less ability than r-selected species do to adapt to changing environments. While K-selected traits were adaptive in pre-human New Zealand, they are clearly disadvantageous in today's much-altered environment.

Breeding biology of New Zealand birds

Most New Zealand birds breed in spring and summer, as would be expected for temperate species. For example, South Island robins at Kowhai Bush begin breeding in August, grey warblers and fantails in September, and brown creepers in October. All four species lay their last eggs in December.[75] A few species deviate from this general pattern. Autumn or winter breeding is characteristic of New Zealand parrots and may correlate with the seasonal abundance of beech seed or podocarp fruits.[76] Alpine-dwelling kea usually begin nesting in winter, although in some years they delay breeding until spring. Kakapo lay eggs in February and chicks leave the nest in midwinter. In the Fiordland beech forests, yellow-crowned parakeets usually breed only once, in late summer, but in beech mast years they breed all year and may raise up to five broods of chicks. Most seabirds breed during summer, but Westland petrel, grey petrel, grey-faced petrel, Fiordland crested penguin and northern populations of little shearwater breed in winter.[77]

A large proportion of New Zealand's endemic birds nest in holes, and this makes the eggs, chicks and incubating adults vulnerable to introduced predators. Unlike most hole-nesting birds elsewhere, the New Zealand species have poorly developed predator-avoidance behaviour.[78] Species such as mohua are especially vulnerable because their nest holes have a single entrance; and furthermore, since only the females incubate, more of them are lost to predators. The chicks are noisy and their incubation and nestling periods are long, so they are vulnerable to predators for an extended period.

All seabirds and most land birds are monogamous and the pair bond is maintained, at least throughout a breeding attempt. Monogamy does not imply that partners are always faithful to one another, and in many species individuals are known to copulate with birds other than their mate. Both parents contribute to the care of the eggs and chicks, but males and females may have different roles; for example, males may maintain the territory while their partner incubates. About 10 per cent of bird species are polygynous, where one male mates with several females, and about one per cent are polyandrous, where one female mates with several males.* Polygamy occurs in species where food is sufficiently abundant that one parent can raise the young without help from its mate. Polygamous individuals may have a stable pair bond with more than one partner, or the birds may merely meet to mate, after which one of them provides all the parental care.

Most New Zealand birds are monogamous: the blue duck is a good example. Pairs live on swift-flowing rivers and maintain a year-round territory that consists of up to two kilometres of river from which they

* Polygamy is a more commonly used term and refers to any situation where an individual mates with more than one partner. Thus polygyny and polyandry are both forms of polygamy. Polygynandry occurs where both sexes have multiple partners.

exclude other blue ducks.[79] Murray Williams found that only pairs with a territory attempt to breed. Some other territorial species defend territories only during the breeding season. Blue ducks usually mate with the same partner in the same territory for several consecutive seasons, although young birds, bereaved birds or non-resident birds (usually males) will challenge a same-sex resident and occasionally usurp it. The two sexes have distinct roles. Only females incubate. While both parents attend to the ducklings, the male is concerned primarily with brood defence, and the female with keeping the brood together, an important task on the fast-flowing rivers on which they live. Nineteen of the 26 ducks that joined Williams's study population were the progeny of resident ducks, and most of these established territories adjacent to that of their parents. Thus neighbours tend to be closely related and birds frequently paired with kin, leading to high levels of inbreeding.[80]

There are some interesting exceptions to the monogamous norm. Several species breed co-operatively, and these are discussed in the next section. The kakapo has the most bizarre breeding system of any New Zealand bird. Both sexes are solitary. Males devote a lot of time and energy to building then maintaining a track-and-bowl system. This consists of one or several areas, called bowls, that are situated against overhanging banks, rock faces or trees on ridges or hillsides.[81] These are cleared of debris and linked by carefully cleared tracks up to 50 metres long and 30–60 centimetres wide. Each male has his own system that he often re-uses year after year. The bowls reflect the males' booming calls out across the valley, where they hope females will be listening. These systems are only used for display. During the booming season (December–March at two- to four-year intervals), males feed and roost within several hundred metres of their bowl. During the non-breeding season, each male has a home range of about 50 hectares, which may include the track-and-bowl system but more often than not is several kilometres away. Males spend six to eight hours a night booming. They inflate their thoracic air sacs to the point where the body almost engulfs their head, the bird assuming the shape of a rugby ball, then it emits a series of low-pitched booms that can be heard up to five kilometres away. Nesting occurs only during booming years. Females move from arena to arena, then choose which of those males they will mate with at a track-and-bowl system. Males have no role in incubation or chick-rearing. At Stewart Island, booming and breeding occurred only in years when podocarps produced large quantities of fruit. However, the birds began displaying long before the appearance of the fruit.

Hihi have the most variable breeding system of any New Zealand bird. Isobel Castro found that two-thirds of nesting attempts were by a monogamous pair, but other nests were by females in polygynous or polyandrous relationships, and one nest was a result of polygynandrous group.[82] Some birds were recorded breeding in more than one mating system, usually with different partners. To provide additional variation, some males visited the nests of non-pair females, suggesting that extra-pair copulations may have occurred. Females could raise their broods without assistance, but males provided food for chicks at some nests. Birds usually copulate by the male perching on the female's back, sperm transfer occurring when their cloacas touch. Hihi mostly copulate in this manner, but males will sometimes force the female to the ground so that she is lying belly-up. With his wings outspread, he holds her on the ground and copulates face to face. Forced copulations occur in other bird species, but face-to-face copulation has not been recorded.

Piopio were allied to the Australian bowerbirds. The males of most bowerbird species build display arenas that they decorate with brightly coloured objects. Females visit and, if the display and the arena impress, they mate with that male. Piopio became extinct before their mating system could be described, but there is no evidence that they built bowers.[83] Observations made by the early naturalists Buller and Potts suggest that the birds were monogamous and both male and female defended the nest.

Kiwi breeding biology

All species of kiwi are monogamous. Both sexes remain in the same territory year-round and the pair bond remains intact from one year to the next.[84] Kiwi eggs weigh up to a fifth of the female's body weight and, in proportion to body size, are among the largest eggs laid by any bird. About 60 per cent of the egg's weight is yolk,[85] so in terms of energetics it is an extremely expensive egg to produce. It takes the female brown kiwi 25–30 days to produce this 430-gram egg, and she may lose up to 200 grams of body weight while doing so. The egg is so large that, in the days before laying, it occupies nearly all her abdominal cavity and she is unable to feed. Each year, North Island brown kiwi usually lay two eggs three weeks apart. If early eggs fail, a female North Island brown kiwi can lay up to five eggs per year. The other species usually lay a single egg each season.[86]

After the egg is laid, the female remains in the nesting burrow with her mate for a day or two, but after that incubation is primarily the responsibility of the male. In the case of North Island brown kiwi and little spotted kiwi, only the male incubates, although females remain close to the nest burrow for a few days after laying and before hatching. The female has made her contribution by the production of that huge egg. Incubation takes 70–85 days, slightly less for the little spotted kiwi, and the incubating male must leave the egg for several hours each night, sometimes all night, to attend to his own needs. Female Stewart Island kiwi (tokoeka) and great spotted kiwi, which lay only one egg each year and occur in cooler parts of the country, assist with incubation.[87] On Stewart Island, birds other than the parents may also assist with incubation. Just over half of the yolk is used during embryonic growth, so the chick hatches at an advanced stage of development with food reserves to tide it through its first week of life.[88]

For kiwi, reproductive investment is in the egg and little effort is invested in chick care. The chick hatches fully feathered and remains in the nest for up to a week before it first goes in search of food. Feeding forays get progressively longer, and after 10 days they are out foraging most of the night. Neither parent accompanies the chicks on these forays.[89] The male broods the chick by day, but after about three or four weeks the chicks are fully independent.

Brown kiwi chicks remain in their natal territory for two or three years and take two years to reach full size. Newly fledged kiwi are vulnerable to introduced predators. Even without predators the reproductive rate would be low; today, pairs fledge on average only one chick every second year.

Sexual dimorphism in huia

> The very development of their mandibles in the two sexes enabled them to perform separate offices. The male always attacked the more decayed portions of the wood, chiselling out his prey after the manner of some woodpeckers, while the female probed with her long, pliant bill the other cells, where the hardness of the surrounding parts resisted the chisel of her mate.
>
> – Sir Walter Buller (1888)

Huia must have been striking birds to watch. Both sexes were glossy black in colour with white-tipped tails, orange wattles and ivory-coloured bills. Male and female huia had very different bills: that of the male was stout and chisel-like, while the female's slender, down-curved bill was a third longer than her mate's. The difference in bill shape and size between the sexes was greater for huia than any other bird. Perhaps this difference evolved to allow the sexes to utilise different foods or, conversely, the ivory-coloured bill, which contrasted with the black plumage, may have been used to attract the opposite sex.[90] In animals that use sexually dimorphic traits to attract mates, the dimorphic feature is often brightly coloured or contrasts with the rest of the body, as it did in the huia.[91]

The male apparently used his bill as an adze to rip into rotting logs, while the female used her long forceps-like bill to extract grubs from those logs. Pairs of huia remained close together when feeding but apparently exploited different foods, thus reducing competition between members of a pair.[92] While their different bills enabled each sex to obtain food unavailable to the other, it is not clear that they shared the food thus obtained, as many accounts have suggested.[93] Huia were apparently monogamous and territorial, and such species do not exhibit extreme sexual dimorphism unless the feature evolved for some other purpose. Other evidence is consistent with the sexes obtaining their food in different ways: there are corresponding differences in the structure and musculature of the head and neck.

There are similar but less extreme examples of dimorphic bills in other wood-excavating birds such as woodpeckers. The females of certain birds of paradise also have longer bills than the males, and this trait is found in those species for which wood-boring insects are a large part of the diet.[94] Birds of paradise are polygynous, so the female has to obtain all the food for her chicks. Clifford Frith suggested that females rather than males evolved the long bill as an adaptation to obtain the protein-rich invertebrate diet required by chicks. He further speculated that the female huia's long bill might suggest that although huia were monogamous, she was still the main food provider for the chicks.

Co-operative breeding

Co-operative (communal) breeding occurs when individuals provide care to young that are not their own. They could be non-breeding adults (helpers) or co-breeders, where two or more same-sex members of the group participate in a breeding attempt. In such cases, two or more females lay in the same nest, or more than one male incubates or feeds the females or chicks. 'Helpers' most often collect food for chicks (or, less often, their parents) but may also defend territory, incubate or repel predators. The use of value-loaded terms such as 'helping' and 'co-operative breeding' should not be taken to suggest the type of motivation they imply when used in a human context. 'Helping' need not be altruistic. It may be a means of acquiring mates or territory by inheriting that position from established breeders.

Until the concept of kin selection was developed in the 1960s, communal breeding appeared rare and quaint.[95] Co-operative breeding is much more common among southern temperate than northern temperate birds, and this meant early observations of such behaviour in pukeko, weka and skuas[96] attracted little attention or even sceptical comment from northern hemisphere biologists. Since then, communal breeding has been found to be much more common than previously imagined. It occurs in a wide variety of both taxonomic groups and habitats, but mostly where resources are patchy.

The reason why individuals forgo an opportunity to breed to 'assist' another are many and varied. In some cases, suitable habitat may be limited or individuals may reproduce more successfully in the long run if they first spend a season as helpers, acquiring skills. Helping at a nest can also provide mating opportunities if a same-sex member dies or an opposite-sex member reaches breeding age.

Communal breeding is most common in southern hemisphere tropical to warm temperate zones, and a disproportionately large number of Australian species breed this way. Hugh Ford[97] has suggested that the unpredictable nature of the Australian environment and its lack of seasonal differentiation may encourage communal breeding, but this does not explain why some New Zealand species breed this way too.

Communal breeding has been reported for eight New Zealand bird species: one seabird (brown skua), three rails (takahe, pukeko and weka), and four insectivorous passerines (whitehead, mohua, brown creeper and rifleman). None of these species always breeds communally; brown creeper and weka probably do it rarely. Pukeko have the most complex social system of the New Zealand communal breeders. Mohua and whiteheads breed in pairs or in groups that consist of a breeding pair and one or more 'helpers'. In whiteheads, the primary males guard their mates during courtship and egg laying, and only the primary female incubates, so it is probable that they are the chicks' genetic parents.[98] Most 'helpers' are males, probably young from a previous brood. They provide food for the chicks and sometimes the breeding female. Groups apparently fledge more young than pairs. Helping probably gives young adult whiteheads the opportunity to remain in a familiar territory until the opportunity to breed arises.

The rifleman has a more complex system, with both regular and casual helpers. Regular helpers remain faithful to a pair throughout the chick-rearing period, while casual 'helpers' make infrequent visits to one or several nests.[99] Greg Sherley found that the casual helpers were usually offspring from a previous clutch. Regular helpers were present at about 15 per cent of nests and were usually single, unrelated males. Only one female helper was seen and she was the mother of the female she associated with. Helpers fed young, removed faecal sacs from the nest and helped defend chicks and nests. Males of pairs receiving no help provided their chicks with more food than the females did. When helpers were present, male parents fed chicks less often, and this suggests they benefited most from the help provided. However, females were more likely to survive if they had received help during the previous breeding season. Some helpers paired with offspring from the nest at which they helped.

Some brown skua territories are defended by trios of adult birds, all of which attend the chicks. On Rangatira Island, in the Chatham group, Evan Young found that trios produced no more eggs than pairs, nor was there any difference in their breeding success.[100] The relationship and role of the third bird is unknown.

The social system of pukeko

Pukeko have a complex social system. Breeding birds live in territorial pairs or in social groups of up to a dozen birds, all of which participate in rearing of chicks and defence of the territory.[101] Non-breeding pukeko live in non-territorial flocks. Within the group, active breeders mate promiscuously with group members of both sexes, and dominant males do not usually prevent subordinate males from copulating. Females have been seen mating with up to three males within just a few minutes. Up to three females may lay in the same nest, but if more eggs are laid than will fit, a second nest is built adjacent to the first and two birds incubate simultaneously.

Group size and social behaviour varies. At one study site John Craig and Ian Jamieson found that pairs were common, whereas at other study sites most birds lived in groups of larger than four. At these sites good-quality nesting areas were in short supply and neighbours or non-breeders would infringe on territories whenever the opportunity arose. Most defence is by males, so groups with extra males can defend larger territories, but there is a trade-off: those extra males have to share in the group's breeding activities. Breeding success was highly variable and was most strongly correlated with habitat quality, not how large the group was. Pukeko bred in pairs wherever habitat quality enabled a single male to defend sufficient resources. Thus, for male pukeko, the optimal size of the breeding group is a trade-off between habitat quality, ability to defend an adequately large territory, and the advantage of increasing territory size by sharing reproductive output with other males.

Chicks hatched in pair territories were expelled from their parents' territory when a year old. However, 91 per cent of chicks born in group territories remained in their natal territories until they were two years old, and 71 per cent spent their reproductive years there. Seldom were outsiders allowed to join these groups. Thus group members were closely related and, with the pukeko's promiscuous mating habits, incest was common. Although these groups were highly inbred, no deleterious results were apparent.

Usually, only pukeko that had copulated incubated, but non-breeders would sometimes take their turn. All members of the group would provide food for the chicks. Craig and Jamieson suggested that helping behaviour was a consequence of communal living. When in habitats that make group-living advantageous, all members of a breeding group are exposed to the demands of hungry chicks or the sight of eggs needing incubation, and they respond with appropriate parental behaviour regardless of their genetic relationship to the young. Craig and Jamieson suggested that helping behaviour is merely parental behaviour stimulated by begging offspring and has no altruistic motive. In the case of the pukeko, where kin alone gain admittance to the group, only related birds get the opportunity to 'help'. In other species where non-kin join the group, they also encounter and respond to the stimulus of a begging chick.[102]

Migration

Terrestrial birds

In the highly seasonal northern hemisphere temperate zone, most birds are migratory, overwintering far south of where they breed. In the less seasonal, more equitable southern hemisphere, few terrestrial birds or waterfowl are migratory. The New Zealand mainland extends over 13 degrees of latitude, from warm temperate climates in the north to cool temperate in the south. Farther south are the subantarctic islands, and to the north the islands of the tropical Pacific. However, the only truly migratory land birds are the two species of cuckoo, although other species, including silvereyes and some introduced finches, make smaller north–south movements within New Zealand. The shining cuckoo breeds in New Zealand, southern Australia, Vanuatu and New Caledonia, and winters in eastern Indonesia, New Guinea and the Solomon Islands. The long-tailed cuckoo breeds only in New Zealand but winters on many islands in Micronesia and Polynesia.[103]

About half of New Zealand's native terrestrial birds have reduced powers of flight, so for them migration is no longer an option, but we do not know why so few others migrate. Perhaps insufficient time has passed since the Pleistocene glaciations, when southern forests were confined to scattered remnants, for birds to evolve migratory behaviour. Perhaps the seasons at opposite ends of the country are not sufficiently different. There would have to be a winter shortage of food in the south when food was abundant in the north. If northern breeders exploited all the food available, or if differences in food availability between north and south were not great, then the energy used in migration could not be recouped.

Trans-equatorial migrants

About 40 species of wading bird that breed in the Arctic during the northern summer visit New Zealand during the southern summer. During the Arctic summer, a flush of food becomes available and vast numbers of wading birds breed on the tundra. As autumn approaches, food supplies dwindle and the birds are forced to migrate south and spend the southern summer on harbours and estuaries in the southern hemisphere. These birds arrive in New Zealand in September and most leave in March or April, although a few non-breeders overwinter here. About 160,000 Arctic waders visit New Zealand each year, 95 per cent of which are bar-tailed godwits or knots.[104] About 24 species are regular visitors, with at least a few individuals arriving each year; of the rest, only small numbers reach New Zealand and only during some summers. Scanning estuaries for rare Arctic migrants is a favourite occupation of many New Zealand birdwatchers.

Another group of trans-equatorial migrants are the petrels and shearwaters that breed in the New Zealand region during the southern summer and spend the northern summer in the Arctic (see Chapter 8). No wading birds breed in New Zealand then overwinter in the northern hemisphere, and the only northern-breeding seabirds that migrate to New Zealand waters are Arctic terns and skuas, which are far less numerous than the migrants from the south. Several species of Charadriiformes breed on the South Island braided rivers and winter in coastal regions, while some of the banded dotterels breeding in such locations migrate to Australia.[105]

Birds of the South Island braided rivers

To the east of the Southern Alps, a bird community unique to New Zealand is found on the braided rivers that begin at the base of the mountains. These shingle-bed rivers have been produced by rapid uplift of mountains combined with equally rapid erosion. The erosion rubble has accumulated in the deeply scoured glacial valleys and extends outwards to form the Canterbury Plains and upper Waitaki basin. The rivers meander over the valley floors and plains to form wide, multi-channelled rivers. These are dynamic systems that are flood-prone, particularly during spring and summer, and channels may change as stones roll downstream and scour out new courses. In the past, periodic floods have restricted vegetation to a few hardy mat plants. The best-developed braided-river systems are in Canterbury and Otago. Elsewhere, rivers are less braided, have more single-channel sections and support fewer birds. Braided rivers are uncommon in the rest of the world.

In summer, these riverbeds can be hot and dusty; in winter, cold and forbidding; yet these habitats support about 20 species of wading and wetland birds that breed on the open banks and islands of the riverbeds. Six endemic birds breed primarily on braided rivers and their associated wetlands, and show specific adaptations to these habitats. Three of these – black stilt, wrybill and black-fronted tern – are threatened or endangered, while even the three more common species – black-billed gull, South Island pied oystercatcher and banded dotterel – have also declined in numbers.[106] The birds of the Cass River have been described by Ray Pierce,[107] who discussed the ecology of braided-river birds.

To the uninitiated, a braided river can appear bleak and monotonous, but for the birds the rivers present a variety of feeding habitats. While most species feed primarily on aquatic insects, they do so in different microhabitats. Deep, swift main channels are used mostly by terns and shags. Small, shallow channels with pools and riffles are favoured by dotterels, stilts, oystercatchers and wrybills. Backwaters and seepages are used mostly by wrybills and dotterels; and pools by stilts, herons, oystercatchers and waterfowl. Nearby swamps are used by waterfowl and some large waders.

The bird most characteristic of the braided rivers is the wrybill, which is the only bird in the world whose bill curves sideways, an adaptation for feeding on mayflies and other insect larvae that cling to the underside of rounded riverbed stones.[108] Mayfly larvae are most numerous in riffles, which are favoured feeding sites for wrybills. The birds nest on shingle spits or islands that are free of vegetation.

In winter, most braided-river birds migrate to estuaries, lagoons and other coastal habitats. Only black stilts and some black-billed gulls remain close to the rivers all winter. The stilts initially use tarns and swamps but move to the river deltas when the tarns freeze. Those gulls that do not disperse to the coast in winter mostly use lakes and fields.[109] Wrybills leave the braided rivers about January and overwinter on the Firth of Thames and other northern estuaries. They return to their breeding grounds in August or September, most breeding within 500 metres of the site they previously used.

Most South Island pied oystercatchers breed at inland South Island locations, many on braided rivers, but some on other habitats and a few breed in the lower North Island. After breeding, they disperse to coastal locations throughout New Zealand. Oystercatchers arrive on estuaries about the time the godwits that fed there during summer depart for the Arctic. The black-billed gull shows both behavioural and physiological adaptations for breeding on braided rivers.[110] These birds must complete their breeding cycle before floods destroy their eggs or chicks. In order to shorten the breeding cycle, black-billed gulls lay fewer eggs than other gulls and much of their courtship is carried out before arriving on the colony. Their chicks are mobile early in life, and if floods wash them downriver, the parents shift camp to where their chicks can scramble ashore.

The braided rivers and their birds are poorly represented in New Zealand's conservation portfolio and all have suffered habitat change during the last century (see Chapter 9).

Ecology of bats

New Zealand had three species of bats, all of which were endemic. Before 1990, they were little studied, but the recent development of tiny radio transmitters and automatic bat detectors has greatly facilitated research into these small nocturnal animals.[111] Two species of short-tailed bat – the greater short-tailed (extinct since 1964) and the lesser short-tailed (threatened) – are currently recognised.[112] The size of greater short-tailed bats varied, with those in the north being largest and size decreasing in the south.[113] Three subspecies of the lesser short-tailed bat are recognised but their taxonomy has not been satisfactorily resolved. A recently discovered population in Fiordland that differs from other populations in both size and vocalisations further complicates the situation.[114]

The taxonomy of the long-tailed bat is similarly confused. The size of this species varies from one part of the country to another, and while only a single taxon is currently recognised, it may in fact comprise several subspecies.[115] The long-tailed bat is a typical insectivorous species. Roosting in hollow trees, rock crevices and caves by day, they are fast-flying, agile insectivores, catching all their food at night on the wing. They were once common throughout New Zealand and remain widely but patchily distributed through much of the North Island (though absent from Wellington and Wanganui regions), in Fiordland and a few localities in Westland, Nelson, Canterbury, Otago and Stewart Island.[116] They also occur on Little Barrier, Great Barrier and Kapiti Islands.

Long-tailed bats can be found in many habitats, most often in native forests, but some populations use exotic forests or farmland. They occur in a variety of native forest types, including kauri, beech, podocarp and regenerating kanuka and manuka. On the Volcanic Plateau, long-tailed bats occur in pine forests, and in South Canterbury they make extensive use of farmland and willow woodlands. They feed near forest edges, in forest gaps, above the forest canopy or along the edges of roads, lakes and streams.[117] Sometimes long-tailed bats travel over 30 kilometres during a night's foraging. Their home ranges are among the largest of any non-migratory bats. They are less active in cold weather than during warm nights, but during winter have been recorded flying in temperatures as low as –1.5°C.

Short-tailed bats are perhaps the most unusual of bats. Just as New Zealand had ground-dwelling birds, we also had two partially terrestrial bats, although unlike the birds, which had reduced powers of flight, both short-tailed bats were still fully flighted.[118] The part of the wing membrane closest to the body of the living species is thick and leathery, and when on the ground the bat can furl its wing membranes against the forearm so that the delicate outer part is protected from harm by the tough inner part. The hind limbs are stout and not enclosed within the wing membrane, as they are in most bats. Lesser short-tailed bats can scurry about on the ground or clamber in trees with mouse-like agility. In the absence of ground-foraging insectivorous mammals, short-tailed bats evolved to exploit tree- and ground-dwelling insects. They even enter burrows and search among leaf litter for food, and appear to spend more time on the ground than in flight, suggesting that aerial foraging is relatively uncommon.[119] The extinct greater short-tailed bat sometimes roosted in petrel burrows. Not only are lesser short-tailed bats adept ground scurryers, but they are capable of rapid flights of up to 17 kilometres, and manoeuvrable flight within cluttered native forests.[120]

Lesser short-tailed bats take both flying and ground-dwelling insects. They feed on the latter during the cooler months and take more flying insects in summer, when these are abundant. Pollen, fruit and nectar are also important components of their diet.[121] They have an extensible brush-tipped tongue for lapping nectar. They are an important pollinator of the threatened wood rose (see Chapter 7), and in northern New Zealand they eat pohutakawa nectar and thus presumably pollinate the trees at the same time. On the islands off Stewart Island, bats (most likely greater short-tailed bats[122]) were recorded as having chewed meat and fat off muttonbirds. Even on Codfish Island, at the southernmost limit of their range, some short-tailed bats are active each night even in mid-winter.[123] During winter, they are active more often and for longer than other cool-climate bats. Jane Sedgeley suggested that this was related to their catholic diet. Even in the depths of winter there are some ground-dwelling insects or fruits available for short-tailed bats to eat, while there are insufficient flying insects to support other bat species.

The short-tailed bat appears to be a lek breeder (where several males perform courtship displays at common or close-by areas but do not contribute further to the reproduction attempt). Males produce a call audible to the human ear from holes in traditional 'singing trees'. Several males display in proximity to one another, presumably to lure females. Females roost in nursery colonies in which they raise their young. Males in the central North Island populations, where females have large and unpredictable home ranges, display outside the maternity roosts.[124]

Today, the lesser short-tailed bat occurs at widely scattered North Island localities (the largest concentrations are in the central North Island), Little Barrier Island, Northwest Nelson, Fiordland and Codfish Island. They occur only in old-growth kauri, podocarp and beech forests, and apparently feed mostly within the forest interior but will commute over open areas when moving between favoured parts of their home range.[125]

Both long-tailed and short-tailed bats shift from one roost to another every few nights, each individual bat making use of several roosts. In the Fiordland beech forests, long-tailed bats usually stayed in a roost only one night and seldom used a roost more than once.[126] Reproductively active females usually roost communally, while males tend to roost alone.

At birth, young bats of both species weigh about a third of their mother's weight, and when females move from roost to roost, the young are carried hanging from their mother's nipple. By the time the baby bat is weaned, it can weigh four-fifths its mother's weight, yet she will still carry it between roosts.[127] Roost sites that are inaccessible to introduced mammals appear to be a prerequisite for the bats' continued survival, at least in some areas. During the course of a night, long-tailed bats have, on average, four feeding bouts and three roosting periods.[128] Night roosts are usually different from those used during the day. At Grand Canyon Cave in the King Country, long-tailed bats roost in the cave between foraging bouts, but by day they settle in nearby native forest. Long-tailed bats show a strong preference for the oldest cavity-bearing trees available. In Fiordland, both bat species roost in old, very large beeches, the very trees sought during selective logging.[129] In the Waitakere Ranges, large kauri trees are used. Long-tailed bats appear to prefer large trees, in open-structured forests, on the valley floor or gentle slopes, the very terrain preferred by loggers.

Short-tailed bats made use of multiple roosts even in midwinter. They roost in large groups using traditional roosts that may be used for decades or even longer.[130] Each population has a number of such sites and, after using one particular roost for a few days or even a few weeks, the group shifts en masse to another of their traditional sites. Short-tailed bats roost in large trees, although the distribution of fossil remains suggests that, in the past, caves were also used. One or both species of short-tailed bat occurred in most forested parts of the country.

Both living species have suffered marked declines in distribution and range, and are now listed as threatened or vulnerable.[131] Both species are susceptible to introduced predators and habitat loss, and short-tailed bats may take baits laid for the control of introduced mammals. However, the reasons for their decline are not known with certainty. Loss of roosting trees and introduced predators are considered to be the main threat to long-tailed bats. Competition for roost sites with introduced mammals, birds and wasps has probably also impacted on New Zealand's bats.

CHAPTER 4

Vertebrate communities in pre-human New Zealand

> With regard to animals, it is a most remarkable fact, that so large an island, extending over more than 700 miles in latitude, and in many parts ninety broad, with varied stations, a fine climate, and land of all heights, from 14,000 feet downwards, with the exception of a small rat, did not possess one indigenous animal. The several species of that gigantic genus of birds, the Deinornis, seem to have replaced mammiferous quadrupeds in the same manner as the reptiles still do on the Galapagos archipelago.
>
> – Charles Darwin (1845)

The pre-human vertebrate fauna

Previous chapters have described how different many New Zealand vertebrates were from species elsewhere. If the animals were peculiar, then so also might be the ecological communities to which they belonged. In most parts of the world, mammals are important members of vertebrate communities, but in their absence here, New Zealand birds adopted certain typically mammalian roles. For example, there was a guild of flightless, foliage-feeding birds and a guild of ground-probing, insectivorous birds (the best-known example of which is the kiwi, though that guild included other, unrelated birds). Madagascar was the only other place where birds were the primary vertebrate herbivores. There, on another Gondwana fragment not reached by browsing mammals, a different group of large ratites – the elephant birds – filled similar ecological roles to moa.

In this chapter I will describe the vertebrate communities of forest and wetland ecosystems in pre-human times. I will also discuss how the species assemblages changed over time as climates warmed and cooled.

This is the most speculative chapter in this book, as knowledge of the ecology of extinct animals can only ever be indirect.

The first moa bones were discovered in 1839. They were considered collectable items, but although vast numbers of them were shipped to Europe and stockpiled in New Zealand museums, only recently has any attempt been made to put these birds into an ecological context. Until the late 1980s, there was even confusion over the number of moa species those bones represented,[1] and work currently under way may again result in changes to moa taxonomy. Bones of smaller birds, bats and reptiles had been found, but the over-riding interest in moa meant that for a long time these 'lesser' creatures were largely overlooked. Only recently have systematic studies of reliably dated material enabled a new breed of biologists, the palaeoecologists, to describe the ecological communities to which these animals belonged. In 1989, Richard Holdaway published a speculative paper on the ecology of the extinct birds. This paper, together with another published two years later by Ian Atkinson and Phil Millener,[2] provides a good introduction to pre-human vertebrate communities and how ecological relationships may be inferred from collections of old bones.

During the last decade there has been increased interest in prehistoric faunas. A series of papers by Trevor Worthy and Richard Holdaway[3] have been of special importance in changing our understanding of New Zealand's pre-human ecosystems. The huge volume of information now available on New Zealand paleoecology has been admirably compiled in their major book *The Lost World of the Moa,* which was published too late for its new findings to be included here but is essential further reading for anyone seriously interested in paleoecology.

An understanding of pre-human ecological communities is of more than just academic interest. In the last 2000 years, 53 species of vertebrates have become extinct from the New Zealand mainland, while others such as the kokako are now so rare that they no longer play a significant role in the ecological community of which they were once part (see Chapter 10). Since first human contact, 14 species of Australian birds have colonised this country, and in the last 150 years 72 species of mammals and birds have been introduced to New Zealand. Some of these recent arrivals, such as the possum and silvereye, now rank among the country's most common vertebrates. As native species were lost and exotic species became established, ecological communities have been profoundly altered. Only by understanding the full range of species present, the ecological roles of those lost, and how these species interacted

Table 4.1 Predators of vertebrates on the New Zealand mainland

Pre-human times	Present day
Haast's eagle*	
Aptornis*	
New Zealand owlet-nightjar*	
New Zealand raven*	
Tuatara*	
Eyles's harrier*	Australasian harrier
New Zealand falcon	New Zealand falcon
Laughing owl*	Little owl
Morepork	Morepork
Weka	Weka
	Cat
	Ferret
	Stoat
	Weasel
	Dog
	Rats (3 species)
	Brushtailed possum

* Species now extinct on the New Zealand mainland (North and South Islands)

Table 4.2 Herbivorous vertebrates on the New Zealand mainland

Pre-human times	Present day
Forests and other lowland habitats	
Moa (10 species)*	Deer (8 species)
Takahe	Goat
Kakapo	Brushtailed possum
Kokako	Kokako
New Zealand goose (2 species)*	Wallaby (2 species)
Finsch's duck*	Pig
New Zealand coot*	Sheep
Kereru	Kereru
Paradise shelduck	Paradise shelduck
Alpine	
Moa (3 species)*	Red deer
Takahe	Takahe
Kea	Kea
Finsch's duck*	Tahr
New Zealand coot*	Chamois
	Hare
	Brushtailed possum

* Extinct

can we grasp the magnitude of these human-induced changes. Tables 4.1 and 4.2 (opposite) show how radically ecological communities have altered.

Forest communities

New Zealand has a variety of forest types, including beech, kauri, podocarp, broadleaf and various combinations of these. They range from sea level to alpine elevations, and from scrub to mature forest. There is a broad regional variation from warm Northland to cool Stewart Island, and from dry Canterbury to Fiordland rainforest. Naturally, the bird species present in each forest type varied.[4] To illustrate the importance of climate and forest type in determining which species were present both during the last glacial advance and during the warmer period shortly before humans discovered this country, I will compare the animals that occurred in Westland with those in Canterbury and Otago.

Extinct predatory birds

Haast's eagle, with a wingspan of three metres and talons like tiger's claws, was the largest eagle known to have occurred anywhere in the world. These birds preyed on large ground-dwelling birds, including moa, probably by pouncing on their prey from a perch overhead. They occurred only in the South Island and were forest-dwelling, but most common where there was a mosaic of open forest, shrubland and grassland. Haast's eagles were apparently absent from rainforests.

Eyles's harrier occurred in forests throughout the North and South Islands but was perhaps most common in dry eastern areas. There has been confusion over the taxonomy of this species, which was apparently related to the harrier hawks but had short, compact wings, which suggests that it behaved like a goshawk and could fly through the forest. Eyle's harriers probably hunted medium-sized birds by waiting at a perch then pursuing or pouncing on birds that passed beneath.

The two species of aptornis, or adzebills, are among the least understood of our birds and very different from the rails, their probable sister group. Their necks and heads were strongly muscled and they were armed with a stout, sharp bill. Perhaps they tore apart rotting logs or dug out large invertebrates, tuatara, lizards, frogs and burrowing petrels. Aptornis were mostly birds of the lowlands, more common in dry, open forests than in rainforests.

Weighing just 200 grams, New Zealand's owlet-nightjar was one of the smallest birds in the pre-human predator guild, even though it was by far the world's largest owlet-nightjar. These birds were nocturnal, flew poorly and probably captured their invertebrate, frog and lizard prey on the ground.[5] At Waitomo and Punakaiki, the location of owlet-nightjar fossils suggest the birds may have roosted in caves.

Laughing owls (see page 92) were widespread in the South Island until the 1880s, after which they declined. The last sighting was in South Canterbury in 1914.

Ravens elsewhere take a wide variety of foods, including small animals, carrion and fruit, so we may assume the New Zealand species behaved in a similar manner. Large insects, lizards, tuatara and petrels were likely prey. New Zealand ravens occurred from the far north to Stewart Island, with a sister species on the Chatham Islands. They were apparently most common at coastal sites, in particular near seal and seabird colonies.

Predators

There is a common misconception that New Zealand animals evolved in the absence of predators. Predatory mammals were absent, but there were predatory birds and reptiles (although the number of species was fewer than on continental land masses). In New Zealand forests, there were six species of birds that preyed mostly or entirely on vertebrates: the New Zealand falcon, Eyles's harrier, Haast's eagle, laughing owl and two species of aptornis. In addition, the owlet-nightjar, long-tailed cuckoo, morepork, kingfisher, tuatara and possibly the giant gecko (kawekaweau) ate small vertebrates and large invertebrates. The New Zealand raven was probably a scavenger that preyed on small animals when given the opportunity.

Today, few species of predatory birds remain, tuatara occur only on some offshore islands, and the kawekaweau is presumably extinct. The native predators have been replaced by introduced mammals, and the Australasian harrier probably became established in New Zealand long after Eyles's harrier became extinct. Insect-eating birds, lizards and frogs were also present. These are predators in the true sense of the word but are discussed in the next section.

Haast's eagle was only found in the South Island, and the North and South Islands each had its own species of aptornis. The other predatory birds and tuatara were widespread on both main islands and, in suitable habitats, all species could co-exist. Today, with virtually all the dryland forest removed, we imagine that the diversity of wildlife was greatest in the rainforests, but for predators this was not so. Fossil deposits suggest that there were more species of predators present in dryer open forests or in areas with a forest/scrub mosaic than in rainforests. Laughing owls favoured rocky areas in open country, dry open forests, open beech forest or forest margins. The remains of laughing owls have been found at Punakaiki, of Haast's eagles at Takaka, of Eyles's harriers at Waitomo, and of aptornis near Karamea. However, these remains date from the last glacial period, when forest was less extensive, and not during later times, when rainforest predominated.[6]

These predators were all more common on the eastern side of the South Island, where a mosaic of forest and scrubland occurred during both epochs.[7] Of the predators, only the owlet-nightjar appears to have been common in western rainforests, and it was also common at Pyramid Valley in Canterbury.[8]

How did these species apportion the food resources among themselves? The species that took mostly vertebrate prey varied in size from the falcon (female average weight 460 grams, males 260) to Haast's eagle (females 13 kilograms, males 10). In addition, there were differences in their hunting methods. Haast's eagles and Eyles's harriers were perch-and-pounce hunters, waiting at perches for prey to pass beneath. The harriers probably engaged more in pursuit than did the eagles, but their main difference was in prey size. Eyles's harriers, being less than a quarter of the eagles' weight, probably took large passerines and birds the size of kaka, kakapo, kereru, kokako and Finsch's duck,[9] while eagles apparently specialised on medium-to-large flightless birds, including moa, geese, takahe, kakapo, Finsch's duck and maybe aptornis.

In pre-human New Zealand, frogs, lizards and tuatara were widespread and probably common in many habitats. The aptornis, owlet-nightjar, laughing owl and raven all appear to have included frogs and reptiles among their prey. Richard Holdaway suggested that aptornis ate a variety of ground-dwelling animals, including tuatara, but they probably also took some plant foods.[10] Aptornis was perhaps the main predator of petrels and their chicks. Today, virtually all petrel colonies are on offshore islands, but in pre-human times petrels bred at many mainland localities, including some far inland, and would have been a ready food source for predators able to exploit them.

Falcons utilise grassland and forest habitats from sea level to the alpine zone. Mostly they hunt in open habitats or along forest edges, taking small prey, mostly vertebrates, by aerial pursuit. They may dive onto prey from high overhead or attack unsuspecting prey while flying low to the ground. In pre-human North Canterbury, parakeets were the most frequently taken vertebrate prey, but they also ate short-tailed bats, lizards, many small birds and occasionally larger birds, including kereru and grey ducks. In South Canterbury, carabid and other large beetles were important falcon food.[11] The morepork, both cuckoo species and the sacred kingfisher also ate invertebrates and small vertebrates. These four species will be discussed in Chapter 7.

In pre-human times tuatara were widespread. They are sit-and-wait hunters; large beetles are their most frequent prey, but weta, spiders, other invertebrates plus reptiles and birds' eggs are also eaten (see Chapter 2).

Scraps from owls' and falcons' tables

We have learned much from laughing owls about the pre-human distribution of reptiles, bats, small birds and large insects. The owls nested in dry caves or crevices, which happened to be suited to the preservation of regurgitated bones or insect exoskeletons. Generations of laughing owls used the same caves and in each generation the remains of their prey accumulated on top of the bones voided by earlier inhabitants.

Laughing owl middens are especially valuable to palaeoecologists because they contain small and flighted animals that are uncommon or poorly represented in human middens or in cave, swamp and dune deposits. Predator Cave, near Takaka, was inhabited by laughing owls for about 10,000 years and the remains were deposited in a stratigraphic sequence that spans the duration of the Holocene.[12] Because the laughing owl survived into European times, its middens document the arrival of kiore, the extinction of many small vertebrates and, eventually, the arrival of European rodents and passerines.

Laughing owls were nocturnal or crepuscular ground-hunting generalist predators, taking animals weighing up to 350 grams. Differences in the composition of the middens between regions and over time presumably reflect differences in the prey faunas and provide insights into the habitat preferences of prey species that are now extinct. At Takaka, prior to the arrival of kiore, short-tailed bats (both species) and geckos each made up a quarter of the owls' prey, with birds constituting the major prey class.[13] Parakeets and New Zealand wrens were the most commonly taken birds. At nearby Gouland Downs, reptiles and frogs were the most important prey items, with small nocturnal geckos making up over half of all individuals present in the middens.[14] Gouland Downs owls also took short-tailed bats and small birds, again preferring parakeets and wrens. In Canterbury, where the climate is much drier, bats, reptiles and wrens were rare in owl middens.[15] Fantails, robins and piopio were common prey in Canterbury, but seldom taken at Takaka. At Hermits Cave on the West Coast, the main prey of laughing owls was petrels, but they also ate other birds, reptiles, frogs, bats and even eels.[16]

In Canterbury, the major invertebrate prey of laughing owls consisted of large, nocturnal, ground-dwelling or trunk-dwelling beetles. They may also have taken soft-bodied invertebrates, which would not be preserved, but notably absent are weta, which would surely have been preserved had they been hunted.[17] Among the insect remains are several large beetles that now occur only on predator-free islands. Owls provide some of the best evidence of the relict nature of those beetles' present distribution.

Soon after kiore first appeared in owl middens, wrens, snipe and some other small birds disappeared. Kiore became the dominant prey until European rats, mice and passerines made their appearance.[18] Soon after this, at the beginning of the twentieth century, the owls themselves became extinct, presumably victims of cats and mustelids.

New Zealand falcons sometimes also nested on rock ledges or caves, where the remains of their prey accumulated, although these have provided less information on prehistoric distributions than have laughing owl middens.[19] Falcons are diurnal predators, so they sample a different range of prey than did laughing owls. Near Waiau in North Canterbury, snipe, a large skink, Duvaucel's gecko and short-tailed bats have been found in falcon middens but not in cave deposits. Fernbird, pipit and New Zealand quail bones first appeared in Polynesian-era falcon middens, reflecting the clearance of forest in the area.

Ground-insectivore guild

A wide variety of insects, earthworms and other invertebrates live in soil and leaf litter, and in most parts of the world these are food for mammals. In the northern hemisphere, moles, shrews and related mammals are the primary ground insectivores; in Australia, small marsupials such as antechinus, bandicoots and the marsupial mole have this job. In New Zealand, birds fill the role, and this ecological guild* was one of the most distinctive features of New Zealand's forest communities.

* An ecological guild is a group of species with closely related ecological niches and that co-exist exploiting similar foods.

The largest members of this guild were kiwi. In addition, two smaller birds, unrelated to kiwi or to each other, also probed leaf litter and soil for invertebrate prey. The New Zealand snipes are related to other snipe but, in appearance and in ecology, resemble tiny kiwi. The snipe-rail had a longer bill than other rails and was quite unlike them in body shape. All of these birds were strictly ground-dwelling. The kiwi and snipe-rail were flightless, and the surviving island races of snipe fly only reluctantly.

The large kiwi (2–2.4 kilograms), little spotted kiwi (1.2 –1.5 kilograms), snipe-rail (275 grams) and New Zealand snipes (90 grams) formed a size series and beak-length series of ecologically similar species that formerly must have co-existed.[20] The large kiwi (great spotted kiwi, brown kiwi, eastern kiwi and tokoeka) had non-overlapping ranges. The little spotted kiwi and snipes were widespread on both main islands. Kiwi were common in rainforests but, together with snipes, were less common in Canterbury and Otago, where in places the eastern kiwi appears to have been the only member of this guild.[21] In the North Island, kiwi were more numerous in the Waitomo region than in drier Hawke's Bay habitats.[22] The snipe-rail is known only from the North Island. The length of the bill limited the depth to which each of these species probed. Females of the large kiwi species had the longest bills, at 130 millimetres (female kiwi have longer bills than males), while snipe bills were about 45 millimetres long.

While kiwi and the living races of snipe obtain some prey from the soil surface and from the topmost layers of the leaf litter, there was a separate group of leaf-litter invertebrate hunters. The surface and leaf-litter guild included three species of flightless wrens and the New Zealand robin. The wrens apparently gleaned insects from leaf litter, decaying logs and undergrowth. Robins take a range of invertebrates, ranging from insects less than 5 millimetres long up to much larger and heavier weta and stick insects.[23] They mostly forage on or close to the ground and catch prey by fossicking through leaf litter or by scanning and then pouncing. The weka is a ground-dwelling omnivore that includes leaf-litter invertebrates in its diet. Weka eat large invertebrates, including large land snails, lizards, fruits and carrion, obtained by a variety of feeding methods. At some sites, all litter-feeding insectivores apparently co-existed. Ground-dwelling insects were once much more abundant than they are today, and only on some mammal-free islands can we appreciate the prior abundance of these invertebrates.

Ecology of the New Zealand wrens

The New Zealand wrens were small, stout-bodied, insectivorous birds. Descended from a common ancestor, each of the six species was adapted for its own distinctive ecological role. All species occurred in the South Island, and fossil deposits suggest that at some localities at least five species co-existed.[24] Four species are now extinct.

Both the diminutive rifleman and the slightly larger bush wren had slender bills and lived in tall forest, in particular beech forests, where they fed on small insects and spiders. The rifleman mostly feeds on tree trunks and large branches, probing into crevices in the bark and searching among epiphytes. In contrast, the now-extinct bush wren obtained most of its food from foliage or outer branches and occasionally from the ground. The rock wren, closely related to the bush wren, inhabits subalpine scrub and open rocky ground. These tiny birds (16 grams) remain above the bushline throughout the winter. They have long legs and long toes, so are accomplished rock climbers, but with their short, rounded wings they do not fly far. Rock wrens find insect food between and under boulders and stunted vegetation, and eat seeds and berries during winter when insects are scarce.[25]

The Stephens Island wren was both discovered then promptly exterminated in 1894, reputedly by the lighthouse keeper's cat, although it is possible that bird collectors took some of the specimens now in museums. Although this wren was confined to Stephens Island at the time of European settlement, fossils have been found on the North and South Islands.[26] The cat's owner, perhaps the only European ever to see the wrens alive, reported that they lived among rocks and ran about like mice.

The stout-legged wren occurred on both the North and South Islands. They were flightless and gleaned insects from leaf litter, decaying logs and undergrowth. The legs of this wren were perhaps used to penetrate thickets and tear rotten logs to expose insects unavailable to other birds. Fernbirds and yellowheads also have stout legs and their lifestyles may suggest ways in which stout-legged wrens fed. Fernbirds live in dense, reedy vegetation, while mohua frequently feed suspended from twigs and branches by their legs.

The long-billed wren had an extraordinarily long downcurved bill, which was perhaps used to probe into deep holes and crevices on tree trunks, thick-barked subalpine shrubs and fallen timber. It, too, was flightless. The long-billed wren was described in 1991 and fossils are known from only a few caves in Nelson, North Westland and Southland.[27]

Tree-dwelling insectivores

The ground-dwelling insectivores and the New Zealand wrens were all specialised insectivores (though even they occasionally ate fruit), and within those ecological guilds each species had quite a distinctive niche, with relatively little overlap. In contrast, the niches were less well defined for members of the arboreal (tree-dwelling) insectivore guild. The diets of these birds tended to be broader, with much overlap between species. To further complicate matters, most arboreal insectivores also ate other foods, and certain herbivorous and predatory species often ate insects. Most members of this guild survive, and the ecology of these is discussed in Chapter 7. In this section I discuss New Zealand's bats and some of its more distinctive insectivorous birds. Although a few species had restricted ranges, most of the insectivores co-existed over large parts of the country.[28]

Huia fed in a manner that separated them from all other species. They extracted huhu and other beetle larvae, weta and other insects from logs. Sir Walter Buller described how pairs fed together, the male using his stout bill to chisel into decaying wood, while the female used her long, curved, forceps-like bill to extract insects from soft wood or crevices.[29] The males could also insert their bill into cracks, then force the bill open to split rotting wood and expose the insects within. Buller noted that huia were denizens of thick forest, where they moved 'by a series of bounds or jumps'. They could not fly far. In pre-human times, huia occurred from North Cape to Wellington, but at the time of European settlement they were restricted to the southern part of the North Island.

Saddlebacks take a range of large invertebrates using their heavy bill to rummage among leaf litter, strip dead bark, break up dead wood and probe into holes in trees, under bark or among leaves.[30] In season they eat fruit and even nectar. Saddlebacks probably prefer dense forest but are adaptable, and on the islands where they now occur they utilise virtually all forest types and even forage around the forest fringe. They feed at all levels, from the ground to the canopy. Saddlebacks are conspicuous, noisy birds whose frenzied activity keeps observers entertained.

The piopio was probably the most omnivorous forest insectivore, eating invertebrates, fruit, seeds and even foliage. It obtained insects by gleaning, turning litter and occasionally hawking, and it fed on the forest floor and in the forest undergrowth. Piopio were found from Northland to Southland, from the coast up into the mountains. Observations by

T. H. Potts[31] suggest that they favoured forest edges, scrub and riverbanks. Fossils indicate that piopio were common in dry open forests but rare in rainforest.[32]

The three species of bats were also insectivores. Short-tailed bats have been more often found as fossils than long-tailed bats, but this could be due to the preference short-tailed bats had for roosting in caves and their greater susceptibility to laughing owl predation. In pre-human times, all three species of bats co-existed at some localities in both the North and South Islands, though some differences in distribution are apparent.[33] For instance, in laughing owl middens at Takaka, lesser short-tailed bats were more common than greater short-tailed bats. Greater short-tails were the more numerous species in most North Canterbury locations, while in South Canterbury (and also at Pyramid Valley) only the larger species was present.[34] Trevor Worthy suggested that this distribution might reflect a preference of the smaller bats for beech forest, or the larger bats for podocarp forest.

Very little is known about the now-extinct greater short-tailed bat. By 1840 the larger species was confined to muttonbird scrub and rata forest on the Big South Cape Islands, atypical of the habitats utilised in prehistoric times. Worthy and co-workers[35] suggested that these bats were even more terrestrial in habit, and targeted larger prey, than their smaller cousins, perhaps including frogs and lizards in their diet. The owlet-nightjar was also a nocturnal insectivore that may have also taken small lizards and frogs.[36]

Browsers and grazers

> It was in the latter extensive open grasslands of the South Island that the moa abounded, particularly the massive forms whose ungainly bulk would probably have caused them to favour the lowlands, leaving the higher tussock and scrub country to the more slender species, e.g., of *Anomalopteryx* and *Megalapteryx*, and probably of the tall *Dinornis* also . . .
>
> – G. E. Archey (1941)

Leaves are the most abundant and most easily obtained forest food, but contain few nutrients and provide little energy. Furthermore, the digestible cell contents are enclosed in cell walls made of cellulose, which vertebrates are unable to digest. In order to fully utilise leaves, birds and mammals must harbour micro-organisms that can digest cellulose, have mechanisms to physically break those cell walls, or resort to quick throughput of huge quantities of food, utilising only the most easily digested material and excreting the bulk.

Herbivorous mammals have teeth, but birds must resort to stones in their muscular gizzards to break down plant material. Takahe, the smallest of the flightless herbivores, utilise the option of rapid throughput and large volume, extracting the juices, which are easily digestible, and voiding the less easily digested fibrous matter.[37] Moa had gizzard stones, and some species may also have had large vat-like stomachs where micro-organisms fermented leaves, as they do in ruminant mammals. The power of flight and adaptation to a heavy, bulky leaf-dominated diet are almost mutually exclusive.

New Zealand once had a diverse guild of herbivorous birds. Moa, geese and takahe were the major vertebrate browsers. The flightless kakapo, Finsch's duck and New Zealand coot, and flighted kokako, kea and kereru took leaves as part of a diet in which fruits, seeds and other more nutritious plant parts were preferred. Of the 11 moa species (Table 4.3), seven occurred in the North Island and nine in the South Island.[38] All nine South Island moa are known to have occurred in Southland, though not all shared the same habitats.[39] Up to four species of moa, plus the takahe and goose, are known to have co-existed in parts of New Zealand.

Until the 1970s, moa were generally considered to be grassland birds, and up to the late 1980s they were usually treated as a single entity, implying that the ecology of all species was similar. However, in recent years we have begun to appreciate the tremendous diversity in the morphology and ecology of moa. They ranged in size from the small upland moa, coastal moa and Mappin's moa, some of which weighed as little as 20 kilograms, to the giant moa, which may have weighed more than 270 kilograms.[40] Six species weighed more than 100 kilograms – the weight of an ostrich, the largest living bird. Bill shapes varied among the six genera and closely related species apparently had different habitat preferences.

All moa were herbivorous, at least as adults (like some other herbivores, young moa probably also ate insects.) The diets of giant moa and slender bush moa are best known because stomach contents of these species have been preserved. They include many twigs and stems, but fewer leaves or seeds. Giant moa are known to have eaten at least 26 species of plants, and the slender bush moa 16 species.[41] Some of the tough old twigs present were neatly sheared through, showing how powerful the bills of these moa were. Some gizzards from dinornithid moa contained more than 5 kilograms of stones, essential to process such poor-quality food. In some places, all three dinornithid moa species have been found together, but it is unlikely that they were all present at the same time. Differences in the height they could feed at was one factor that allowed them to co-exist. Giant moa may have occurred mostly in forest margins and open habitats, while the other dinornithid moa

Table 4.3 The taxonomy, distribution and habitats of moa

ORDER DINORNITHIFORMES, moa

FAMILY DINORNITHIDAE, dinornithid moa
Giant moa *Dinornis giganteus.*
An uncommon species found in North Island and eastern South Island lowlands, usually in scrub and open forests.
Large bush moa *D. novaezealandiae*
North and South Islands; wet dense lowland forests and upland areas.
Slender bush moa *D. struthoides*
Widespread in North and South Islands from lowland to subalpine habitats, most common in forests. The smallest dinornithid.

FAMILY EMEIDAE, emeid moa

SUBFAMILY EMEINAE
Stout-legged moa *Euryapteryx geranoides*
North and South Islands, dry inland forests and scrublands; also swamplands but rare in wet climates.
Coastal moa *E. curtus*
North Island, open areas preferred, mostly coastal; rare in wet lowland forests.
Eastern moa *Emeus crassus*
South Island, dry scrub and open forests in lowland eastern areas; in swamplands but absent from wet western areas.

SUBFAMILY ANOMALOPTERYGINAE

Upland moa *Megalapteryx didinus*
South Island, montane to subalpine forest, scrub and tussock; rare in lowlands.
Little bush moa *Anomalopteryx didiformis*
North and South Island, wet lowland podocarp and broadleaf forests.
Mappin's moa *Pachyornis mappini*
North Island, coastal and wet lowland shrub/forest mosaics; common only in swampy locations.
Heavy-footed moa *P. elephantopus*
South Island, dry eastern forests and grasslands, present on the west coast during glacial periods when shrublands prevailed. Absent from wet, tall forests.
Crested moa *P. australis*
South Island, western montane forests and subalpine areas; rare in wet lowlands.

Primary sources for this table were Worthy 1990, 1991b, 1998b and Holdaway & Worthy 1991. Current research is likely to result in significant changes to moa taxonomy.

are believed to have been forest-dwellers. The dinornithid moa were taller and more slender than other species.

Moa of the subfamily Emeinae all had short, rounded bills and weak head and jaw musculature, suggesting they ate less-fibrous foods such as new leaves, buds and fruit.[42] The stout-legged and coastal moa are known to have carried only a few small gizzard stones. There was a fourfold difference in size between the stout-legged and coastal moa, both of which were in the same genus and co-existed in some places. Upland moa also had weak bills, limiting them to softer foods. In contrast, all other moa in the subfamily Anomalopteryginae had sharp, robust bills, apparently adapted to cutting fibrous leaves. The heavy-footed moa included flax and other fibrous leaves in its diet, suggesting that it frequented swamp margins, but its bones have also been found in forested areas. Presumably these robust-billed birds and the dinornithid moa had micro-organisms in their guts that further broke down fibrous foods.

Trevor Worthy[43] recognised three broad assemblages of moa, each associated with different habitat types. In tall, dense, wet forests, the little bush moa, large bush moa and slender bush moa were the most common. Species of *Pachyornis* and *Euryapteryx* were present though rare. In the dense forests at Waitomo, large bush moa and little bush moa were most common.

A different group lived in North Island coastal dunelands and in the eastern South Island lowlands, areas that had a mosaic of forest, shrubs and grasslands. In the North Island, coastal moa were most common. The stout-legged, crested, Mappin's, giant and slender bush moa were sometimes also present. Mappin's moa was associated with wetland margins or shrublands.[44] In the eastern South Island lowlands, heavy-footed moa, eastern moa, stout-legged moa and giant moa were present.[45]

Although moa were primarily birds of the lowland forests, there was a third group that used shrubland and grassland habitats in montane and subalpine zones. The upland moa was the most common species in all upland areas, but crested moa and (less often) large bush moa or slender bush moa were sometimes also present. Upland moa had feathered tarsi (lower legs), presumably an adaptation to living in cold, sometimes snowy habitats. Their long toes and light build may have facilitated walking on soft surfaces such as snow and the moss carpets typical of montane and subalpine habitats. Moa that inhabited open forests and scrublands – for example, species of *Euryapteryx* and *Pachyornis* – were generally stouter and heavier than moa that lived in dense forests.

Worthy concluded that each family or subfamily of moa had a distinctive bill morphology, which enabled its members to exploit quite different foods. Differences in the height, size and habitat preferences of closely related birds would have minimised competition among them. Closely related birds of a similar size were not found in the same location during the same era.

Takahe were formerly widespread, but the North Island species is now extinct and the South Island takahe endangered. Without human intervention, takahe are now confined to subalpine grasslands in Fiordland, where they have a highly specialised diet: most of their food consists of three species of snow tussock and one species of mountain daisy. Takahe only eat the basal portions of tussock tillers and the leaf bases of the daisies, and they even select individual plants high in nitrogen and phosphorus.[46] Jim Mills suggested that takahe are a relict of the Pleistocene era, especially adapted to the subalpine grasslands. However, the bones of takahe have been found in many parts of New Zealand, including forested areas, suggesting that in the past they must have eaten a more diverse diet in a wider range of habitats.[47] Their preferred habitat may have been the ecotone between forest and shrubland,[48] in which case their present specialised feeding ecology and restricted montane distribution give a misleading impression of the species' former ecological role.

Some waterfowl and rails were terrestrial herbivores. Goose bones have most commonly been recovered from places that supported grass or shrub vegetation. Geese were common at low-altitude inland localities but rare near the coast, where paradise shelducks utilised both wetland and terrestrial habitats.[49] Finsch's duck occurred in shrublands, grasslands and open forests, making little use of wetlands, and presumably ate mostly invertebrates and fallen fruit. During the Holocene, South Island geese and Finsch's duck were more common in the forest mosaics of Canterbury and Otago than in Westland rainforests. In wetter regions or in closed-canopy forest, Finsch's duck were replaced by brown teal,[50] which are now almost solely restricted by day to slow-moving waterways with overhanging vegetation, although at night the Little Barrier Island birds feed ashore. Fossil evidence and the bird's relatively large legs and pelvis suggest that in wet, forested parts of the country brown teal utilised both wetlands and forests. The New Zealand coot was also primarily terrestrial and, while most common near wetlands, also utilised dry shrublands.[51] Hodgen's rail was probably omnivorous.

Adaptations of plants to browsing by moa

Over time, plants evolve defences against the herbivores with which they co-exist. New Zealand plants exhibit several unusual features that some scientists believe evolved in response to moa browsing.[52] The best known are the divaricating plants, which have tightly intertangled, spreading twiggy branches with few small leaves and no spines. The leaves of some divaricating plants are found only beneath the tough, interlaced, often springy surface layers.

New Zealand has 53 species of divaricating plants, belonging to 20 genera from 16 families of flowering plants. Most of these genera – for example, *Coprosma* – have both divaricating and non-divaricating species. Nearly 10 per cent of native woody plants show divarication. A few plants with a divaricating habit do occur elsewhere in the world, but most are spiny or have larger leaves than the New Zealand species.

Divaricating species occur in forests, scrublands and grasslands, in the lowlands and in the mountains, in both dry and wet locations, but are more common on high-fertility than low-fertility soils. There are few divaricating epiphytes.

A further nine species have a divaricating habit when young but, when taller, adopt an upright growth form and produce larger leaves. Moa did eat at least some species of divaricating plants,[53] but Ian Atkinson and R. M. Greenwood believe that divaricating plants co-evolved with moa, their divaricating habit giving them a selective advantage.

Not everyone agrees. Matt McGlone and Colin Webb[54] suggested that the divaricating growth form evolved during the Pleistocene, and the tangled outer branches protected the growing points and leaves from damage by wind, frost and desiccation. Atkinson and Greenwood defend their hypothesis by pointing out that in New Zealand, divaricating plants are found in cold and warm, wet and dry habitats, and that elsewhere in the world divaricating species are rare. Moa only occurred in New Zealand, but cold, dry, windy conditions existed in other countries as well. Perhaps the debate will be resolved by the current genetic studies that should indicate how long plants have been divaricating. If divarication evolved in response to moa, the trait will be ancient; if it evolved in response to climate, the divaricating habit should have evolved within the last two and a half million years.

New Zealand plants show certain other peculiarities that may also be related to moa browsing.[55] The tough, fibrous leaves of flax and cabbage tree would provide resistance to all but the most determined herbivores.

The long, stiff, downward-pointing brown leaves of juvenile lancewood must have been less conspicuous to moa than were the more elliptical, green leaves produced by the adult tree. Most large mainland species of *Aciphylla* have rosettes of needle-sharp leaves that protect the sugar-rich leaf bases; however, the two species from the Chatham Islands, where moa never roamed, have soft leaves. It may also be significant that New Zealand has a relatively high number of plant species with purplish-black or bronze-coloured leaves. If, like other ratites, moa had poor colour vision, these leaves would be difficult for them to see.

Tree-dwelling eaters of plants

> . . . the wooded valley of the Mangaone, in which we have been camped for the night is ringing with delightful music – The silvery notes of the bellbird, the bolder song of the Tui, the loud continuous strain of the native Robin, the joyous chirping of a flock of whiteheads, and the whistling cry of the Piopio – all these voices of the forest are blended together in wild harmony. And the music is occasionally varied by the harsh scream of a Kaka passing overhead, or the noisy chattering of a pair of Parakeets, – at regular intervals the far off cry of the Long-tailed Cuckoo and the whistling call of its bronze-winged congener; while on every hand may be heard the soft trilling notes of Myiomoira toitoi. For more than an hour after this concert had ceased, and the sylvan choristers had dispersed – one species continued to enliven the valley with his musical notes. This bird was the Piopio – unquestionably the best of the native songsters.
>
> – Sir Walter Buller (1888)

In New Zealand's evergreen forests, leaves are the most abundant plant food and are available year-round. Seeds, fruits, nectar, rhizomes and flowers are more nutritious but are only available for part of the year. Most tree-dwelling herbivorous birds have broad diets comprising leaves, fruits, seeds, flowers, rhizomes, nectar and insects in varying ratios. Most browsers and insectivores also eat fruit during the months they are available. Diets vary not just between species but for each species seasonally, from year to year and between different forest types. These tree-dwelling herbivorous birds include the parrots, kereru, kokako and honeyeaters. A thousand years ago most of these species probably co-existed in many parts of the country. By and large, information on the foods and foraging of birds in this guild is derived from observations made in today's fragmented and depleted forests. The ecology of some species may have been quite different in pre-human ecosystems.

We can use the parrots to show how foods, foraging behaviour and habitat use varied among herbivorous birds. The giant flightless kakapo is the only parrot that is exclusively vegetarian. Kakapo eat a wide variety of plants, ranging from herbs, ferns and tussocks to shrubs and trees, but they favour storage organs such as rhizomes and bulbs, or new growth including leaves, flowers, fruits or the cambian tissue that lies under bark.[56] The kakapo's bill has a series of ridges on the upper mandible that enables the bird to grind plant material and extract the nutritious sap, leaving behind indigestible fibrous material. Kakapo may be flightless, but they do climb trees and thus feed from ground level up to the forest canopy. Fossil remains, and their distribution in the nineteenth century, suggest that kakapo were the most common parrots in wet, western forests of both islands, but they were rare in seasonally dry eastern zones of the South Island, where rimu, an important food, was scarce. However, they were apparently common in Hawke's Bay, in the eastern North Island.[57]

Kaka eat fruits, nectar, pollen, sap, seeds and insects. They also dig into live wood to obtain sap exudate or insect larvae.[58] In beech forests, honeydew is now their main energy source, but they also extract longhorn beetles from trees to obtain the proteins and fats necessary for a balanced diet.[59] Kaka take nectar with their brush-tipped tongues and, in Westland during spring and summer, this can comprise up to a quarter of their diet, while at other times insects are a vital food.[60] Although their diet includes many different foods, kaka tend to concentrate on particular foods at different times of the year.

Most people think of kea as alpine birds, but they also utilise forests, in particular beech forests, mostly near the treeline but also at lower altitudes. Fossils indicate that they used forested habitats in pre-human times. They are known to take over a hundred species of plants, and from these eat buds, fruits, roots, leaves, seeds, nectar, shoots and leaf bases. Their favoured food plants include alpine daisies (*Celmisia* spp.), spaniards (*Aciphylla* spp.), shrubs such as snow totara and coprosmas, and trees, including beech and rata.[61] Insects appear to be a minor component of their diet. Kea will scavenge and sometimes kill larger animals. They are best described as opportunists that prefer fruits, berries or seeds. Food resources in the kea's montane and subalpine environment vary greatly between seasons and years, while those exploited by the closely related forest-dwelling kaka are less variable. Kea are more inquisitive than kaka, a feature that Judy Diamond and Alan

Bond[62] suggest evolved as an adaptation to their less predictable food supply.

The main islands have two species of parakeets that were once widespread throughout New Zealand. Both take a wide variety of plant foods plus some insects, their diets varying from place to place and between seasons. Yellow-crowned parakeets eat more insects than do their red-crowned cousins. Rowley Taylor[63] found that mainland yellow-crowned parakeets favoured large areas of tall forest, whereas red-crowns are more often seen near forest edges, in scrub and at lower altitudes. A third species, the orange-fronted parakeet, was present in Canterbury and Nelson, and possibly elsewhere in the South Island. The ecology of orange-fronted parakeets is little studied. Taylor did not find any significant differences in morphology, foods eaten or habitats used, between it and the yellow-crowned parakeet. However, studies currently under way do indicate real differences in the ecology of the two species.

Kereru, kokako and kakapo eat lots of leaves but prefer fruit, and even appear to require it for successful breeding. For kokako, fruit is an important food in summer and autumn, leaves predominate in winter and invertebrates are important during breeding and chick-rearing.[64] In central North Island forests, kokako ate the leaves of about 50 plant species, and foliage comprised about a third of their diet. They relied on leaves from trees, epiphytes and shrubs, but also ate mosses, ferns, orchids, vines and shrubs. The sixpenny scale insect, which lives on the leaves of broadleaf trees, is also an important food.

Kereru have a broad diet and eat leaves and fruits from a large number of native trees and shrubs. Fruits are preferred when available. They usually feed alone or in pairs, but when fruit is abundant, flocks of kereru may gather in a single tree. Despite the wide range of plants taken, they often specialise on one species at a time and the browsing pressure on favoured plants can be intense. Mick Clout and Rod Hay[65] estimated that one kereru could eat 2300 kowhai leaflets, 2050 *Coprosma areolata* leaves or 1200 *Parsonsia heterophylla* leaves in a day. Since even today's depleted kereru populations can defoliate favoured plants, the larger populations in earlier times must have exerted considerable pressure on some forest trees.

Kereru fossils are much more common in eastern areas than in the western rainforests[66] – the reverse of their present distribution, which probably reflects the loss of eastern forests rather than a preference for rainforest. Kereru remain present, sometimes common, in forest remnants in eastern parts of the country.

New Zealand has three species of honeyeaters. Unlike some Australian species, none of the New Zealand birds relies solely on nectar but also feeds on both fruits and insects to varying degrees. Their diets vary seasonally, from year to year and between localities. The tui, the largest of the three honeyeaters, is the most nectivorous, the bellbird the most insectivorous, and the small hihi eats the most fruit. Tui and bellbirds occurred throughout New Zealand, while the hihi was restricted to the North Island. On Little Barrier Island, one of few places where all three currently co-exist, all eat insects or fruits in winter when nectar is scarce. The tui and hihi have similar diets, but the tui feeds mostly in the canopy, and the hihi in lower forest tiers.[67] In South Westland, tui fed in higher forest tiers than do bellbirds.[68] When the species compete for resources, tui are dominant over bellbirds, which in turn dominate hihi.

Birds present in North Westland during the Otiran glaciation and the Holocene

The extensive bone deposits recovered from caves near Karamea and at Punakaiki, both on the West Coast, span the last 25,000 years.[69] The oldest of these deposits accumulated during the most recent of the Pleistocene glacial advances, the Otiran glaciation, when the cave entrances were surrounded by subalpine scrub and tussockland interspersed with patches of beech, podocarp or New Zealand cedar forest. The climate began warming and glaciers retreating about 14,000 years ago. By 10,000 years ago the climate was warm and wet, and dense podocarp/beech rainforest surrounded the caves. The forest cover and animals present probably changed little during the Holocene (the last 10,000 years), until the arrival of kiore 2000 years ago and Maori about 800 years ago. Habitats were probably little altered until European settlement. Plates 1 and 2 show reconstructed scenes with animals that would have been present in the Oparara Valley, near Karamea, during the Otiran glaciation, then a short time before kiore and humans arrived. Plate 4 shows the same location in the late twentieth century.

Although the vegetation was very different between the cool, dry Otiran period and the warm, wet Holocene, many of the same birds were present during both periods (Table 4.4). Certain species, however, were present only in one period or the other. During the Pleistocene, upland and crested were the most numerous moa in the Oparara Valley, which at that time supported a mosaic of open understorey forest, scrub, tussock and herbfields. Closer to the coast, in the more heavily forested

Table 4.4 Terrestrial birds of North Westland

	Otiran glaciation	Holocene	Late twentieth century
Browsers			
Little bush moa		•••	
Upland moa	•••	•	
Heavy-footed moa	••		
Crested moa	••		
Stout-legged moa	••		
Slender moa	•	•	
Large bush moa		••	
Giant moa		•	
Takahe	•		
South Island goose	•		
New Zealand coot	•		
Other primarily herbivorous species			
Kakapo		•••	
Kaka		•	•
Kea	••	•	•
Parakeets	•	•	•
Bellbird		•	•••
Tui	•	•	••
Kokako	•	•••	
Kereru	•	•	•
Song thrush			•
Blackbird			••
Silver-eye			•••
Chaffinch			••
Redpoll			•
Goldfinch			•
Greenfinch			•
Omnivores			
Finsch's duck	•••		
New Zealand quail	•	•	
Weka	•	••	•
Hodgen's rail		•	
Piopio	•	•	
New Zealand raven	•		
Predators			
Haast's eagle	•		
Eyles's harrier	•	•	
Australasian harrier			•
New Zealand falcon	•		•
Aptornis	••		
Laughing owl	•	•	

	Otiran glaciation	Holocene	Late twentieth century
Morepork		•	••
Owlet-nightjar	•	•	
Shining cuckoo			•
Kingfisher			•
Insectivores			
Great spotted kiwi	•	•	•
Little spotted kiwi		•	
Snipe	•	••	
Rifleman	•	••	••
Bush/rock wren	•	•••	
Stephens Island wren	•	••	
Stout legged wren	•	•••	
Long-billed wren		•	
New Zealand pipit	•	•	•
Brown creeper		•	•
Mohua	•	•	
Grey warbler		•	••
Fantail		•	••
South Island tomtit	•	••	•••
South Island robin	•	•••	•
Saddleback		••	
Dunnock			•
Long-tailed bat		?	
Greater short-tailed bat	•		

• •• ••• relative abundance

Sources: Worthy & Mildenhall 1989, Worthy & Holdaway 1993, Onley 1980, 1983, and personal observations

Punakaiki region, the heavy footed-moa and stout-legged moa were the most common species.[70] The upland moa was the only species of moa present at Punakaiki in both the Pleistocene and the Holocene, but during the Holocene those birds in the rainforests were markedly smaller than the Pleistocene upland moa. As the climate warmed and rainforest replaced the montane forests, the little bush moa and large bush moa replaced the upland and crested moa.

Of the predators, Haast's eagle, aptornis and laughing owl were probably more common during the Pleistocene than during the later, warmer period. The bones of bats, reptiles, frogs and small birds are poorly preserved in the fossil record, so more species were probably present than Table 4.4 shows. Kea would have been common in the Pleistocene but were also present after the climate had warmed. Kakapo and kokako were common in the Holocene rainforests and are but two of many species that became locally extinct after human contact.

During the Otiran glaciation, a mosaic of tussock and scrub with patches of beech/podocarp forest dominated the area. During the Holocene, North Westland supported dense podocarp/beech rainforest. In the last century, much of the coastal and lowland area has been cleared, but forest cover is still more extensive than in most other parts of New Zealand.

A comparison with Canterbury and Otago

To the east of the Southern Alps, Pleistocene and the Holocene vegetation, and consequently bird faunas, differed to a much lesser degree than in Westland. During both eras they were similar to the Westland fauna of the Otiran glaciation.[71] Although eastern vegetation changed little between the two eras, there was much greater spatial variability than in Westland. At Waikari in North Canterbury, the climate was relatively dry during both periods, so giant moa, eastern moa, stout-legged moa and heavy-footed moa were present throughout. The relative proportions of these species did differ, the heavy-footed moa being more numerous during the Otiran than the Holocene. At Mt Cookson, closer to the mountains and with a slightly wetter climate and colder winters, the heavy-footed moa was the most common species during the Pleistocene, but during the Holocene, when contiguous beech forest clothed the area, little bush moa and large bush moa were present.[72]

During the Otiran, the vegetation at both North Canterbury study areas was probably grass, herbfield and open shrubland. Although forest or tall shrubland dominated Canterbury during parts of the Holocene, the forest was drier and more open than in Westland. Unlike western areas, it was still interspersed with patches of scrub and grassland even during the warmest periods. Trevor Worthy[73] suggested that extensive forest was a significant feature of the Canterbury environment for only 4000 of the last 70,000 years. In Canterbury and Otago, the relative abundance of moa species varied over time and from place to place, presumably reflecting local differences in topography and vegetation. For instance, the heavy-footed moa was common whenever or wherever there was open scrub, and the stout-legged moa and giant moa where or when there was tall forest.[74] In Otago, upland moa, little bush moa and crested moa occurred near Wanaka, where it was cool and wet; while in the warmer, more varied habitats of coastal North Otago, heavy-footed moa, eastern moa, stout-legged moa, slender bush moa and giant moa were present.[75]

Quail and pipits, species characteristic of open habitats, were common during the Otiran but declined whenever forest became more extensive. During the Holocene, the Canterbury and Otago bird faunas included most forest-dwelling species as well as species characteristic of more open habitats; thus eastern bird faunas were more diverse than in the west. Piopio, fantails, kereru, mohua, Hodgen's rail, paradise shelduck and snipe were more common in Canterbury and Otago than in wetter climates in Westland and at Takaka Hill.[76] Conversely, weka, kakapo, kiwi, wrens, owlet-nightjar, parakeets and bellbirds were less numerous in lowland Canterbury and Otago than in wetter regions. Laughing owl, aptornis, coot, South Island goose, Finsch's duck and Haast's eagle were common in Canterbury, but on the West Coast they have been recorded only during the Otiran period. At Mt Cookson, forest birds were prevalent up to 800 years ago, but as the forest was cleared following Maori settlement, pipits, quail and fernbirds reappeared in the fossil record.[77]

Holocene faunas in the North Island are less well documented, but there too the species distributions were strongly related to vegetation type.[78] The most common bird species found in eastern areas such as Hawke's Bay and Martinborough were also common in Canterbury, whereas in the wetter Waitomo forests the fauna more closely resembled that of Westland.

Seabird colonies

Seabirds can play an important role in terrestrial ecosystems. Sometimes they breed in high densities and transfer nutrients from the sea on to the land in the form of faeces, dead birds and eggs. On islands where seabirds are still numerous, invertebrate and reptile densities can be very high, and presumably densities were once similarly high at and around mainland seabird colonies. These localities would have provided rich pickings for aptornis and perhaps other predators. Today, very few petrels breed on the mainland. Most species are confined to islands free from predatory mammals (see Chapter 8), but Trevor Worthy and Richard Holdaway have shown that burrow-breeding petrels were once present at many locations around New Zealand. For instance, Cook's petrel, now confined to two offshore islands, appears in Holocene deposits scattered throughout the country. Grey-backed and black-bellied storm petrels, which now breed only on the Chathams or subantarctic islands, apparently once bred on the South Island mainland. Petrel colonies were most common in coastal locations, but four species apparently bred near Mt Cookson, more than 25 kilometres from the coast.[79]

Wetlands and waterfowl

Waterfowl (ducks, swans and geese) and certain other wetland birds were vulnerable to human hunters.* Fourteen species have become extinct since people first colonised New Zealand, and during the European era wetland habitats have been reduced in area by 90 per cent. In this section I describe the wetland birds that once occurred in New Zealand and discuss some aspects of their ecology.

Wetland habitats are diverse and include estuaries, brackish water lagoons, freshwater swamps, lakes and rivers. Naturally, the birds utilising each of these habitat types differed, but space prohibits a detailed discussion of the different habitat types.

Extinct waterfowl, rails and other birds of freshwater habitats

Pelicans were rare in New Zealand and fossils have only been found at Lake Waikaremoana and four coastal locations. These birds were thought to be larger than Australian pelicans and have traditionally been considered a distinct species. However, Trevor Worthy[80] considers the New Zealand specimens to be vagrants from Australia and no longer recognises the local birds as a separate species.

The New Zealand swan occurred on the mainland and at Stewart and the Chatham Islands. It was similar (but was possibly larger and with stouter legs) to the Australian black swan, which became established here in the nineteenth century. Worthy does not consider that the New Zealand birds differed significantly from Australian swans and suggests they are best treated as members of that species. If he is correct, then the species became locally extinct in New Zealand, later to be re-introduced.

The South Island goose and the smaller North Island species were more primitive than modern geese and may have been different enough from other waterfowl to warrant their own family. The New Zealand geese were terrestrial and lived in grasslands and scrublands.

---

* The term waterfowl includes all swans, ducks and geese, not all of which lived in wetlands.

Bones of Finsch's duck have been commonly found throughout New Zealand. This was a heavily built goose-like duck that appears to have lost the ability to fly during the last 12,000 years. By the time humans arrived, it was a terrestrial ground-foraging bird that seldom used wetlands.[81] Finsch's duck was common wherever or whenever there was a mosaic of forest, scrub and grassland. It was particularly common in Canterbury and Hawke's Bay, but rare in closed-canopy, wet forests. Finsch's duck is not closely related to other ducks.

Scarlett's duck was closely related to but larger than the Australian pink-eared duck. It is known from bones found in Marlborough, Can-

Table 4.5 Birds found on New Zealand lakes, ponds and swamps in pre-human times (see also Plate 3)

Herbivores

Black swan
Paradise shelduck
Grey duck
New Zealand scaup
New Zealand coot*
Hodgen's rail*

Eaters of crustaceans, insects and other invertebrates

Dabchick
Brown teal
Grey teal
Scarlett's duck*
New Zealand scaup
De Lautour's duck*
Black stilt
Banded dotterel
Little bittern*
Marsh crake
Spotless crake
Fernbird

Eaters of fish and invertebrates

Auckland Island merganser*
Crested grebe
Black shag
Little shag
White heron
Little bittern*
Black-fronted tern
Black-billed gull
Sacred kingfisher

* Extinct

In compiling this table I follow Holdaway et al. (2001) in assuming that shoveler, pied stilt, Australasian harrier, Australasian bittern, little black shag and pukeko were rare or absent in pre-human times.

terbury, Otago, Hawke's Bay and the Chatham Islands. It was a filter-feeding duck, probably a denizen of swamps and shallow lakes.[82]

De Lautour's duck was related to the musk duck of Australia, which uses deep water in lakes, swamps and estuaries. It was probably rare and fossils have been found at just four sites, two from each main island.

For a description of the Auckland Island merganser see page 117.

The large, flightless New Zealand and Chatham Island coots were probably descended from early colonisations by the Australian coot. The New Zealand coot was a long-legged species that was more like the pukeko than its ancestral coot. It was common in wetlands adjacent to lakes but also occurred in shrublands far away from lakes or swamps. Both these endemic species became extinct in pre-European times, but during the twentieth century the Australian coot once again colonised New Zealand.

The flightless Chatham Island duck is known only from those islands.

Hodgen's rail was a flightless moorhen similar to the Australian black-tailed native hen. It occurred in both the North and South Islands, and its habitat preferences appear similar to those of Finsch's duck. The Australian species is common, utilises many wetland habitats and offers no insight into why the New Zealand species became extinct.

New Zealand little bitterns are known only from specimens collected in Westland last century and fossils from the North Island and the Chathams. They probably used wooded margins of lagoons and creeks.

Herbivores

The swan, grey duck, coot and Hodgen's rail probably grazed on aquatic plants and other plants growing near the water's edge. The swan and grey duck also fed while floating in shallow water, reaching for food on the lake bottom. The swan could feed in deeper water than the duck and was probably most common on large lakes and lagoons. Today, the grey duck occurs in many freshwater and estuarine habitats. They also eat some invertebrates.

The New Zealand coot was very different from its Australian counterpart, which has recently become established here. The latter feeds on plants and invertebrates on pond and lake bottoms, but the New Zealand species had long legs suited to walking on land. It was common in wetlands surrounding lakes but also occurred in shrublands well away from water.[83] Plants and insects probably made up the bulk of its diet. If we assume that Hodgen's rail was similar to the related Australian black-tailed native-hen, it would have been primarily herbivorous but perhaps also ate some insects. Trevor Worthy and Richard Holdaway[84] suggested that the coot and Hodgen's rail were most common along stream margins.

Paradise shelduck, Finsch's duck, the native geese and brown teal were all herbivorous. The geese and Finsch's duck were terrestrial and not members of the wetland community. Although paradise shelduck are now common inland, fossils are rare at inland localities but common near coastal wetlands. Worthy[85] suggested that paradise shelducks and Finsch's duck had mutually exclusive ranges and, since the extinction of the latter, the shelduck has expanded into the interior.

Eaters of insects and crustaceans

Wetland consumers of insects, crustaceans and other invertebrates formed the largest and most complex of the aquatic guilds. Few of these species were specialists, and many also ate plants. Each species of water bird tends to prefer certain aquatic habitats over others, so on that basis we can subdivide this guild into the following broad groups:

- Birds that normally feed in water while swimming – some ducks.
- Birds that fed in emergent vegetation – fernbirds, bitterns and crakes.
- Birds that fed along the shoreline and in shallow waters – herons, New Zealand coot and wading birds.

Doubtless there is overlap between these species in the foods they ate and the habitats they used, but we can divide those birds that fed in water along two further resource gradients:

- Those that occured in small waterways such as ponds and swamps, and those that frequented large lakes and lagoons.
- Those that fed mostly on insects and crustaceans, and those for whom these were minor components of the diet.

Scarlett's duck probably inhabited swamps and large shallow lakes,[86] while scaup preferred large, clear, deep lakes. Brown teal and de Lautour's duck would have usually been found on water bodies of intermediate characteristics. The grey duck is a generalist that occurred in most freshwater habitats. Assuming that the extinct species ate similar foods to their Australian analogue species, brown teal, Scarlett's duck and scaup probably ate mostly invertebrates but some plant material, and the grey duck mostly plant material. De Lautour's duck probably preyed on crayfish, large insects, molluscs and fish – larger prey than those taken by the other ducks.

Fossils suggest that in wetter parts of the country or in closed-canopy forest where Finsch's duck were absent, brown teal were both forest-floor and stream-dwelling omnivores.[87] Today, blue duck mostly inhabit fast-flowing rivers. Occasionally they are seen in forests some distance from running water, and at Takaka their bones have been recovered far from rivers, suggesting that in pre-human times they may have also foraged in terrestrial habitats.

Many ducks filter small organisms from the water, but the shoveler and Scarlett's duck, having spoon-shaped bills with extensive lamellae, were adapted to take planktonic animals that were too small for other ducks to catch. The scaup and de Lautour's duck probably took most of their food by diving. The brown teal, grey duck and grey teal would have fed by dabbling and pecking at prey on the water's surface. These three species ate both plant material and invertebrates, the two teal probably relying more on invertebrates than did the duck. A few grey teal fossils have been found near swamps in Canterbury and Marlborough, suggesting they were present but rare during the pre-human era. Trevor Worthy has suggested that the decline of brown teal enabled the recent expansion in grey teal numbers. Shoveler are most likely recent (though still pre-European) additions to the New Zealand fauna. Perhaps they colonised New Zealand after the extinction of Scarlett's duck, or after human-induced habitat changes created the larger open-water wetlands they prefer.[88]

In the thick emergent vegetation that in the past fringed many aquatic habitats, and in swamps and marshes, there lived a different group of birds that fed on insects and crustaceans. These were the bitterns, crakes, rails and fernbird. Birds of the emergent vegetation are skulkers. Hidden in the thick vegetation, they are seldom seen and less often studied. Bitterns wait motionless, then lunge at passing prey or, with stealth, stalk it in the dense vegetation. Virtually nothing is known about the extinct little bittern but, assuming it was similar to the Australian little bittern, it would mostly have eaten crustaceans and insects.

Both species of crake occur in swamps and marshes. They take a variety of invertebrates and plant material. Spotless crakes utilise forests on some islands, but the extent to which they would have used drier ground in the past is unknown. No banded rail fossils have been found, but no fossil sites sample coastal marshes, where banded rail are common today.

The fernbird is the only passerine in the aquatic community and had a distinctive ecological role: it was the only bird in this community to mostly hunt above ground level, gleaning insects from wetland vegetation. Some of these birds would have visited the lake and terrestrial habitats between which they were sandwiched.

Piscivores

Birds use a variety of methods to catch fish. Shags, grebes and the merganser take them by underwater pursuit; herons and bitterns wade in the shallows then lunge at prey; and black-billed gulls and black-fronted terns capture fish (and other animals) while floating on the water or by alighting briefly to seize prey swimming close to the surface.

The four species of freshwater and estuarine shags all dive from the surface to pursue prey underwater. Today, all eat mostly fish or crustaceans. Presumably their modern diets largely reflect those of the past, although the black shag, which today is partial to trout, did not enjoy that delicacy in earlier times. The two large shags (black and pied) generally take larger prey and feed in deeper water, with longer dive times than the little shags. The pied shag is primarily a bird of sheltered marine habitats, although it also occurs in coastal lagoons and lakes. The black shag is mostly found in lakes, rivers and estuaries. The little shag occurs throughout the country in a wide range of aquatic habitats, from sheltered marine areas to high-country lakes.

The crested grebe now occurs only on a few high-country lakes in Canterbury and coastal lakes in Westland. The dabchick's present range is similarly restricted, mostly to shallow freshwater lakes and lagoons in parts of the North Island. In the past, these birds used a wider range of freshwater habitats on both the North and South Islands.[89] The larger crested grebe takes mostly fish and some insects, while invertebrates predominate in the diet of the small grebe. The Auckland Island merganser would have taken fish and crustaceans by pursuit-diving. It appears primarily to have been a marine duck but also utilised coastal lakes.

Herons were probably rare. The white-faced species that is now common only colonised New Zealand in the mid-twentieth century. In prehuman New Zealand, the white heron (kotuku) was probably the only freshwater heron, but Maori legend and subfossil remains suggest that it was uncommon. The sacred kingfisher occasionally eats lizards and small birds but more often takes fish, large insects and crustaceans.[90] This species spots prey from a perch then swoops down to capture it. The kingfisher occurs in many wetland and coastal habitats. The black-billed gull and black-fronted tern (see Chapter 3) occur on some larger lakes and on the South Island braided rivers.

The Auckland Island merganser

By the time it was discovered in 1840, the Auckland Island merganser was restricted to that group of islands. Over the next 60 years it was seen by several naturalists who considered it more important to collect museum specimens than to observe live birds. Today, these birds can be found in museums in New Zealand, Britain, United States, France, Austria and Germany, but none has been seen at the Auckland Islands since 1902.

Auckland Island mergansers were most often observed on rivers, inlets and harbours. They dived for food and ate mostly fish and crustaceans, but a mollusc and a marine worm were present in the only gut that has been preserved. These birds had longer, more slender bills, smaller bodies and proportionately shorter wings than other mergansers. Like many other island waterfowl, they had reduced powers of flight. They lived in pairs, at least during their summer breeding season. The only clutch recorded comprised four ducklings.

Merganser bones have been found on the North, South, Stewart and Chatham Islands, some in early Polynesian middens. The mergansers seen on the Auckland Islands were extraordinarily tame, so they must have been easy prey for hungry Maori, just as they were for European collectors.

The merganser's last refuge was on predator-free Adams Island, the southernmost of the Auckland group, suggesting that the birds on other islands were exterminated by cats, pigs and dogs. There are few suitable streams and harbours on the Auckland Islands, so they probably never supported more than a few hundred mergansers. The ecology of the Auckland Island merganser has been reviewed by Janet Kear and Ron Scarlett,[91] and their evolutionary history by Bradley Livezey.[92]

CHAPTER 5

Extinctions of New Zealand vertebrates

> Ko te huna i te moa (Destroyed like the moa).
>
> – Maori proverb

> In New Zealand today we are not studying an avifauna but the wreckage of an avifauna.
>
> – Jared Diamond (1984)

During the last 2000 years, 40 per cent of the terrestrial and freshwater bird species native to the New Zealand mainland, five marine birds, three frogs, a bat, at least three lizards, one freshwater fish, some plants and unknown species of invertebrates have become extinct (Table 5.1). Of the 53 vertebrates lost from the North and South Islands, all but two (the black swan, which was subsequently re-introduced) and the shore plover (which survives on Rangatira Island in the Chatham group) occurred nowhere else. An additional 13 bird species endemic to the Chatham Islands have been lost (Table 5.6), bringing the total number of vertebrate species lost from the New Zealand region to 64. Other species are now extinct on the mainland but survive as refugees on small islands where they have escaped the onslaught of hunters and predators. Another few species persist in remote corners of the mainland where the full impact of ecological change has yet to be felt.

Although climate and vegetation change have caused the demise of localised populations, all vertebrate species known to be present during the Pleistocene survived somewhere in New Zealand until human contact.[1] During the last two millennia, species have become extinct as they were faced by a series of increasingly demanding environmental changes brought about initially by kiore and subsequently by Polynesian and then European immigrants.[2]

The rate at which species are becoming extinct is faster now than at any time in the last 65 million years. In order to tackle the current global biodiversity crisis, we need to understand what makes species extinct and be better able to predict which of them will be vulnerable to further environmental change. New Zealand is one of the best places to study human-induced extinctions. It was the last land mass settled by people, so the record of ecological change is more intact than for countries discovered earlier. During prehistoric times, there were only three introduced predators: kiore, people and kuri (Maori dogs), and during the historic period the dates of other human-induced ecological changes (including the arrival of further predators) are well documented. The New Zealand vertebrate fauna is comparatively small and the post-Pleistocene fossil deposits are extensive, little disturbed and well studied.

Populations become extinct when the last individual in that population dies, yet the species may survive, even thrive, elsewhere. It is important to distinguish between the demise of a population (commonly termed 'local extinction') and 'global extinction' (when the last individual from the last remaining population dies). Table 5.1 (page 120) lists New Zealand species that are extinct on the mainland but still survive on islands. For the elephant seal and brown skua, extinction on the New Zealand mainland was a relatively trivial event because both species still occur on islands throughout the Southern Ocean, and the New Zealand mainland populations were relatively small. For the endemic reptiles, frogs and terrestrial birds, loss of the mainland populations meant that each species had become extinct throughout most of its range and only remnant populations survived on small islands.

The Stephens Island wren illustrates the need to distinguish between local and global extinction. The species became extinct through virtually all its range soon after the arrival of kiore, but centuries later a single cat destroyed the last small population. Kiore caused extinction over the vast bulk of the wren's range long before 1800, yet the species is listed in Table 5.1 as a post-1800 extinction. We commonly blame the cat for the wren's extinction, but in reality that predator merely administered the *coup de grâce*.

New Zealand's extinction cascade has not finished, and only by intensive management of endangered species have almost 40 years gone without a global extinction of a native vertebrate. However, the mainland populations of various species continue to decline; for instance, during the twentieth century there has been a retrenchment of brown teal, with the species becoming extinct from Stewart Island, the Waikato and Coromandel, while the only two remaining mainland populations,

Table 5.1 Vertebrate species that have become extinct on the New Zealand mainland

	Extinct pre-1800	Extinct 1800–present	Extinct on mainland but survive on offshore islands	Total number of native species
Freshwater fish	0	1	0	c. 30
Frogs	3	0	2	7
Tuatara	0	0	2	2
Geckos	0	1	3	32
Skinks	2	0	11	32
Bats	0	1	0	3
Seals	0	0	1	3
Moa	11	0	0	11
Kiwi	1	0	1	5
Petrels	2	0	11[a]	20
Bitterns	0	1	0	2
Waterfowl	7[b]	0	0	14
Raptors	2	0	0	4
Quail	0	1	0	1
Rails	4	1 subsp.[c]	0	9
Aptornis	2	0	0	2
Waders	0	1	0	8
New Zealand snipe	0	2	0	2
Skuas	0	0	1	1
Parrots	0	0	1	6
Owls	0	1	0	2
Owlet-nightjar	1	0	0	1
New Zealand wrens	2	2	0	6
Passerines	1	4	3	22
Total	38	15[d]	36	

Note that the offshore islands are those small islands close to the North, South and Stewart Islands. The table does not include 13 now-extinct birds endemic to the Chatham Islands, nor those species that became extinct on the main islands in the Kermadec, Chatham, Auckland and Campbell groups but survive on small islands in those groups.

(a) The number of petrel species that bred on the mainland is not known and likely exceeds the 11 species listed by Holdaway 1999.

(b) Worthy 1998c suggested that the New Zealand was the same species as that found in Australia. Thus, the swan listed here as extinct before 1800 re-established in New Zealand during the nineteenth century. The Auckland Island merganser became extinct on the New Zealand mainland prior to 1800 but lingered on at the Auckland Islands until 1902.

(c) The buff weka is extinct in its native mainland range but survives on the Chatham Islands, where it was introduced.

(d) Plus one subspecies.

in Northland and Fiordland, continue to decline.[3] Extinction of these mainland populations is likely and the few islands on which they are found do not offer enduring sanctuary.

This chapter will review the impact predators and people have had on native vertebrates, and identify the groups of animals vulnerable to human-induced environmental change. The objective of this chapter is not to apportion blame. The first priorities of any colonist are food and shelter. New Zealand's animals and ecosystems were very different from those in the Pacific and European homelands from which our forebears came. Unlike the present generation, our ancestors had little prior experience in sustainable harvest of such resources, nor the luxury of being able to empathise with the strange creatures they discovered. We can only prevent more species becoming extinct if we appreciate the huge ecological changes that have occurred, understand the factors that caused these changes, and identify the species vulnerable to further change. That is the objective of this chapter.

Extinction scenarios elsewhere

Animals, mostly large mammals and birds, became extinct in many parts of the world following colonisation by people.[4] The Aborigines reached Australia more than 40,000 years ago, and a number of species became extinct during this early phase of human occupation. These were mostly large animals, including kangaroos, echidnas, mound-building birds, emu-like birds, terrestrial crocodiles, monitor lizards and a hundred-kilogram snake. For millennia Aborigines managed the land by frequent burning to encourage new growth that attracted game or encouraged the regrowth of favoured food plants. Tim Flannery[5] suggested that, over time, a new ecological balance was struck between the now-impoverished biota, the Aborigines and their land-management techniques. This balance was subsequently disrupted when the European colonists failed to appreciate the subtleties of Aboriginal land-management techniques and triggered a new set of ecological changes and a second extinction cascade.

In Europe, Asia and the Americas, the spread of people coincided with the Pleistocene glaciations, and in the first two regions it is virtually impossible to separate human-caused extinctions from those caused by climate change. Evidence suggests that, at the very least, humans caused the ranges of some large mammals to contract, but their impact was probably very much greater.[6] In the Americas, many large mammals, including horses, mammoths and sabre-toothed tigers, became extinct 12,000–10,000 years ago. This coincided with two events: retreating glaciers and the arrival of people. These animals had survived previous glacial cycles, but never before had human hunters confronted them.

It seems incredible that stone-age people could exterminate so many large mammals from two continents in just a few thousand years, yet Jared Diamond[7] presents a convincing scenario to support this contention.

Extinction rates are far higher on islands than on continents, and thousands of species have become extinct since humans dispersed across the world's oceans.[8] Islands in the tropical Pacific may seem like paradise to the unknowing tourist, yet during the few millenia people have lived on them countless species have become extinct. The Pacific Islands, each of which now has only a few species, once had among them as many bird species as the South American continent. David Steadman[9] suggested that 2000 species of rails, plus countless other birds, may have been lost from the Pacific Islands alone.

In Hawaii, at least 45 species of birds, over half of all the native non-marine species, became extinct following the arrival of Polynesians, and 16 more have been lost since Captain Cook began the European onslaught.[10] The threats continue to intensify and only 10 of the Hawaiian Islands' non-marine native birds are not currently threatened with extinction. On Henderson Island in the Pitcairn group, over half of all land birds became extinct and most ground-nesting seabirds became locally extinct during the first few hundred years of human occupation.[11] After the bird and fish populations had been overexploited, the Henderson people themselves became extinct, a salutary lesson for societies that continue to overexploit their resource base. On most other oceanic islands the fossil record is less well studied, yet evidence suggests that losses of this magnitude were neither unusual nor the worst on record.[12] On Easter Island, for example, all terrestrial birds and almost all seabirds became extinct.

Is it really possible that all these extinctions were caused by people, their camp-followers and the habitat changes they induced? Climates also changed and such natural perturbations must surely have had some role in sealing the fates of some species. The roles of climate and habitat change have been hotly debated,[13] but as the evidence accumulates, the case against the people becomes ever more convincing. If extinctions were a result of climate change, we would expect them to be closely synchronised in different parts of the world. However, in Australia they occurred more than 40,000 years ago; in the Americas between 12,000 and 10,000 years ago; and in the Pacific between 3000 and 1000 years ago. These times coincide with only one thing: the arrival of humans. In addition, the bones of most extinct taxa occur in middens associated with archaeological sites. The evidence may be largely circumstantial, but the case seems proven beyond reasonable doubt.

The sixth major mass extinction

Extinction, like death, is inevitable; but like death it is best postponed. All species will eventually become extinct, as surely as new species will evolve. The natural or background rate at which species become extinct is low, and Edward Wilson[14] has estimated that without human intervention it can be expected that between one in a million and one in 10 million species would become extinct each year. Assuming there are about 12.5 million species on Earth,[15] this means that each year we can expect the natural extinction of up to 12 species. We have no idea just how many species actually do become extinct each year, but Wilson estimates that about 17,000 species are lost from tropical rainforests annually. If he is correct, the rate at which species are lost from tropical rainforests alone is 1000–10,000 times higher than we might expect for the entire world. Subsequent analyses of extinction rates suggest that the higher figure is closer to the mark.[16]

Extinction rates are highly variable, with long periods of low rates being infrequently interrupted by periods of high rates. In the history of life there have been five major mass-extinction events, in each of which more than half the world's species were lost. The most recent of these was the Cretaceous/Tertiary event that killed not just the dinosaurs but about two-thirds of all animal species. Extinction events of this magnitude occur only about once every hundred million years. New species evolve, and after each of these events it apparently took five to ten million years until there were as many species as there had been before the extinction event.

The extinctions that have occurred, and will continue to occur, in New Zealand are part of a global phenomenon. We are currently witnessing a worldwide loss of species that may equal the major mass-extinction events of the past. We have no reliable estimate of the number of species that may become extinct, but several leading biologists have suggested that over half of all species living at present may be doomed unless people drastically reduce their demands on this finite planet.[17] Most of those species will be invertebrates, so their loss will be less conspicuous than that of moa or dodos. The vast majority of those species are not yet known to science, so their passing will go unnoticed and unlamented. This is the greatest biological catastrophe for 65 million years and it is caused by humans, yet we lack the fundamental knowledge needed to assess the magnitude of the problem, let alone remedy it.

What makes species become extinct?

Species decline towards extinction when more individuals die than are produced. For some species, extinction may take many generations and the rate of decline is so slow that it is not detectable during a human lifetime. Rare, unforeseeable events may cause the extinction of well-adapted, 'evolutionarily fit' species. Species evolve in response to the prevailing ecological conditions, and each has some ability to adapt as conditions change. In New Zealand, change was rapid, massive and of a kind the animals had never before been confronted with.

Throughout the Pleistocene, moa and other species were confronted with environmental change as climates warmed then cooled and glaciers shrank then grew. Thus the area covered by forest changed and the vegetation was in a state of flux. The distribution of the various moa species changed with the vegetation, but all species of moa known to have existed before or during the Pleistocene were alive and well at the time Polynesians settled this country.[18] Moa were not antiquated relics surviving in a benign environment; they were well attuned to slowly changing ecological conditions but ill-adapted to the tumultuous changes about to confront them.

Kiore: New Zealand's first alien invaders

The kiore is a small rat that originated in Southeast Asia and was carried by Pacific peoples to virtually all the islands in that great ocean. By 900 BC, kiore had reached Samoa, Fiji, Tonga, New Caledonia and many other islands in between those countries and the rodent's Asian homeland.[19] Subsequently, the Polynesians carried kiore, either intentionally for food or accidentally as stowaways, to most of the islands in Polynesia, including Hawaii, Easter Island, Henderson Island (where they outlasted the humans that took them there), New Zealand and the Chatham Islands. Before Europeans and their rats began to spread around the globe in the fifteenth century, the kiore was the most widespread of those rodent species that live alongside people.

Gunther's tuatara. When rediscovered in 1990 this species was confined to tiny, windswept North Brothers Island.
Photograph by Kerry-Jayne Wilson

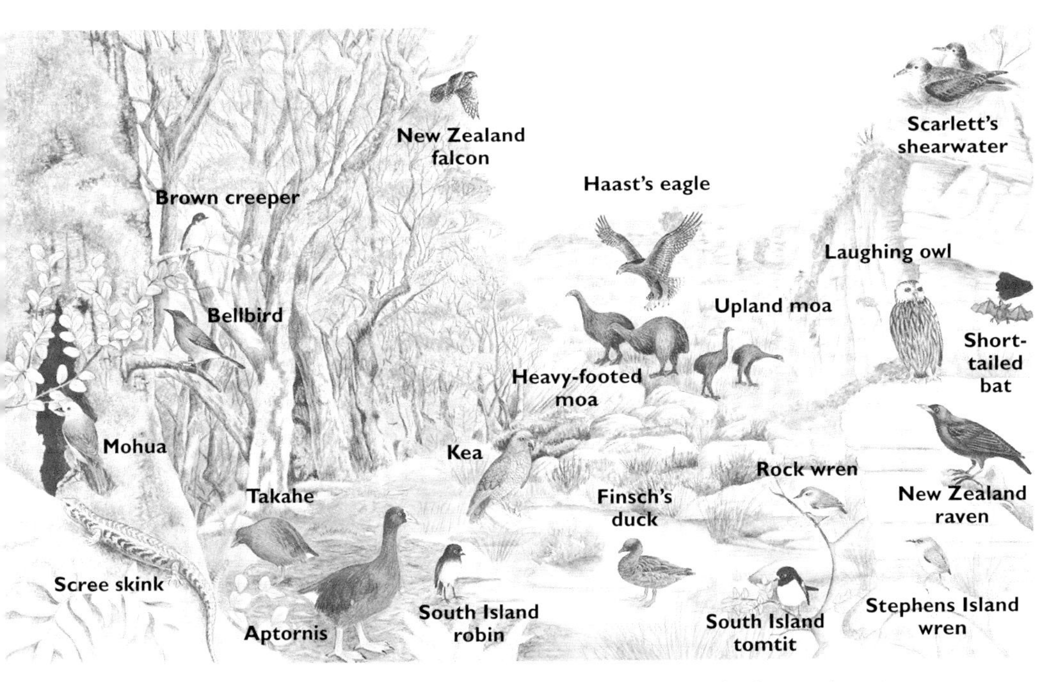

Key to Plate 1 (overleaf): The Oparara Valley, Karamea, during the Otiran glaciation, 14,000 years ago, when beech forest, tussock grasslands and subalpine scrub dominated the area.
Painting by Pauline Morse

Archey's frog. This Jurassic survivor is one of the world's most primitive frogs.
It is a terrestrial species and has webless feet.

Photograph by John Marris

Duvaucel's gecko. Bones of this species have been found in many localities, some as far south as Otago. Today, the species survives only on islands in Cook Strait and off the North Island.

Photograph by John Marris

Until recently it had been assumed that kiore had accompanied Maori to New Zealand about 800 years ago. However, in 1996 Richard Holdaway, using radiocarbon and other dating methods, showed that kiore had been in this country for almost 2000 years – much longer than the Maori.[20] Kiore certainly did not get here without human assistance, and genetic studies indicate that the New Zealand populations were derived from animals in the southern Cook Islands and the Society Islands.[21] Holdaway concluded that the most likely date for the arrival of kiore was between AD 50 and 150. He suggested that the people who first brought kiore to New Zealand were explorers who either returned to their land of origin or, if they settled here, did not endure. Kiore remains of this age have been found in the North and South Islands, so these sailors must have visited both.

Holdaway's findings were, and remain, controversial. Kiore remains have now been reliably dated using several methods, so his early date of arrival appears robust,[22] but not everyone is convinced. A review by Ian Atkinson and David Towns[23] presents both sides of the debate. Further introductions of kiore occurred when the Polynesian settlers arrived a millenium later.

Kiore normally weigh 60–80 grams (maximum 180),[24] only half the size of the familiar ship rat. They are omnivores that most often eat insects, seeds or fruits, but all rats are opportunists and will eat almost anything they can catch. This ostensibly benign little creature, once set loose in an environment where no predatory mammal had gone before, proved an insidious threat to large invertebrates and small vertebrates. Kiore regularly kill birds up to their own body weight in size and break open eggs up to 68 millimetres long.[25] On occasions they have killed seabirds many times their own body weight. Kiore posed a significant threat to petrels up to 250 grams in body weight, snipe, small ducks and rails, flightless or poorly flighted wrens, ground-dwelling reptiles, frogs and large invertebrates. Although kiore are agile climbers, it was the ground-dwelling birds that proved most vulnerable to predation.

Laughing owl middens provide good evidence of the impact kiore had on the native fauna. These birds were generalist predators that ate kiore, other small vertebrates and large invertebrates. Generations of owls ate their prey in the same caves, and the prey remains accumulated on top of bones voided by earlier inhabitants (see Chapter 4). Owl middens document the arrival of kiore and the subsequent local extinction of bats, reptiles, small birds and large insects.[26]

Small burrow-breeding petrels such as diving petrels and storm petrels are absent or uncommon on islands with kiore, and these rats have been shown to take little shearwater eggs,[27] indicating that the decline of mainland-breeding petrels began with their arrival. The breeding success of little shearwater, Cook's petrel and Pycroft's petrel all increased following the eradication of kiore from islands inhabited by those birds.[28] Kiore possibly caused the extinction of Scarlett's shearwater. The extinction of mainland petrel colonies has continued throughout the past 2000 years, and today only the largest and most robust petrels remain on the North and South Islands. The decline has not yet finished: the remaining mainland sooty shearwater numbers decrease year by year as stoats, rats and cats eat eggs, chicks and adults.[29]

Unlike the native species, which were K-selected (see Chapter 3), kiore could respond quickly to the abundance of food, and no doubt quickly became numerous. They had spread 80 kilometres inland by the time of the Taupo eruption about 1800 years ago.[30] We can imagine a blitzkrieg of kiore sweeping across the land, exerting a multi-pronged attack on the natives. They not only preyed on adults, eggs and young of ground-dwelling animals, but also competed with them for food. The reptiles, frogs and birds could at best produce two small broods each year, whereas kiore could have multiple broods with large litters. Some of those young could themselves breed during the same season they were born. Their impact on reptiles has been discussed in Chapter 2. On the mainland, among the first birds to succumb were snipe, the larger wrens and snipe-rail. In South Canterbury, kiore are implicated in the local extinction of about 30 species of small birds, the greater short-tailed bat, tuatara and Duvaucel's gecko.[31]

The time taken from the arrival of a new predator to extinction of a vulnerable prey species depends on the population dynamics of prey and predator. Their relative rates of reproduction are obviously important, but so is the generation time of the prey and the stage in the life cycle it is taken. The ground-dwelling wrens had shorter generation times than the long-lived kakapo, and kiore probably ate both their chicks and their eggs. Thus the wrens would have become extinct soon after the rats arrived, but kakapo, with their long generation time and only chicks being vulnerable to kiore, survived until the European predators arrived.

Extinctions during the Polynesian period

> The two domestic quadrupeds are the hog and the dog. The Society Isles alone are fortunate enough to possess them both, New Zeeland and the low islands must be content with dogs alone.
>
> – Johann F. Forster (1778)

When Polynesians first arrived in New Zealand about 800 years ago,* they had to rely on hunting and fishing for sustenance. The climate was too cool for all but a few of the tropical crops they brought with them, and the only native plant suitable for extensive cultivation was bracken, whose root became a staple food. Meat, on the other hand, was easily obtained. During the settlement phase, the main animal foods were seals, moa and large waterfowl – easy prey for nimble-footed hunters. At the time of Polynesian settlement, fur seals and, less commonly, sea lions occurred as far north as Northland, and these were hunted to local extinction very quickly.[32] As the larger game was depleted, smaller game was hunted, and midden deposits dating from after AD 1500 contain mostly fish and shellfish.[33]

Some of the first birds to become extinct during the Polynesian period were large meaty species such as moa and those found in estuaries and lagoons such as the swan, de Lautour's duck and the Auckland Island merganser.[34] Species most common in dry eastern habitats such as geese and aptornis were other early casualties. In southern New Zealand, the swan, goose, Finsch's duck, Eyles's harrier, coot and New Zealand raven probably became extinct during or soon after the twelfth century, and Hodgen's rail by 1300.[35] Most of these species were flightless and were vulnerable for at least one other reason. Some moa and aptornis probably lived at low population densities and frequented open habitats. Moa were extinct in coastal areas and the eastern South Island interior by the late fifteenth century. A preference for the habitats least used by Maori enabled the little bush moa, the large bush moa (inhabitants of tall wet forests) and the upland moa (an alpine and subalpine species) to survive a little longer than the others, but still not long enough to be seen by Europeans. Takahe also favoured edge habitats, where they were vulnerable to hunting, and they apparently became extinct over

* McGlone & Wilmshurst (1999) conclude that the first Polynesian-induced fires to cause vegetation change occurred between 750 and 550 years ago and that initial settlement was unlikely to have occurred significantly earlier than AD 1200. Holdaway and Jacomb (2000) and Higham et al. (1999) suggest settlement occurred no earlier than 750 years before present, i.e. about AD 1250.

most of their range early in the Polynesian period.[36] They survived into the twentieth century thanks to their ability to persist in Fiordland tussocklands that were marginal for both takahe and their hunters. Seabirds were important foods, and the distributions of many became restricted during the Maori era.[37]

Polynesians introduced the kuri, their breed of domestic dog. Dogs are carnivores and surely this was a much more ferocious predator than a small rat to enter an arena filled with flightless prey. However, kuri appear to have remained domesticated animals, raised for food and skins. They were prized because, apart from seals, they were the only fur-bearing mammals in a climate where this would have been a desirable commodity. Doubtless kuri did accompany people on hunting expeditions, but their part in the extinction scenario appears to have been minor. Had feral populations been common (no doubt there were some), the story might well have been different. Pigs and chickens, the other Polynesian camp-followers, either did not accompany the first settlers to New Zealand or, if they were brought here, did not survive.

Extinctions were caused not just by overhunting but also by habitat loss. Polynesian-induced fires began soon after settlement. By 1500, there had been extensive loss of the eastern South Island lowland podocarp forests and beech forests further inland.[38] In these dry regions, regeneration was slow and the tussocklands that dominated the plains and foothills of Canterbury, Otago and Marlborough at the time of European settlement were an artefact of Maori fires. Extensive areas of North Island forest were also burned, but the climate was milder and damper, enabling bracken to colonise cleared areas. This was repeatedly burned to enable the roots to be harvested and to encourage regrowth. Forest and scrub were also burned to facilitate access through densely vegetated areas. Some forest types – for instance, the dry podocarp/broadleaf forest – were virtually eliminated and others substantially reduced in area prior to European settlement.[39] The coastal lowland forest, especially in dry eastern regions where deforestation was most extensive, was the prime habitat for some moa, Haast's eagle, Finsch's duck, geese, Eyles's harrier and New Zealand raven. Trevor Worthy[40] suggested that, at Marfells Beach in Marlborough, hunting caused the local extinction of those species that are now globally extinct, whereas habitat change was probably of greater significance in the local extinction of bird species that survive elsewhere.

When the Polynesians arrived in New Zealand, over 85 per cent of the country was forested.[41] By 1840, that had fallen to 53 per cent, and most of that clearing had occured before 1600. The loss of a third of the forest cover had been challenge enough for those birds that survived the first 650 years of human occupation, but a new era of more rapid and even more extensive habitat loss was to follow.

Overkill of moa

> We are looking here, I suggest, at the depredations of an efficient, but inexperienced predator which was set loose in a resource environment where the narrowness of the choices and the naivety of the prey ensured that the first swing of the predator-prey pendulum pulled the clock off the wall.
>
> – Athol Anderson (1983)

How could 115,000 Maori armed with stone-age weaponry drive all 11 species of moa to extinction?* Surprisingly easily, it seems: it now even appears the moa were gone long before the human population grew to anywhere approaching this number.

Anderson[42] discussed the history of moa hunting in detail. He suggested that most dated moa-hunting sites in both the North and South Islands were deposited between 800 and 500 years ago, with the period of most intense moa hunting occurring 700–650 years ago. However, subsequent evidence suggests moa hunting may have only begun 700 years ago and lasted little over a century.[43] In Anderson's scenario, South Island middens with moa tend to be older than in North Island localities, and coastal moa-hunting sites tend to be older than those inland. At any particular location, moa hunting would have lasted only a short time. The odds were staked heavily in favour of the hunter. Estimates of the total number of moa, rough and ready though they may be, suggest that there were probably fewer than 100,000 moa divided among 11 species. They were flightless, probably slow-moving and had no prior experience of terrestrial predators, so they lacked appropriate anti-predator responses. Moa probably laid a single egg for which there was a long incubation period. Adult birds were apparently the preferred prey, but eggs were also gathered. Moa probably lived in localised populations separated from one another by ranges and rivers, so there was limited opportunity

* The pre-European population of New Zealand is estimated by various archaeologists and historians to have been between 100,000 and 200,000. I have used the figure of 115,000 given by McGlone et al. (1994).

for adjacent populations to recolonise depleted areas before they were themselves exploited.

Anderson also discussed the mechanics of moa overkill. He suggested that 800 years ago, following an initial period of exploration, a series of coastal moa-hunting sites had been established along the east coast of the South Island, with another near Foxton in the southern part of the North Island. Most of these sites are near rocky coasts, where fur seals were numerous, or at river mouths, where moa and waterfowl would have been easy to hunt. With such abundant food, these would have been prime sites for first settlement. Seals were even more prone than moa to overharvesting and became extinct in most places north of Otago early in Polynesian history.[44] These sites were not occupied for long before the hunters moved on in search of new game.[45] Anderson suggested that, after overharvesting local seal and moa stocks, Maori radiated out from these sites, extinguishing moa in one catchment then moving on to the next. He pictured a series of local extinctions occurring as moa hunters advanced both along the coast and inland on a number of fronts. As the dependence on small prey increased, moa hunting may have continued with seasonal expeditions to the more remote areas.

The eastern South Island would have been an easy area for settlement to begin. The dry, open forests and shrublands held the greatest abundance of moa, birds were abundant in coastal lagoons and estuaries, and the braided rivers, stocked with birds and fish, were obvious highways into the interior.[46] In contrast, Anderson suggested, the densely forested North Island held fewer moa and the resources there would have been more difficult to exploit until a specialised knowledge of the habitats and the quarry had been gained.

The extinction of moa from the easily burned eastern forests appears feasible, but it is harder to imagine that moa could be rendered extinct in the interior and from rainforest. One scenario[47] suggests that hunting flightless birds in rough, forested country may not have been as difficult as we imagine. The birds may have been confined to, or at least largely dependent on, high-quality browse along rivers, in valley bottoms and along swamp margins. Intensive hunting of such sites, which were attractive to the birds and accessible to the hunters, may have been sufficient to bring moa populations to the brink of extinction.

Predator/prey dynamics depend on the relative numbers of prey and predators. Anderson attempted to calculate standing stocks and the numbers of moa killed. Despite their limitations, these calculations indicate that overhunting was feasible. He suggested that there were tens of thousands of moa on each of the two main islands, with about twice as many on the South Island as the North. The southern South Island has 120 known moa-hunting sites, containing the remains of 100,000–500,000 moa. This is likely to be a conservative estimate of numbers actually killed, as some known sites have been reduced in area by erosion and it is unlikely all moa-hunter sites are known.

Anderson initially assumed that these bones had accumulated over a 400-year period and suggested that up to 3.2 per cent of the population would have been killed each year, a level of harvest that might have been sustainable had the destruction of the moa's favoured habitats not also taken place. Anderson suggested[48] that in most parts of the North Island moa hunting was never a mainstay of Maori, but as the numbers of people increased, continued casual hunting was enough to drive the birds to extinction. In the South Island, the Maori population probably plateaued quickly and the greater abundance of moa encouraged a more systematic assault.

Richard Holdaway and Chris Jacomb[49] carefully dated middens with and without moa and combined that evidence with mathematical models to suggest that all 11 species of moa may have become extinct within a century of Polynesian colonisation. They suggest that hunting alone may have driven moa to extinction, and that in parts of the North Island where moa populations were low, the time to extinction could have been less than 20 years. Scientists will continue to debate just how long it took for moa to become extinct and whether hunting alone was sufficient to cause their extinction.

Other possible causes of extinctions

In the preceding sections I have attributed the extinction of moa and other birds to hunting, human-induced habitat change and kiore, but is it possible that natural factors could have been responsible? Probably not. Bones from 10 moa species plus most other large birds have been found in archaeological sites, proving that their extinction occurred after humans arrived in New Zealand. Moa bones are only found in older middens, with more recent ones reflecting an ever-increasing reliance on fish, shellfish and plants. Surely Maori would have continued to hunt moa and other large birds had they been available. Similarly, most species of small birds are known to have lived up to the time kiore arrived in New Zealand.

Climate and vegetation did change during those crucial 2000 years, but the birds had previously survived both warmer and cooler periods. Temperatures were apparently warmest about AD 1200 and coolest during a minor glacial advance around 1650.[50] Moa had already proven their ability to cope with climate-induced habitat change. At the end of the Pleistocene, tall forest replaced the forest/grassland mosaic and the ranges of open-country birds contracted, while forest-dwelling birds expanded their range. Kea and Haast's eagle apparently became extinct on the North Island long before human contact, but they survived in cooler southern locations.[51]

Trevor Worthy and Richard Holdaway[52] compared faunal change at Takaka, near Nelson, at the end of the Pleistocene and following the arrival of kiore and humans. At the end of the Pleistocene, some birds apparently became extinct locally, but they all survived elsewhere in New Zealand. Since the arrival of kiore and humans, 24 species of birds have become extinct at Takaka, and 13 of these are now globally extinct. Extinctions still continue; for example, weka numbers have declined since the mid-1980s and these birds are now almost extinct in the Takaka area.[53] Worthy and Holdaway concluded that 'the impact of man was considerably worse than that of the major climate change from the Otiran to the Holocene'. Elsewhere in the world the story is similar: climate change did cause redistribution of species and a few extinctions, but humans caused the loss of many more.[54]

The most extreme habitat changes were the result of fire, yet natural fires had occurred in New Zealand long before people arrived. Why then were the animals unable to adapt? In pre-human New Zealand, fire was a rare and localised event. Only in semi-arid inland parts of Otago and Canterbury did lightning-induced fire cause the permanent reduction of forest cover, and even there patches of forest survived.[55] Volcanic eruptions are locally destructive, yet even they do not cause long-term deforestation. Tongariro National Park contains two active volcanoes yet it has extensive areas of indigenous forest. Forests recovered even after the huge Taupo eruption about 1850 years ago.

In Polynesian times, fires became much more frequent and more extensive. In European times, fires became even more frequent and even more extensive. Unpalatable as it may be, there is no alternative explanation: these extinctions were caused by kiore, by hunting and by human-induced environmental change.

The European period

> Everything indigenous to N.Z. is dying out at such a rapid rate from the introduction of British and foreign plants and birds and beasts and especially blights that before long birds, &c. which are now fairly common will be totally extinct.
>
> – Edward A. Wilson (Diary of Travels in New Zealand, 1904)

The European colonists brought with them a host of formidable predators plus new animals that competed with native species for food. This new wave of agriculturally based settlers, and their determined efforts to change the landscape and biota to resemble that of their homeland, caused much faster and more extensive loss of forest and wetland habitats than had the Polynesians. After 200 years of European impact, about 16 more species of bird plus one bat and possibly one lizard have become globally extinct. This is far fewer than during the Maori era but, at an average of one species per decade since the signing of the Treaty of Waitangi, it represents more species per century than had occurred in previous eras (Table 5.2, page 134). These figures still mask the real magnitude of European impact.

After only 200 years of Pakeha impact, and just 120 years with mustelids present, the European and mustelid extinction events are still far from finished. Various species survive only on small predator-free islands, and most (maybe all) currently threatened or endangered species will become extinct during the present century unless we continue to intervene. Pakeha colonisation would certainly have caused many more extinctions had Maori and kiore not already removed the most susceptible species.

For some of the post-1800 extinctions, the European settlers only dealt the *coup de grâce*. The initial reduction in the ranges of huia, takahe, little spotted kiwi and kakapo had been a consequence of Maori hunting. Once the Europeans and their camp-followers joined the fray, the huia was soon extinct, takahe and kakapo were critically endangered and the little spotted kiwi was an island refugee. By the time of European settlement, the Auckland Island merganser was extinct on the mainland and the only remaining population was lost soon after people, pigs, cats and rats got to the Auckland Islands. Long before 1800, snipe and Stephens Island wrens, once common on the North and South Islands, survived only on small islands, their extinction from the mainland a consequence of kiore predation.

Table 5.2 Birds and bats that have become extinct in New Zealand since European contact

		Approx. date of extinction	Likely cause
Dieffenbach's rail	Chatham Islands	1840	Rats, cats
NZ little bittern	North Island	Pre-1800	
	South Island	1868–70	
NZ quail	North, South Islands	1870-75	Disease? hunting, ship rats?
North Island snipe	North Island	pre-1200	Kiore
	Little Barrier Island	1870	Kiore
Stephens Island wren	North, South Islands	pre-1200	Kiore
	Stephens Island	1894	A cat, collectors?
Chatham Island rail	Chatham Islands	1900	Cats, rats
Chatham Island fernbird	Chatham Islands	1900	Cats
North Island piopio	North Island	1900	Predators
South Island piopio	South Island	1900	Predators
Auckland Islands merganser	North, South Islands	pre-1800	Hunting
	Auckland Islands	1902	Hunting, predators
Chatham Island bellbird	Chatham Islands	1906	Cats
Huia	North Island	1907	Collectors, predators? habitat change?
Laughing owl	Stewart Island	1880s	
	North Island	1892	Mustelids
	South Island	1914	Mustelids
Stewart Island snipe	Stewart, South Islands	pre-1200	Kiore
	Big South Cape Island	1964	Ship rats
Greater short-tailed bat	North, South Islands	pre-1800?	
	Big South Cape Island	1964	Ship rats
Bush wren	North Island	1880	
	South Island	1900?	
	Big South Cape Island	1965	Ship rats
South Island kokako (possibly survives)	South I	1960?	Predators, competitors

Some species that became extinct on the mainland (including Chatham Island) since 1800 but survive on offshore islands

Kakapo	North Island	1906	
	South Island	1980s	
	Stewart Island	1990s	Predators, transferred to offshore islands

Mottled petrel	South Island	1910s?	
	North Island	1960s?	Predators, survives on offshore islands
Cook's petrel	North Island	1930s?	
	South Island	1970s?	Predators, survives on offshore islands.
South Island saddleback	South Island	1940s	Predators, survives on offshore islands
	Big South Cape Island	1964	Ship rats
Shore plover	North, South Islands	1880	Predators, survives on South East Island, Chatham Islands
Little spotted kiwi	North Island	1890	
	South Island	1970?	Predators, translocated to offshore islands
Buff weka	Eastern South Island	1917	Disease? predators? introduced to the Chatham Islands

Note: All species that have become globally extinct since 1800 are listed, plus species that have become extinct on the North, South or Stewart Islands or the Chatham Island mainland since 1800 (even if they survive on smaller islands). Dates of extinction from the North and South Islands are given even if they were before 1800. Holdaway (1999b) gives the full list of extinctions and is the primary source of the information in this table.

During the last 200 years, the pace of change has been so rapid that it is difficult to unravel the roles of predators, habitat loss, habitat change, competitors, hunting and disease in the extinction process. The vast tracts of lowland forest that remained in 1800 were broken into isolated remnants. Today, on the mainland, small forest patches support only the most widespread and common bush birds, and rare species survive only in large tracts of forest.[56] However, predators proved a more immediate threat than habitat change, as indicated by the large number of populations lost in so short a time and the fact that the species lost had attributes such as limited powers of flight and hole-nesting habits that rendered them vulnerable.[57] Most bird species that on the mainland survive only in large tracts of forest remain on small predator-free islands, suggesting that predators were a more immediate threat than habitat fragmentation. Habitat loss is an insidious and little-understood threat, and if animals can cope with the predators present, they may persist in remnants for decades after forest fragmentation. While I argue that predators were the immediate threat, this is not to diminish the importance of habitat loss.

Native birds may have been exposed to diseases brought in with exotic birds, though this appears to have been a minor factor in New Zealand. Circumstantial evidence suggests that disease may have contributed to the demise of the New Zealand quail and buff weka,[58] but this is unlikely to be the sole reason for the decline of either. Weka appear susceptible to introduced diseases or parasites, but other factors including predation and changes in land use may also have contributed to their decline.[59]

A new wave of predators

The first Norway rats and house cats to reach New Zealand arrived with Captain Cook. Thanks to explorers and sealers, by the early nineteenth century Norway rats were advancing through the country from a number of liberation points. Ship rats followed, spreading through the North Island in the 1860s and the South Island 30 years later. Stoats, ferrets and weasels were released in the 1880s.[60] Hedgehogs, dogs, pigs and possums also kill native vertebrates or eat their eggs, but, compared with rats, cats and mustelids, they have made only a minor contribution to the extinction story.

This new suite of predators posed far greater threats to native animals than kiore ever had. The new rats were larger and there were four species of true carnivores that had evolved specifically to hunt small vertebrates. It would be useful if we could unravel the impact of each species, but only on some islands where just one or two introduced predators are present is that possible. The three mustelids were introduced within a decade of one another and spread through the country very soon after cats became widespread.[61] On the South Island, these predators arrived at the same time as ship rats. It was also the time when habitat destruction was most rapid and when deer, possums and other browsing mammals were dispersing across the country.

Although all rats pose broadly similar threats to native animals, the ecology of the three species differs, so that cumulatively they put more native species at risk (Figure 5.1). Kiore are mostly terrestrial, avoid water, do not burrow but can climb trees. They are a particular threat to small ground-dwelling animals. The Norway rat is larger than the others, is ground-dwelling, prefers wet habitats, readily enters burrows but seldom climbs trees. It is a serious threat to ground-nesting or ground-dwelling species, including some large enough to be safe from kiore or ship rats. Norway rats are a particular threat to burrow-breeding seabirds.

Ship rats, being arboreal, threatened tree-nesting birds that had been safe from the earlier invaders. Ship rats are implicated in the decline of mainland populations of bellbird, robin, hihi, saddleback, piopio, mohua, kokako, brown teal and possibly New Zealand quail.[62] The impact of rats on Big South Cape Island is discussed on page 148.

The threat rats posed to native birds, reptiles, and invertebrates is hard to imagine now that both prey and predator are less numerous and birds make up only a small part of rats' diets (Chapter 7). Because rats are more numerous and more fecund than native birds, even occasional predation can be significant. Owing to differences in the ecology of the three rats, it is important to prevent further species from invading islands even where other rats are already present.

	Kiore	Ship rat	Norway rat
Body weight	60–80 g	120–160 g	200–300 g
Date of liberation			
North Island	AD 50–150	1860s	c. 1800
South Island	AD 50–150	1890s	Late 18th century
Tree climbing	✓	✓✓✓	
Ground hunting	✓✓	✓✓	✓✓
Use of wet habitats			✓✓✓
Burrow hunting			✓✓✓
Main species threatened	Small flightless or ground-dwelling birds, small petrels, tuatara, lizards and large frogs, beetles and land snails	Tree-nesting birds up to the size of kokako, large invertebrates	Small to medium-sized ground-nesting birds, burrow-breeding petrels, large beetles and land snails

Figure 5.1 Comparative ecology of the three species of rats and the threats posed by each to native animals

Cats have become feral in many parts of the world and pose a serious threat to native wildlife wherever they become naturalised.[63] Although the first of the carnivores to reach New Zealand, they did not become widespread until the mid-nineteenth century. Cats are the largest feral carnivores in this country. They can take prey on the ground or in trees, and readily desert humans for a feral existence. Today, they are widespread on the North, South and Stewart Islands and on the larger islands in the Kermadec, Auckland and Chatham groups. They also occur on about 20 smaller islands to which they have been introduced by lighthouse keepers, muttonbirders or settlers.[64]

On the mainland, cats are just one of the threats faced by native animals, but on certain islands their threat is more obvious. The expatriation of 12 bird species from Mangere Island in the Chatham group is attributed to cats (see page 147). Cats were introduced to Herekopere Island, near Stewart Island, in the 1920s, and by 1932 at least eight species of birds were locally extinct.[65]

The most infamous of all cats was owned by a Stephens Island lighthouse keeper: with perhaps some assistance from human collectors, it caused the final demise of the Stephens Island wren.[66] On cat-infested Raoul Island, Kermadec and black-winged petrels, wedge-tailed shearwaters and parakeets are now extinct, but all these species remain common on the adjacent, cat-free Meyer Islets. Hihi, parakeets and robins increased in numbers and saddlebacks were successfully re-introduced after cats had been eradicated from Little Barrier Island. Kakapo might have become extinct had the last remaining birds not been relocated from cat-infested Stewart Island to smaller, cat-free islands. Cats are one of the principal threats faced by the endangered black stilt. They also kill bats, lizards and invertebrates, but their impact on non-feathered animals is less well documented.

On Herekopare Island, those species that fed or bred on the ground proved most susceptible to cat predation.[67] Of the six forest birds lost from that island, only one, the brown creeper, did not habitually use the ground or low vegetation. Two species of burrow-breeding petrels became extinct, while two species survived. The diving petrel was susceptible for two reasons: it was the smallest, and the adults used to return nightly to the island to relieve their incubating mate or feed their chicks. Other petrels visit their breeding colonies less often. Diving petrels and broad-billed prions, the two species expatriated, breed earlier in the season than fairy prions and sooty shearwaters, which survived there. Weka were present on the island for some years and may have contributed to the loss of some species. The cats were exterminated in 1970.

Kim King[68] suggested that in New Zealand cats were implicated in the extinction of six endemic species and in some 70 local extinctions – 26 times as many as all three mustelid species combined. Dick Veitch[69] listed the islands in the New Zealand region to which cats have been introduced, and the bird species they have extinguished. The evidence is damning, yet pet cats are still kept near parks and reserves where their presence threatens vulnerable native wildlife. Even pet cats kill native birds, lizards and large insects.

The mustelids were the last predatory mammals to arrive, and potentially the most destructive. However, in most parts of the country they arrived too late to inflict the full damage of which they were capable.[70] The only parts of New Zealand where mustelids played a significant role in the extinction of native birds were South Westland and Fiordland, but even there ship rats and cats arrived at about the same time. In the 20 years following the arrival of these predators, piopio, laughing owl, bush wren, saddleback and kokako became regionally extinct, and the numbers of kakapo, kiwi, weka, blue duck and most other forest birds declined markedly. King suggested that even in Fiordland and Westland, ship rats alone would probably have caused the decline of most, possibly all, of these species if stoats had not been present to speed up their demise. She showed that 'barriers of time or geography prevented stoats from even meeting 137 of the 153 distinct local populations of birds that have drastically diminished or disappeared'. The very rapid decline of almost all bird species in South Westland and Fiordland – parts of the country where habitat loss was least significant – illustrates the threat mustelids would have posed had they arrived earlier. The species that had survived kiore and Norway rats but succumbed to ship rats and stoats either had small clutches (for example, piopio), nested in holes (huia, mohua and kaka) or lacked predator-distraction displays (kokako).[71] The survivors were species such as robins and tomtits, which retained the predator-distraction displays used by their Australian congeners and had larger clutches that could quickly be replaced if lost to predators.

King[72] suggested that the damage predators would inflict on native birds had then been done and only the hardier species remained. She considered that, except for the protection of a few vulnerable species such as black stilt, kakapo, takahe and kokako, attempting to control predators on the mainland was not generally a wise use of scarce conservation dollars. Subsequent research has shown mustelids to have a greater impact than King believed. Stoats are a major factor in the continuing decline of mohua, the birds being especially vulnerable after beech mast years (see Chapter 7). Stoats appear to be the most significant threat faced by kaka.[73] On Stewart Island, where rats and cats are present but mustelids are absent, kiwi, weka and kaka remain common and kakapo survived longer than on the South Island, where mustelids are also present. Unless controlled, stoats may cause the extinction of the remaining North and South Island kaka and kiwi populations. Weka are abundant on mustelid-free Chatham Island, yet are declining on the North and South Islands. Ferrets have caused a dramatic decline in the number of white-flippered penguins on Banks Peninsula.[74] Predators, probably mustelids, are causing declines in South Island colonies of sooty shearwaters, whose extinction is inevitable unless the colonies are protected.[75]

Collectors are not normally classified as predators, but for our purposes it makes no difference whether a bird is killed and eaten by a stoat or killed and stuffed by a person. During the nineenth century, collecting birds was a popular pursuit and both museums and private collectors had created a global demand for the hitherto unknown species. Collectors particularly sought out those species too rare to be of interest to other predators. There was a widespread assumption that New Zealand's endemic species were doomed, so it was considered important that a good range of specimens be collected and preserved before the species became extinct. Besides, there was good money to be made from the sale of huia, kiwi and other such rarities.

Andreas Reischek was the most infamous of the nineteenth-century collectors. He took 150 hihi from Little Barrier Island at a time when the species was already extinct elsewhere. Even on that island they were so rare that on his first visit he saw none, and on his second visit he searched for 10 days before finding a bird to shoot.[76] Comparative ecology of the introduced carnivores and collectors and the threats posed by each to native animals. The extinction of the huia and possibly the final demise of the Stephens Island wren were hastened, if not caused, by collectors. Thirteen bird species, including piopio, kokako, saddleback and Stephens Island wren, disappeared from Stephens Island in just a few years, their demise hastened by scientific collectors, forest clearance and that infamous cat.

Kiore, although the least damaging of the predatory mammals, got the best pickings and perhaps caused the extinction of more species than all the other rodents and carnivores combined. Most vulnerable

	Weasel	Stoat	Ferret	Cat	Human collectors
Body weight	F 57 g M 126 g	F 207 g M 324 g	F 600 g M 1200 g	F 2.7 kg M 3.7 kg	>50 kg
Date of liberation	1880s	1880s	1880s	c. 1800, main spread	1772, main spread
Ground hunting		✓✓	✓✓✓	✓✓✓	✓✓✓
Arborial		✓✓✓		✓	✓✓
Forest		✓✓✓	✓	✓✓	✓✓✓
Open country	✓	✓	✓✓✓	✓✓	✓✓✓
Abundance	Rare	Common	Common	Low density but roams widely	Once common, now controlled
Main species threatened	Poses little threat in the presence of rats and larger mustelids	Forest-dwelling, hole-nesting birds up to the size of kaka or kiwi	Ground-nesting and braided-river birds up to the size of blue penguins	Small to medium-sized ground-nesting and ground-foraging species	Species too rare to be of interest to other predators

Figure 5.2 Comparative ecology of the introduced carnivores and collectors and the threats posed by each to native animals

species were extinct by the time mustelids, the most potentially devastating predators, arrived. The story is similar with humans. Europeans, with their greater firepower, could have caused the extinction of the moa, geese and aptornis, but the opportunity was denied them.

The lost birds of Banks Peninsula

About 12 million years ago, the volcanic complex that was to become Banks Peninsula first rose above sea level.[77] At that time this new island was 60 kilometres seaward of the nascent Southern Alps, but as these mountains were raised by tectonic uplift, their erosion caused the

Canterbury Plains to creep eastward until they joined 'Banks Island' to the mainland about 20,000 years ago. Before then, plants and animals could only colonise the island if they were blown over from the mainland or rafted across the open ocean. Sea levels rose and fell as climates cooled and warmed, so the island may have been temporarily joined to the mainland at some prior date, in which case the connection with the mainland was certainly severed several times during the Pleistocene. Flightless birds could only colonise Banks Peninsula while the land bridge was in place.

The volcanic rocks of the peninsula and the gravels of the Canterbury Plains are poor substrates for the preservation of bone, so knowledge of the birds present before 1800 is incomplete. Bones have been preserved in the peninsula's loess soils, but because these mostly accumulated during the cool Otiran period, the bird species represented were probably different from those present in the dense forest that clothed the peninsula at the time of Polynesian settlement.[78] The remains of four species of moa, eastern kiwi, South Island aptornis, South Island goose, black swan, Auckland Island merganser, a crested penguin and tuatara have been found in Banks Peninsula archaeological sites, but these were all locally extinct before 1800.[79] Remains of two other species of moa were found at Redcliffs, where the plains meet the peninsula, and they may also have occurred on Banks Peninsula.

The loss of forests and birds during the European era is better documented for Banks Peninsula than for most other parts of New Zealand. When the first European settlers arrived in 1840, about two-thirds of the peninsula was still bush-clad and forest birds were abundant. By 1900, only 1.2 per cent of the peninsula was still in forest,[80] and only small remnants survived into the twentieth century. Until the 1980s, the largest reserve on the peninsula was 240 hectares. There was only one other reserve larger than a hundred hectares, and all reserves exceeding 40 hectares were on hilltops, ridges or in steep valley heads. Only minute remnants remained in the lowlands, where bird densities and diversity had been greatest.[81] In recent decades, many landowners have allowed or even encouraged the regeneration of native scrub and forest, so that today about a quarter of the Banks Peninsula is reverting back to native forest.

Europeans did not record takahe or kakapo on the peninsula,[82] so if those flightless species were ever present, they were locally extinct before 1800. However, all flighted forest birds present on the South Island in 1840 were also present on Banks Peninsula. It is possible that the New Zealand raven may have survived into the 1840s.* By 1870, piopio, kokako, saddleback and probably bush wrens were extinct or very rare on Banks Peninsula;[83] seven further species were lost by 1900, and the last kaka and parakeets were seen about 1912.[84]

It is possible to identify factors that contributed to the demise of some of these species. Large areas of forest remained on Banks Peninsula in 1870, so the extinctions that occurred at that time were probably not caused, but may have been accentuated, by habitat loss. At that time cats and rats were the only European-introduced predators likely to affect piopio and kokako. Piopio were extinct in most parts of New Zealand by the turn of the century, and Sir Walter Buller attributed their decline in forested areas to cats.[85] By 1870, Norway rats and possibly ship rats were common on Banks Peninsula, and it is likely that they caused the demise of saddlebacks and bush wrens. The nationwide decline of both species began after the arrival of European rats but before mustelids were introduced.[86] Ship rats caused the extinction of both these species on Big South Cape Island (see page 148) and saddlebacks failed to establish on Kapiti Island while Norway rats were present.

Mustelids were introduced in the 1880s and may have contributed to the loss of some remaining species on Banks Peninsula, but the story is complicated by the concurrent fragmentation of forests. Elsewhere, mohua and parakeets remain only in large areas of little-modified forest, but surviving populations are declining, owing to predatory mammals.

Factors contributing to the local extinction of robins and falcons are less obvious, but because these species survive elsewhere on the mainland and use a variety of habitats, it seems probable that both habitat loss and predators were implicated. Fernbirds are now virtually restricted to swamps but may once have used a wider range of habitats. Banks Peninsula itself had few wetlands, although neighbouring Lake Ellesmere did support wetland habitats suited to fernbirds. The buff weka was apparently extinct on Banks Peninsula by 1900, when it was still common elsewhere in Canterbury. Disease has been blamed for its sudden demise, but it is unlikely that this alone was the cause.

---

* The New Zealand raven is thought to have become extinct prior to European settlement. However, a description is given by James Hay (1915) of 'a large black crow which was very rare . . . The bird was larger than a wild pigeon, and smaller than a fowl. It was glossy black, with a strong beak like a fowl's. It had poor flight and generally frequented the same part of bush.' This provides tantalising evidence for the survival of the species into European times. Three were seen by Hay, sought by ornithologist John Healy, then shot and eaten by others at Pigeon Bay in 1847 or 1848. These birds could not have been rooks, which were not introduced to Canterbury until 1871 (Lamb 1964).

Today, kaka retain only a tenuous clawhold on the mainland. Edgar Stead[87] suggested that possums denned in the tree hollows used by nesting kaka, thus contributing to the that bird's demise. Stoats are now the most serious threat to mainland kaka,[88] which remain only in large tracts of forest. Habitat loss, stoats and possums may all have contributed to the kaka's demise. Red-crowned parakeets were locally extinct by about

Table 5.3 The changing status of terrestrial birds native to Banks Peninsula

NOW EXTINCT

Pre-European

Little bush moa
Stout-legged moa
Eastern moa
Large bush moa
Possibly other moa species
Brown kiwi
Aptornis
South Island goose
New Zealand coot

Extinct 1840s

New Zealand crow?

Extinct c. 1860

New Zealand quail

Extinct c. 1870

Piopio
Kokako
Saddleback
Bush wren
Laughing owl

Extinct c. 1900

Mohua
Yellow-crowned parakeet
Weka
New Zealand falcon
Fernbird

Extinct c. 1912

Kaka
Red-crowned parakeet

Extinct after 1920

Robin

STILL SURVIVE

Now rare

Tui
Long-tailed cuckoo
Morepork

Now common

Rifleman
Bellbird
Brown creeper
Tomtit
Grey warbler
Fantail
Shining cuckoo

Arrived since 1850

Silvereye
Welcome swallow

Sources: Potts 1873, 1874; Hay 1915, Stead 1927, Dawson & Cresswell 1949, Trotter 1975, McCulloch 1987, Challis 1995

1912, despite the huge flocks that used to raid Canterbury orchards in the late nineteenth century.[89]

The decline of native birds continued throughout the twentieth century. The robin was rare by 1927, but the last population may have persisted until the 1960s.[90] By the end of the twentieth century, the last few Banks Peninsula tui were restricted to Akaroa, yet in 1949 they were common at Governors Bay, and not uncommon in Christchurch and on Banks Peninsula in the 1970s.[91] Their decline was rapid and remains unexplained. In the 1960s, moreporks occurred in indigenous and exotic habitats, but a decade later were recorded at only one Banks Peninsula location.[92] In the 1920s, tomtits were seen in Christchurch gardens, and in the 1970s they were present in the Hunter Reserve at Church Bay,[93] but they do not occur at either place now, though they are present in many scenic reserves on the peninsula.

Birds of the open country have generally fared well, though the New Zealand quail, which was abundant on the plains and grasslands of Canterbury when settlers arrived, was locally extinct by about 1860.[94] Extensive burning of the grasslands, coupled with hunting, may have been the cause.

Coastal and marine birds have also suffered. Crested penguins, black swans, small petrels and king shags were locally extinct prior to 1800, although the swans were subsequently re-introduced.[95] Less than a century ago, sooty shearwaters and fairy prions nested on many peninsula headlands, and mottled petrels on the Canterbury foothills and perhaps also Banks Peninsula.[96] Today, very few shearwaters remain, the prions nest only on offshore islets, and the mottled petrels are long gone from Canterbury. White-flippered penguins remained common on the peninsula until about 1980, but since then the population has almost halved, mainly because of predation by ferrets.[97] The number of these predators on Banks Peninsula may have increased since rabbit control was relaxed in the 1980s.

Tuatara became extinct on Banks Peninsula some time before 1800, and short-tailed bats during the nineteenth century.

Extinct birds of the Chatham Islands

The Chatham Islands consist of two large and numerous small islands 800 kilometres east of mainland New Zealand.[98] They were settled about 600 years ago by Moriori, a Polynesian race who bought with them kiore but not kuri. Early in the nineteenth century, the islands were settled by Europeans, and then by Maori from the North Island. Most Chatham Island birds are sister species to those on the mainland, or subspecies descended from ancestors blown over from there. The Chathams have not had a land connection with New Zealand since the Cretaceous, so flightless and poorly flighted groups, including kiwi, moa, New Zealand wrens, wattlebirds, piopio, frogs, tuatara and geckos, were not present on the islands.

Species are more prone to extinction on small than on large islands – unless of course the small islands remain free of introduced predators – so an even greater proportion of the Chatham Island bird fauna has become extinct. Prior to human contact, about 36 species of terrestrial and freshwater birds bred at the Chatham Islands. Twenty are now locally extinct (Table 5.4). Of those, six also occurred on the New Zealand mainland, and two of these are extinct there as well. Within the Chatham archipelago, certain species survived on larger islands but became extinct when predators reached Mangere Island, further illustrating the greater probability of extinction on small islands.

As on the New Zealand mainland, waterfowl and rails were especially vulnerable. Of the eight native waterfowl, only one, the grey duck, survives at the Chathams; five species were extinct before 1800. Four rails have been lost, two before 1800 and two since European settlement.[99] As on the mainland, few seabirds became extinct, most species being saved by populations on small islands. A crested penguin and possibly a petrel in the genus *Pterodroma* (both undescribed species) are the only seabirds known to have become extinct. However, bone deposits suggest that at least 10 species of albatrosses and petrels bred on Chatham and Pitt Islands before the Moriori arrived. Only three of these remain on those two islands, one of which is the critically endangered taiko.

Extinctions probably happened more quickly on the Chatham Islands than on the New Zealand mainland, with all 13 prehistoric extinctions apparently occurring within 150 years of Moriori settlement.[100] Most of the seabirds lost from Chatham and Pitt Islands probably became extinct during that period. A further seven species (four of which survive on the mainland) have became extinct on the Chatham Islands since 1800.

Certain species survived only on Rangatira Island or on other small islands that remained free of rodents and other predatory mammals. Although Rangatira was farmed, only domestic stock were introduced and the last of these were removed in 1961. About a third of the island

Table 5.4 Extinct birds of Chatham Islands and Mangere Island

	Chatham Islands	Mangere Island
Diving petrel		*r
A gadfly petrel	pre-1800	*
Grey backed storm petrel		*r
White-faced storm petrel		*r
Chatham crested penguin	pre-1800	*
New Zealand little bittern^	pre-1800	
Black swan^	pre-1800	
Chatham Is shelduck	pre-1800	?
Brown teal^	1915	
New Zealand shoveler^	1925	
Chatham duck	pre-1800	
New Zealand scaup^	pre-1800	
Auckland Islands merganser^	pre-1800	
New Zealand falcon^	1900	
Dieffenbach's rail	1840	*
Chatham Islands rail	1900	*
Giant Chatham rail	pre-1800	
Chatham Islands coot	pre-1800	
Chatham Islands oystercatcher		*r
Shore plover		*
Chatham Islands snipe		*r
Extinct Chatham Islands snipe	pre-1800	*
Parea		*
Chatham Islands kaka^	pre-1800	*
Red-crowned parakeet		*r
Forbes' parakeet		*r
Chatham Islands fernbird	1900	*
Chatham Islands fantail		*
Chatham Islands tomtit		*
Black robin		*r
Chatham Islands bellbird	1906	*
Chatham Islands raven	pre-1800	

* Lost from Mangere Isalnd

^ Species that became extinct on the Chatham Islands that also occurred, or still do occur, on the New Zealand mainland.

r Species that became extinct on Mangere Island that survived on other islands and subsequently recolonised Mangere, with or without human assistance.

Note that the Chatham Islands sea-eagle was described from four bones allegedly collected at the islands in the 1890s. No sea-eagle bones have been found since and it is now assumed that the bones were collected elsewhere and wrongly labelled.

Adapted from Millener 1996 and Tennyson & Millener 1994

remained forested and three species – Chatham petrel, Chatham Island snipe and shore plover – survived there but nowhere else. The snipe and plover have since been translocated to other islands. Thanks to the absence of predatory mammals, Rangatira is an ark: a home to species that are now rare or extinct on the larger islands where rats, cats, dogs and weka have been introduced.

Moriori probably only visited Mangere Island to hunt birds or seals, and kiore never reached there, so most extinctions that occurred on Mangere happened during the last 120 years.[101] When farming began in the early 1890s, the island was still largely forested. A decade later, the island was virtually denuded and sheep, goats, cats and rabbits were present. By 1950, the cats and rabbits had died out, and in 1968 the last sheep were removed, leaving the island once again free of introduced mammals. About 18 species of terrestrial and wading birds formerly bred on Mangere Island, and all but two of these – pipit and Chatham warbler – were expatriated from the island. About five species of seabird also became locally extinct.

About 12 of these species were apparently exterminated by cats,[102] while for the rest extinction was a result of habitat loss and hunting by humans. Since the cats on Mangere died out, eight of the species lost from there but surviving on other islands have recolonised, some, such as the black robin, with human assistance.

The Big South Cape Islands rat invasion

> Kotiwhenu (Solomon Island) fostered an avifauna and flora not affected by the changes that have elsewhere transformed New Zealand into a second England. Its surface conditions remained as they had been centuries before, as they had been anterior to the last great migration of the Maori race.
>
> – W. H. Guthrie-Smith (1925)

The Big South Cape Islands are a group of three islands – Big South Cape, Solomon and Pukeweka – near the southern tip of Stewart Island. Each autumn they are visited by Maori who harvest muttonbirds (sooty shearwater fledglings), an important traditional food. Prior to 1964, these islands were the last refuges of the greater short-tailed bat, Stead's bush wren, South Island saddleback and Stewart Island snipe. Most other native birds were more numerous on Big South Cape than on neighbouring Stewart Island, while, conversely, introduced species were rare. In 1913, blackbirds were the only introduced species Guthrie-Smith found breeding on Solomon Island; in 1961, redpolls were the only introduced bird common on Big South Cape.[103] Weka had been introduced some time between 1930 and 1950, but the islands had no introduced mammals.

In March 1964, muttonbirders discovered ship rats on these islands. When and how they got to the islands is unknown. One had been caught on the large island in 1955, but none was recorded during two subsequent visits by Wildlife officers or by the muttonbirders during their annual sojourns.[104] By April 1964, the rats had reached plague numbers in the northern part of Big South Cape Island and were spreading southwards. Saddlebacks, wrens, fernbirds, robins, snipe and short-tailed bats had already disappeared from the northern end of the island, and few bellbirds and parakeets were seen where rats were abundant. As the rats advanced south, the birds at the far end of the island were under threat.

Not just birds and bats were affected (Table 5.5). Insects, in particular weta and other large flightless species that were common before

Table 5.5 The Big South Cape Islands rat invasion

Cast in approximate order of disappearance

Stewart Island robin	Also occurred elsewhere, locally extinct
Stewart Island fernbird	Also occurred elsewhere, locally extinct
South Island saddleback	Last surviving population, saved
Stead's bush wren	Last surviving population, extinct
Stewart Island snipe	Last surviving population, extinct
Greater short tailed bat	Last surviving population, extinct
Hadramphus stilbocarpae	Also occurred elsewhere, locally extinct
Red-crowned parakeet	Survived but numbers reduced
Yellow-crowned parakeet	Survived but numbers reduced
Bellbird	Survived but numbers reduced
Weta	Survived but numbers reduced
Dorcus helmsi	Survived but numbers reduced
Other large flightless insects	Survived but numbers reduced
Stilbocarpa	Survived but numbers reduced
Tomtit	Increased with loss of the robin
Blackbird	Increased
Chaffinch	Increased
Dunnock	Increased

Adapted from Bell 1978 and Watt 1975

rats arrived, soon became rare. One large flightless weevil (*Hadramphus stilbocarpae*) became locally extinct.[105] Some plants, in particular *Pseudopanax* spp. and *Stilbocarpa lyalli*, a giant fleshy herb and host to that weevil, were eaten out, although they did recover once the rat population declined after the initially abundant food was eaten out. Of the survivors, tui were unaffected and may now be more common than previously. Tomtits were uncommon before the rat plague, but once the robins were gone they increased in number. Bellbirds and parakeets, once numerous, stabilised at reduced numbers. The introduced blackbird, chaffinch and dunnock are now much more common than they were before rats changed the island ecosystem.

A Wildlife Service team was dispatched to the Big South Cape Islands in August 1964 and made a valiant attempt to save those species for which the islands were the last refuge. They successfully transferred saddlebacks to two nearby rat-free islands, but their attempts to translocate wrens and snipe failed, and nothing could be done to save the bats. These three species were soon extinct. The rescue operation is discussed in Chapter 9.

It is unfortunate that the dynamics of the rat invasion and the way rats changed the ecosystem was not described in greater detail. The small Wildlife team was too busy saving species to undertake research that could help us better understand the dynamics of rat invasions and how best to manage such events in the future. Brian Bell[106] described the rat invasion and the extinctions that followed. He noted that what happened in the course of just a few years on the Big South Cape Islands mirrored the changes in bird faunas that occurred over a longer period on the mainland.

Why was the New Zealand fauna vulnerable?

Species have become extinct whenever and wherever people or alien animals have colonised new lands, and further species have been lost following the arrival of each wave of new settlers and with each new suite of land-altering technologies. Although extinctions have occurred everywhere, the proportion of the fauna lost has varied from place to place. It has been greatest on oceanic islands and on long-isolated fragments of former continents. The species that were most prone to extinction exhibited one or more of the following characteristics:

- they were archaic
- they evolved in the absence of mammalian predators

- they were flightless or poorly flighted birds
- they had ecological roles similar to animals subsequently introduced, especially when ecological conditions were changed to better suit the introduced species
- they were K-selected species with little ability to cope with rapid environmental change
- they had long incubation periods and slow growth rates, so they were vulnerable to introduced predators during a greater proportion of their life cycle.

The New Zealand fauna included a host of archaic and K-selected species. Many of the birds had reduced powers of flight or were ground-dwelling with predator-defence strategies that were ineffective against introduced predatory mammals. Virtually all of New Zealand's archaic animals – for example, the native frogs and tuatara, flightless and K-selected birds such as kakapo – mammal-niche species (in particular the moa), and species belonging to endemic families or orders (including kiwi and New Zealand wrens), are now extinct or endangered.[107] Extinctions were inevitable once humans, no matter who they were, discovered and settled this country.

Madagascar is another long-isolated continental fragment with a human history and extinction history similar to New Zealand's. That island, in the Indian Ocean off the coast of Africa, was first settled by the Malagasy people about 1500 years ago. Following settlement, all species of elephant birds (large, flightless, moa-like ratites), giant tortoises and about 30 species of mammals, including 12 species of lemurs, a small hippopotamus, a small aardvark and a large 'mongoose', became extinct.[108] Other than the elephant birds, few other birds are known to have become extinct. Madagascar already had predatory mammals so, unlike New Zealand, its small ground-dwelling birds and reptiles were apparently little affected.

Flightless birds, rails, waterfowl (especially terrestrial species), birds of prey, pigeons, parrots and crows were also vulnerable in Hawaii and other Pacific Islands.[109] Rails once occurred throughout the Pacific, with numerous species being endemic to one or few islands. Today, few species survive, and those that do are mostly small and widely dispersed. Gone too are most of the ground-dwelling flightless birds, and of the few birds endemic to Pacific Islands that still survive, most are endangered. As with New Zealand, the species surviving today give little indication of the diverse and fascinating faunas that once occurred on these oceanic islands.

In New Zealand and on islands elsewhere, the loss of major predators, most browsing birds and the seabirds that played a key role in transferring nutrients from sea to shore must have greatly changed the ecosystem dynamics. A large proportion of the fruit-eating birds became extinct in New Zealand and elsewhere in the Pacific.[110] With most of those seed-dispersing birds now gone, the long-term survival of some plant species must now be in question. The ability of the forests to persist now so many species are extinct will be discussed in Chapter 7.

CHAPTER 6

Acclimatisation

> So with zeal unfettered by scientific knowledge, they proceeded to endeavour to reproduce – the best remembered and most cherished features of the country from which they came. – No biological considerations ever disturbed their dreams, nor indeed did they ever enter into their considerations.
>
> – G. M. Thomson (1922)

Whenever people have settled new lands they have carried useful plants and animals with them, and explorers have also brought back useful and interesting species from the lands they have visited. By the time of European settlement of New Zealand, a few species had already been introduced. The first kiore to arrive presumably accompanied travellers 2000 years ago.[1] The Polynesian settlers successfully established kuri, kumara, taro, yam, bottle gourd, paper mulberry and ti (a species of cabbage tree).[2] They probably also attempted to introduce the pig, domestic fowl, coconut and banana, which they carried with them whenever they settled new islands.

By the start of the colonial era, the Europeans had a long history of biotic exchange. They had carried crops, farm animals, rats, cats and other animals to many parts of the world. Tomatoes, potatoes and maize from the Americas and grains from the Middle East had been introduced to Europe and, at the same time that European species were being bought to New Zealand, curious species from here were taken to Europe.

While movement of useful plants and animals from place to place had a long history, special efforts were made to introduce plants and animals to New Zealand. The settlements were planned with the expressed aim of bringing to this country everything required to establish British colonies in a distant land, and that included British animals. The four ships that left England in 1850 to found the Canterbury settlement carried not just settlers and their tools of trade, but all the impedimenta required to establish an English colony in the antipodes. Also on board were some British birds, reminders of a homeland few settlers would see again. Only two pheasants and a partridge survived the voyage, and dogs soon killed them. The Canterbury settlers' first attempt at acclimatisation failed, but by the century's end many species of exotic birds and mammals had been successfully introduced.

The latter half of the nineteenth century was the age of acclimatisation, when the 'enrichment' of biotic communities was considered a highly desirable public service. The formation of acclimatisation societies for the express purpose of introducing foreign biota began in Europe. The first such society, La Societé Zoologique d'Acclimatation, was founded in France in 1854 to assess the value of new species for importation into that country and its colonies. Within a decade the idea had spread to Britain and thence to Australia, North America, South Africa, New Zealand and elsewhere. Acclimatisation societies sprang up in countries as diverse as Egypt, India, China and Russia.[3] The heyday of acclimatisation was contemporaneous in the United States and New Zealand, despite the colonial period occurring much earlier in the former. The reasons for acclimatisation were similar in all these nations: utilitarian purposes, nostalgia, amenity values, sport and biological control. Although part of a global trend, the New Zealand societies were longer-lived and more influential than those in most other countries.

The Canterbury Horticultural and Acclimatisation Society (later to become the North Canterbury Acclimatisation Society) was established in 1864, only 14 years after the colony was founded. During that time private individuals had attempted to introduce bees, gamebirds, rabbits and fish, but with little success. The acclimatisation society was formed to ensure future attempts at the naturalisation of exotic species were more successful. The society had the backing of the provincial government and its inaugural committee included Dr Julius von Haast, the district geologist; W. T. L. Travers and T. H. Potts, both prominent naturalists; the archdeacon and other leading citizens. Two years earlier, Haast had suggested an acclimatisation society be set up as a branch of the Philosophical Institute, the province's society for science and learning.[4] At that time acclimatisation was considered a branch of science.

Not only the Canterbury society enjoyed the support of scientists and leading citizens. Sir Walter Buller, New Zealand's leading nineteenth-century ornithologist, served as secretary of the Wanganui Acclimatisation Society.[5] Societies were formed in all districts in New Zealand with the purpose of liberating the widest possible array of useful and interesting animals and plants. Foreign biota were not just introduced for food and other utilitarian reasons, but also for recreational and other less tangible needs. With a century or more of retrospective wisdom we may question the settlers' motives and we are painfully aware of the disastrous impacts certain introduced species have had. However, the early settlers viewed the New Zealand environment and biota very differently.

Animals and plants have been introduced to virtually all parts of the world, but it was on isolated archipelagos such as New Zealand and Hawaii that their impact on the native biota was most marked. In New Zealand, introduced species have profoundly changed ecological communities. Competition and predation by introduced animals, particularly mammals, has contributed to the extinction of many native species, and threatens others. These concerns are discussed elsewhere in this book. In this chapter I review the history of acclimatisation, describe the range of species introduced (both deliberately and accidentally) and discuss how or why they were bought here. I also consider how the way people value certain introduced species has changed during the last 150 years. I confine my discussion to species that arrived in New Zealand with direct assistance from people. Not included are species that have colonised the country in historic times through natural dispersal but whose successful establishment was made possible by human-induced habitat changes. In Chapter 10, both introduced species and natural colonisers are treated together as part of the 'cosmopolitanisation' of the New Zealand biota.

Explorers, sealers and missionaries, the pre-acclimatisation era

> For a sly cat on board had no sooner perceived so excellent an opportunity of obtaining delicious meals, than she regularly took a walk in the woods every morning and made great havock among the little birds, that were not aware of such an insidious enemy.
>
> – George Forster (1777)

Rats and cats were two of the first European imports. The Norway rat was probably the ship-dwelling rat of the eighteenth century and it would have been this species that first arrived with Captain Cook.[6]

The explorers, sealers, whalers, traders and missionaries that followed Cook unwittingly set rats ashore at numerous locations around New Zealand. The more destructive ship rat became widespread in the North Island in the 1860s and the South Island about 30 years later. Ships of the time were infested with rats, and cats were carried on board to keep them under some semblance of control. Cats are more skilful predators than rats, and their larger size enables them to prey on animals immune to rat predation. When Cook visited Dusky Sound in 1773, George Forster, one of the naturalists, recorded the exploits of one ship's cat in the quotation heading this section. That marauding moggy apparently rejoined the ship, but others that followed settled down to life ashore. During this early period, cats appear to have remained close to human settlements and there were few truly feral populations.[7]

While the rats that deserted Cook's vessels did so without his leave, during his three visits to New Zealand he deliberately released geese, sheep, pigs, goats and domestic fowls at Dusky Sound, Queen Charlotte Sound, Cape Kidnappers and elsewhere. Cook and other explorers released animals and planted vegetables to provide food for explorers and traders who would follow them, and as gifts to Maori. Wild pigs were present in the North Island by 1810, but few other plants or animals left by the explorers established wild populations without further human assistance. Habitat modification was needed before most introduced species could compete with the established native species.

The sealers not only spread rats but, like the Maori before them, took weka, native to the mainland, to offshore islands where they extinguished populations of small ground-dwelling birds and large invertebrates.[8] Whalers, flax and timber traders and missionaries provided many more opportunities for the spread of exotic plants and animals. Missionary activity began in 1814 when the Reverend Samuel Marsden founded the first mission station in the Bay of Islands. More animals, farmed and feral, followed as whalers, missionaries and traders established bases in New Zealand. Unlike the explorers, who merely sowed seeds and released animals then left, whalers, traders and missionaries cultivated land and burned and felled forests. This broke the tenacious hold native species had on the land and created habitats in which the introduced species were competitive or even advantaged.

Settlers, the acclimatisation era begins

> We don't go to New Zealand with pick and pan, to snatch dear-won nuggets, gulp gallons of rum, and then rich or ragged hurry home. We go to the 'Britain of the South' to create an estate, raise a home wherein to anchor fast and plant our household goods.
>
> – Charles Hursthouse (1857)

The rock wren is one of only two extant species of New Zealand wren. These birds diverged from the other passerines 80 million years ago.

Photograph by Kerry-Jayne Wilson

Although the South Island robin weighs only 35 grams, it is a biological giant, being more than twice the weight of any non-New Zealand member of its genus.

Photograph by Kerry-Jayne Wilson

Key to Plate 2 (overleaf): The Oparara Valley, Karamea during the Holocene, 2,000–10,000 years ago, when podocarp-beech rainforest clothed the valley.

Painting by Pauline Morse

The kakapo has been described as 'the most wonderful perhaps of all living birds'. This outsize nocturnal parrot has a most bizarre breeding system.

Photograph by Kerry-Jayne Wilson

The flighted pukeko (left) has colonised New Zealand on at least three occasions, one of the early events giving rise to the giant, flightless South Island takahe (right).

Photographs by Kerry-Jayne Wilson

The year 1839 was a turning point in New Zealand's history, with far-reaching consequences for native wildlife. Until then, areas of European influence were restricted mostly to the coast, and of the animals introduced, only Norway rats, pigs and goats had established truly feral populations. In 1839 the first permanent settlement was founded at Port Nicholson (Wellington). Others soon followed: the Auckland and Taranaki settlements in 1840, Nelson in 1841, Otago in 1848 and Canterbury in 1850.

Few of the previous European visitors had intended settling in New Zealand: most were transients, here to get rich or save souls, then return home. But this new wave of Europeans was different. Settlement sites had been reconnoitred while settlement companies had recruited a balanced mixture of the upper class, professionals, tradesmen and workers with all the impedimenta needed to establish a British colony. From the beginning they set out to bring with them livestock, game animals, plants and animals reminiscent of home. This desire was reinforced when the colonists found that their adopted country had a strange fauna with few species of useful or familiar animals, an unfortunate state of affairs they sought to rectify. Because few native birds could live in areas cleared for farmland or settlement, the new land must have seemed strangely silent. Acclimatisation was the obvious solution. At that time knowledge of how biological systems functioned was rudimentary and the acclimatisers were unable to predict the consequences of their actions.

Today's urban and rural landscapes, where introduced plants and animals dominate and there are few native species, give a misleading impression of the fragility of New Zealand's indigenous ecosystems and the competitive edge those introduced species had. By 1840, few introduced plants or animals had become established, and most of these were confined to places where people had carved out farms or villages from the native bush. These new settlements presented a multi-pronged ecological onslaught. Not only were new animals introduced, but so were the plants that had co-existed alongside them. The foreign species were liberated into a landscape that the new settlers were deliberately changing to resemble the homelands of the species introduced. Few introduced animals would have successfully established in New Zealand if such inordinate efforts had not been made to Europeanise the land. Alfred Crosby[9] discusses the social and ecological imperialism of the nineteenth-century settlement of New Zealand and other temperate lands colonised by the Europeans.

The New Zealand settlements were to be more egalitarian than contemporary Britain with its rigid class structure. Opportunities in Britain were limited for those with no family fortune to inherit, whereas it was possible for even poor men to thrive in the colonies. For many settlers, New Zealand represented an escape from the oppression and poverty of industrial Britain. Those of upper-class birth wished to partake in the pursuits they had enjoyed at home, and so did those from less privileged origins. In New Zealand, deer, game birds, trout and salmon were to be accessible to all. Ironically, for some lower-class settlers banishment to the Australasian colonies was punishment for hunting the self-same species in Britain.

Dr Julius von Haast encapsulated the prevailing views on wildlife in his inaugural address to the Philosophical Institute of Canterbury

> We should like to see the hare and the partridge in our fields, the stately deer, the roe, and the pheasant occupying our hills and our forests, whilst our Alpine rivers are well calculated for the propagation of the salmon and the trout. The most rugged of our mountain summits might become the venue of the chamois, and offer not only to us but to future generations, the exciting pleasure and manly exercise of the chase.

The first attempts at acclimatisation were haphazard before regional societies were formed in the 1860s. The objective of these was to bring order to the introduction of 'all innoxious animals, birds, fish, insects, and vegetables whether useful or ornamental' and 'the spread of indigenous animals from parts of the colonies where they are already known to other localities where they are not known'. Acclimatisation societies played an important role in the introduction of exotic species to New Zealand and in the subsequent management of exotic and native game animals. Bob McDowall[10] has documented the history of these societies and their role in wildlife management.

Once formed in 1864, the Canterbury Acclimatisation Society set to work with diligence and enthusiasm, building enclosures for birds and mammals and rearing ponds for fish in what is now Hagley Park. These facilities were primarily to rear animals for release into the wild, but they became a popular venue for Sunday outings, and at various times even had emus, kangaroos, a bear, a Tasmanian devil, a lemur, a llama and a tortoise, along with the more familiar deer, rabbits, game birds and ferrets.[11] Not all animals held there were intended for release.

The acclimatisation of small British birds was of particular interest to the settlers. In 1867, visitors came to Hagley Park to see the first dunnock to arrive in Canterbury, and in 1872 a lone English robin drew crowds. When the Canterbury society's single nightingale died, 'deep regret was felt far and wide'.[12] Enormous efforts were made to obtain birds from Britain. In 1864, Charles Prince left England for Lyttelton with some 300 birds in his care, but only 30 pigeons survived. Two years later, a Christchurch resident visiting England was sent £150 to procure birds, including insectivorous species that would control caterpillar plagues. He shipped 444 birds, of which 7 pheasants, 11 partridges, 46 blackbirds, 36 thrushes, 21 linnets, 13 skylarks, 11 chaffinches, 20 starlings and the aforementioned dunnock were landed safely at Lyttelton. These were then sold subject to the condition that the buyer would turn them loose once he was home. Further such shipments followed.

Richard Bills became a professional bird collector. In 1870, he left London bound for Otago with more than a thousand birds, most of which arrived at their destination alive and well. A few months later he returned on the first of three collecting trips to the 'Old Country' in the employment of the Canterbury Acclimatisation Society. He was paid a salary of £9 per month, plus a bonus of one shilling for each small bird and two shillings for each large bird landed at Lyttelton. Bills returned to England the following year, taking with him kiwi, blue duck and kea, plus bird skins and moa bones, to exchange for British animals with the London Zoological Society.

Other acclimatisation societies went to equal lengths to procure exotic animals. During its founding year the Auckland society introduced more than 30 species and devoted more than £1000 to acclimatisation.[13] In the following decades the provincial societies exchanged species with one another, or with sister societies in Australia, to enhance the spread of exotic species. During this period many seafarers also carried birds to New Zealand for sale to the settlers. If the birds survived, they provided a useful supplement to the seaman's regular income: in 1864 a pair of grouse was worth 10 guineas in Canterbury; thrushes, blackbirds and skylarks £2 a pair; and sparrows and linnets 15 shillings.[14]

In all, 130 bird species are known to have been bought to New Zealand, but not all were released into the wild.[15] Other introductions no doubt went unrecorded. Many were European passerines, but other unusual species included emu, Solomon's cassowary, Java sparrow and some type of seagull. Only 39 of those 130 species still have self-sustaining wild populations. In their first 20 years the acclimatisation societies had been quite unselective in what they introduced, but by the 1880s the desire for unrestricted introductions seemed to have been satisfied. They narrowed their focus to game animals and fur-bearing mammals such as possums. Later still, government agencies usurped the role of the societies so that official agencies were responsible for most twentieth-century introductions.

Game birds were sought from the beginning of European settlement, and 25 species of waterfowl are known to have been liberated.[16] Only four of these – black swan, mute swan, Canada goose and mallard – became successfully established. Some of the first waterfowl introduced were black swans, which were set free in the Avon River in Christchurch to control watercress. They quickly left the city and proliferated, along with others of their kind liberated elsewhere (and perhaps others that colonised by trans-Tasman dispersal). By the early twentieth century, black swans had acquired nuisance status on Lake Ellesmere and the Waikato lakes.[17]

Pheasants, partridges, quail and their kin, the so-called upland game birds, had been traditional quarry of the upper class in Britain. The first pheasants arrived as early as 1842, and over the next century there were many further releases by most, if not all, of the regional acclimatisation societies. Many of the societies had farms where game birds (mostly pheasants) were bred for release in an attempt to establish new populations, to boost existing populations or to enhance shooting opportunities. The Auckland society was especially assiduous in its efforts to enhance game-bird populations. In total, about 28 species of upland game birds were introduced,[18] two early species being the grey partridge (1842) and California quail (1864). The most recent species to be introduced was the red-legged partridge (1980). Despite the vast sums of money invested and the inordinate efforts made to get them established, upland game birds provide limited opportunities for hunters.

The first pests

. . . a pert, mischievous and immensely reproductive little bird.
– W. J. Steward, MHR for Waimate (1882)

The first acclimatised vertebrates to be accorded pest status were not possums, rabbits or deer, but house sparrows. These, and a selection of other small birds, had been introduced to control caterpillar plagues in several parts of the country. In some areas sparrows were selected because they were prolific and, since they ate both plant material and insects, it was argued they could sustain themselves through the winter when insects were scarce.[19] Elsewhere it was insectivorous hedge sparrows (dunnocks) that were sought but seed-eating house sparrows that arrived. These settled in quickly and multiplied rapidly. Late in 1868, the very year they were first introduced to the city, house sparrows stripped a Christchurch cherry tree of its fruit. Seven years later the Canterbury Provincial Government allowed sparrows to be killed at will. In Nelson, and elsewhere soon after, the sparrow menace was rife and clubs were formed to deal with the vermin.[20] In 1881, the Papanui Sparrow Club destroyed 237 dozen eggs and 347 dozen young. In 1882, they laid 24 bushels of poisoned grain.[21] Poisoned grain was scattered from the guards vans of trains as they steamed through sparrow-infested countryside. Acclimatisation societies were quick to disassociate themselves from the introduction of these suddenly unpopular immigrants.

Not only sparrows proved troublesome. Blackbirds and thrushes were welcomed ashore in the 1860s, then, 15 years later, condemned for destroying fruit crops.[22] Blackbirds, song thrushes, house sparrows, finches, starlings (which are a mixed blessing in that they help control grassland pests), mynas and other introduced birds, plus the native silvereye, still cause local damage in orchards, vineyards and to other horticultural crops. A survey suggested that blackbirds, song thrushes, silvereyes and house sparrows were the most destructive species, and berryfruits, grapes, cherries, pears and apples the crops most affected.[23] The economic losses caused by birds is poorly quantified, but it has been estimated that up to 30 per cent of a wheat crop and a fifth of grape crops can be eaten by birds.[24] Today, however, any nuisance caused by small birds pales into insignificance when compared with damage done by mammals.

Mammals, the real troubles begin

> . . . a grant of money for the purpose of introducing weasels, as a natural check – would (if the object were attained) be of very great service.
>
> – Select Committee inquiring into the Rabbit Nuisance (1876)

European rabbits have accompanied people to many parts of the world. They are easily domesticated, can be kept in confined conditions on board ships and are renowned for their fecundity. They are adaptable animals that produce fur and meat, they make good pets and are popular small game animals. Until about 2300 years ago, the species was restricted to its native range, the Iberian Peninsula. Phoenician and

Roman traders dispersed them around the Mediterranean. By the twelfth century they were present in Britain and subsequently, through human-assisted spread, came to occupy much of Europe.[25] During the golden age of exploration European explorers, and subsequently settlers, carried rabbits to most places they went, so that now they also occur in Australia, North and South America, South Africa and on over 800 islands around the world.

The first rabbits introduced to New Zealand were a pair Captain Cook released on Motuara Island in the Marlborough Sounds in 1777. Domestic breeds were released on Mana Island in 1834, in Southland in 1838, the Bay of Islands in 1838, Nelson in 1842 and the Wairarapa in 1847, but none of these releases was very successful. Wild-type rabbits were introduced around 1850, and efforts to get them established were stepped up in the 1860s. Success quickly followed because land opened up for grazing provided good rabbit habitat.[26] By the 1870s, rabbits were widespread and 'grey plagues' swept over the land, destroying grass intended for sheep. On Burwood Station in Otago, rabbits reduced the sheep-carrying capacity of the farm from 110,000 to 30,000 head.[27] Large areas of grazing land were abandoned and runholders walked off their properties in despair. Rabbit numbers peaked in the 1890s and, while these animals are no longer a major threat to agriculture, pastoralists still regard them as a major culprit in overgrazing of semi-arid grasslands.[28]

In the late 1870s, the loss of grazing caused by rabbits was so severe that runholders urged the introduction of predators to control them. Initially, pastoralists in rabbit-infested areas sought cats to release in the country as rabbit killers, and there are stories of enterprising city lads capturing cats for sale to rabbit-plagued farmers. After further debate, three species of mustelid – ferret, stoat and weasel – were introduced in an attempt to control 'the rabbit menace'. These new invaders were real predators: carnivores evolved specifically to prey on birds and mammals, not omnivorous rodents that would catch only the small and unwary. For the fourth time in 2000 years, birds, bats and reptiles that had evolved in the absence of mammalian predators were faced with new enemies they were ill-equipped to contend with.

Cats have contributed to the extinction of native animals in many parts of the world, and island birds that evolved in the absence of mammalian carnivores have proven especially susceptible to cat predation.[29] Although they arrived in New Zealand with the first Europeans, until 1870 cats were localised, usually close to human settlements, but they appear to have became widespread in the wild about the same time as the mustelids.

The three species of introduced mustelids are all slim and long-bodied, adapted for preying on small animals in constricted spaces. They form a size series (see page 185), the smallest being the weasel, then the stoat, and the largest the ferret. All had a reputation for ferocity and cunning, and there was debate concerning the wisdom of their introduction. Buller, Hutton and Reischek in New Zealand, and in Britain Professor Newton, were among the leading zoologists who predicted that the introduction of these new predators would result in the further loss of native birds.[30] Even some of the acclimatisation societies, still enthusiastic liberators of new species, argued against their release. However, farmers wanted a 'quick fix' for a pressing economic problem, and then (as now) they had greater political sway than biologists.

The first ferrets were imported in 1879 and releases continued until 1912, initially from wild British animals and later from captive-bred stock. Stoats and weasels were first liberated in 1884. They were eagerly sought-after: in 1885 stoats cost £5 each, and in 1888 alone, William Acton-Adams, the runholder at Molesworth, Marlborough, invested £800 in stoats and weasels.[31]

By the turn of the century, it was obvious that the mustelids had failed to control the rabbit plague. Those acclimatisation societies that had not too publicly supported the introduction of mustelids attributed a decline in introduced game birds and native birds to these new immigrant killers.[32] In 1903, the law was changed to allow mustelids to be killed at will, but no attempt at control has ever done more than crop the population.

The effectiveness of predators in controlling rabbit populations depends on habitat. In most parts of New Zealand, predators and diseases, with assistance from pest controllers, keep rabbit numbers in check and they are no longer a major threat to agriculture. Rabbits remain a problem in the semi-arid tussocklands of the South Island, where in some years rabbits reach very high densities and neither predators nor other controls are effective. In these drier areas, the infestation of rabbits, the degradation of the grasslands and the invasions of *Hieracium* and other weeds are now considered symptoms of 150 years of overgrazing by rabbits and sheep.[33]

Quick fixes are still sought. In August 1997, rabbit haemorrhagic (calicivirus) disease (RHD) was illegally introduced to New Zealand by South Island high-country farmers in another ill-conceived attempt to control rabbits. Unlike the mustelids, RHD probably had the potential to reduce rabbit populations and prevent them from again reaching such high densities. However, insufficient research had been done to know when and how to introduce the disease in order to achieve the best results while minimising the risk to native species. The illegal release and the factors leading up to it have been analysed by the Parliamentary Commissioner for the Environment,[34] who concluded that while only a small number of farmers were badly affected, the impact rabbits had on them had been financially crippling, particularly since 1979, when the government opted out of rabbit control and the costs were primarily borne by the landowners. Research had suggested that RHD might well be effective in controlling rabbits, but the public perceived the risks of introducing a new organism to New Zealand to be unacceptable. When, in July 1997, the Ministry of Agriculture declined requests for legal import and release of RHD, affected farmers believed that the government had favoured the whims of city-dwellers over their own financial strife. Out of frustration, a normally law-abiding group of citizens committed a serious breach of New Zealand's biosecurity. The Parliamentary Commissioner's report is a revealing analysis of how different stakeholders view the risks and costs of pests and their control agents. RHD did reduce rabbit abundance in most parts of New Zealand, with varying results,[35] and epidemics have recurred naturally since then, exerting continuing control.

Mustelids, especially stoats, spread rapidly, far beyond the rabbit-infested farmlands. They reached Tutira Station in Hawke's Bay before the rabbits.[36] Stoats spread quickly even to the most remote parts of Fiordland, assisted by a release on the western shores of Lake Manapouri, where rabbits were absent but kakapo still common. Stoats are now the most widespread and most common of the mustelids. They occur throughout both main islands in virtually all habitats, from coastal dunelands to subalpine scrub.[37] Stoats are able swimmers and have been recorded on 26 islands, all within 1100 metres of the nearest stoat-infested landmass.[38] There are no mustelids on Stewart Island, Great Barrier Island or on the Chathams, Kermadecs and subantarctic islands.

The introduction of mustelids and their impact on native animals has been so well documented by Kim King in her book *Immigrant Killers* that it is not necessary to repeat the details here. King concluded that although mustelids were capable of causing the extinction of many species of native animals, not just birds, they arrived too late to exercise their prowess or have failed to gain access to the last island refuges of vulnerable species. As shown in Chapter 5, rats, cats or habitat destruction had already done most of the damage that could be done. It remains crucial that stoats and other predators be prevented from gaining access to the remote islands that are the last refuges of so many native species. The threat mustelids pose to native wildlife is further discussed in Chapter 7.

Hares were also introduced. New Zealand's first hare jumped through a ship's porthole and swam ashore at Lyttelton in 1851.[39] Between 1867 and 1876, further releases were made in Otago, Southland, Canterbury, Nelson, Wellington and Auckland. Hares adapted well and, by 1876, were causing sufficient damage that severely affected runholders were allowed to shoot them, but only on their own land. Meanwhile, other farmers continued purchasing hares for release on their own properties. In Britain, hare coursing – the hunting of hares using greyhounds – was a traditional sport, and hares were introduced to New Zealand with this in mind. Hare coursing was popular for a time, but by the early decades of the twentieth century hares had become so troublesome that the noble sport was replaced by the colonial alternative, the hare drive – mass slaughter by any means practicable.

Sir George Grey, acclimatiser extraordinaire

Sir George Grey had an illustrious career as an explorer and colonial administrator. He spent several years in Australia, a term as Lieutenant-Governor of New Zealand, then in 1854 became Governor of the Cape Colony (South Africa). He returned to New Zealand in 1861 and between then and 1879, when he retired to England, he held office as Governor, a member of Parliament and eventually Premier. Grey had an abiding interest in animals and plants. During his second term in New Zealand, he purchased Kawau Island in the Hauraki Gulf, where he built the famous Mansion House and in its grounds planted trees from all the lands he had visited. He then stocked his gentleman's estate with free-ranging animals, liberating kangaroos, wallabies, possums, emus and kookaburras from Australia, monkeys, zebras, antelopes, Egyptian geese, Cape doves, gnu and guinea fowl from Africa, plus a miscellany of other animals, including deer, cassowary and peacocks. Fortunately not all survived: one of the two zebras was shot; the bull wapiti proved dangerous and had to be put down; and the cassowaries and emus soon died of unknown causes.

A few species did establish wild populations, and within a few years wallabies were eating out the carefully nurtured exotic plants and inhibiting regeneration of the native forest. In 1965, a Parma wallaby, a species believed extinct, was shot on Kawau Island, after which the species was given protected status – the only protected introduced mammal. Between 1967 and 1975, 736 Parmas were sent to zoos for captive breeding or were repatriated to Australia until small populations were rediscovered in New South Wales,[40] so the traffic ceased. In 1984, protection of the Kawau Parmas was revoked, changing this once officially extinct species back to a noxious animal to be shot on sight.

One of the other wallabies on the island is the brushtailed rock wallaby, listed as a vulnerable species in its native range while it remains an introduced pest on Kawau. Today, attractive plants from all parts of the world grace the grounds of Mansion House, four species of wallabies graze the lawns and kookaburras mock Grey's folly.

Deer and other herbivores

Six herbivorous marsupials, rabbits, hares, eight species of deer, tahr, chamois, goats and pigs, plus feral horses, sheep and cattle (See Appendix 1 for a full species list) have successfully established wild populations in New Zealand. Herbivorous mammals from four continents now run wild here, and species from two other continents were offered residence in this land where plant-eating mammals had never before set hoof or claw. Some came from noble homes. Many deer came from British game parks such as the Duke of Bedford's estate at Woburn Abbey or the Prince Consort's herds at Windsor Great Park. New Zealand's chamois were a gift from Emperor Franz Josef of Austria, while the wapiti and moose were presented by US President Theodore Roosevelt.

Herbivores threaten native plants and animals by damaging and modifying indigenous habitats. The damage they do may be no less severe but more subtle and happen more slowly than that caused by predators, hence many decades elapsed before their damage was acknowledged. For example, the possum is now New Zealand's most serious mainland vertebrate pest. In the early twentieth century, possums were known to damage orchards and other cultivated lands, but in the 1920s Professor H. B. Kirk and Leonard Cockayne[41] (an early proponent of native-forest conservation) considered the damage done to forests so slight that further releases would be beneficial. Fur-bearing possums were one way native forests could produce an economic return. For the next 25 years, possums were harvested under licence, but by 1946 the damage they were doing was acknowledged and protection lifted. Ever since, increasing sums of money have been spent in forlorn attempts to control their numbers. Possums have the dubious distinction of being classed as a pest in virtually all habitats, from cities to remote forests.

Deer, in particular red deer, perhaps play a more important part in the New Zealand psyche than any other introduced animal, and for that reason the history of these animals is discussed here in some detail. Over the last 150 years, red deer have been a protected species, a managed game animal and a feral pest. They have supported a multi-million-dollar wild meat industry, have been the subject of commercial safari hunting and are now farmed. Kiwi deerstalkers have enjoyed hunting opportunities unequalled anywhere else in the world. The history of red deer in New Zealand and people's changing attitudes to them has been described in Graeme Caughley's entertaining and informative book *The Deer Wars*. Caughley was a deer culler who later became a wildlife scientist. His perspective on deer, ecology and New Zealanders is illuminating, witty and sometimes controversial. *Deer: The New Zealand Story*, by David Yerex, offers a rather different viewpont.

Red deer were first introduced to Nelson in 1854. That first release failed, but many further introductions followed. At first they were protected to allow the herds to increase and animals were transferred from areas where deer had established to places where their presence was desired. Caughley estimated that there were around 300 liberations, setting free more than a thousand red deer in sites scattered over almost the entire country. By the late 1940s, they had found their way into most suitable areas, and by 1990 colonisation of the last remaining deer-free areas appeared likely.[42] During the 1990s there was a dramatic increase in the spread of deer, with 166 new populations established.[43] Most of these were the result of farm escape or illegal liberations, and natural spread accounted for only five per cent of those new populations. Half of the newly established populations were red deer, with fallow and sika accounting for most of the rest.

Once herds were large enough to support a harvest, the acclimatisation societies allowed limited hunting. The first red deer to be legally hunted were shot in the Nelson district in 1882. During that first season at least 14 red stags and one fallow were shot and the best trophy was proudly exhibited in a Nelson shop.[44] Soon other provinces were able to allow deer hunting. For the next 40 years red deer were managed as game animals. There were strict bag limits, only stags could be shot, and then only by licensed hunters during a closely regulated season.[45]

Whenever a new population of herbivorous mammal became established, the sequence of events was similar. Failure to establish was likely and, at best, the initial increase in numbers was slow. Once they were established, the increase was rapid and the animals spread out from the point of liberation. Food was abundant and only the most favoured food plants were eaten; animals were in prime condition, hinds bred precociously and survival was high. After 10 to 30 years the herds could be huge, but the condition of both animals and forest had deteriorated. The plants they preferred to eat had become rare and species previously spurned were now eaten. The population that a decade earlier had produced trophy animals now produced stags with poor antlers, many hinds failed to rear fawns and mortality, even of adults, was high. The forest had become overbrowsed, undergrowth was scant and regeneration impaired. Subsequently the herbivore population would fall to a low density. Eventually forest health would recover and the herbivore population would 'stabilise' in a now-altered forest habitat. The food plants the deer had favoured had become scarce, replaced by less palatable species. Neither forest nor herbivore would ever again attain the previous levels of health.[46] From each of many liberation points red deer and other herbivores would spread in roughly concentric waves of population irruption, crash, then stabilise. Even after that 'stable' state was reached, deer would continue to slowly change forest composition by browsing seedlings and preventing regeneration. In areas where food quality is poor – for example, D'Urville Island – stags produce small antlers even when the population density is low.

In 1909, deer incurred the wrath of the Forest Service for damage caused to pine seedlings. The following year farmers were annoyed when deer ate pasture grass and destroyed crops. A decade later, deer were blamed for eating out five-finger, tree fuchsia and other native forest plants. Soon deer and other ungulates were being blamed for destroying the vegetation cover and speeding up high-country erosion. In a forlorn attempt to reduce deer numbers, all restrictions were removed. Licences were no longer required and hunters were allowed to shoot as many deer as they wished. By 1924, a bounty scheme had been instigated, with a payment for each deer tail turned in.

In 1930, government cullers began an all-out campaign to rid the country of 'the deer menace'. Deer, tahr and chamois were systematically slaughtered in catchments where their populations were high. Once numbers were reduced, the cullers shifted to other catchments. Later, cullers were strategically located in mountains whose rapid erosion would endanger the cities and farmlands below. By the 1950s, the dogma that ungulates ate out the vegetation, thus were responsible for high erosion rates, was firmly entrenched, and only much later questioned and found wanting. The deer campaign continued despite research by Thane Riney that showed culling was not focused on the catchments where deer damage was most serious, and that it was ineffective at achieving significant long-term reductions in deer numbers.[47]

We now know that erosion rates are naturally high in this rain-drenched geologically active landscape, and that introduced ungulates are of little geomorphological influence.[48] Some statistics help put this into perspective. Uplift in the New Zealand mountains is unusually rapid, in parts of the Southern Alps exceeding a metre a century.[49] Erosion rates on these over-steep slopes are primarily determined by rainfall, and in the South Island, west of the Main Divide, annual rainfall can exceed 10 metres. In the Hokitika catchment, erosion rates are estimated to be about 17,000 tonnes per square kilometre each year; in the Rakaia catchment, on the dry side of the mountains, the corresponding figure is a mere 1800 tonnes.[50] Deer may not enhance erosion, but they do damage indigenous forests. They have eaten out forest understoreys, inhibited regeneration and contributed to changes in the composition of native forests, tussock grasslands and alpine areas.[51]

The deer-culling days came to an end in the 1960s when private hunters supplying the emerging venison market usurped the government shooters. The demand for venison, in Germany in particular, was insatiable and prices rose sufficiently to entice hunters into the commercial sector. Foot-based hunting was hard work and few men could obtain enough deer sufficiently close to road ends to make this pusuit lucrative. Boats, four-wheel-drive vehicles and light aircraft were used to ferry hunters in and meat out, but commercial hunting was impractical over much of the country.

In 1963, helicopters began to be used for venison recovery, initially to position hunters in remote areas and to recover their kill, thus opening up vast, previously unshot areas. A year later, helicopters were used as shooting platforms, and so began a short but colourful episode in New Zealand's wildlife history. Helicopter hunting was capital-intensive, dangerous, but exciting and potentially lucrative. Chris Challies and Graham Caughley have documented the history of the game-meat industry. The number of deer exported reached 50,000 in 1965 and over 100,000 in 1969, but after 1973 began to decline. In total, nearly two million deer carcasses were processed for export.[52]

In some catchments, commercial hunting reduced deer numbers by more than 80 per cent, although the reduction was less in densely forested areas.[53] As deer became scarce, the ever-enterprising hunters turned to live capture in order to stock an emerging deer-farming industry. Initially, deer were captured by 'bulldogging' – jumping from a helicopter on to the deer, restraining and trussing it up for transport to a staging point. Tranquilising dart guns and net guns soon made the hazardous operation more efficient and slightly safer, but within a few years farms could be more cheaply stocked using farm-bred rather than wild-caught animals.

Commercial deer hunting is now on a much smaller scale: Challies estimated that in the 1987–88 season about 2300 deer, mostly reds, were sold by ground-based commercial hunters and about 17,000 by helicopter hunters. Only a third of these were taken alive. Since then the annual commercial harvest of deer has fluctuated from 12,800 to 31,900, the number taken being an indicator of the price of venison rather than deer numbers.[54] Commercial venison hunting seems likely to continue, although today it is mostly a part-time occupation. Deer farming is well established, with over 400,000 farmed animals supplying meat to Europe and antler velvet to the pharmacies of eastern Asia.

During the venison boom a pest that hunters had been encouraged to shoot on sight became a multi-million-dollar export commodity. A threat to the conservation estate that once appeared uncontrollable was reduced to low numbers by commercial operators, at minimal cost to the taxpayer. Kim King[55] compared this exploitive boom-and-bust industry with sealing in the early nineteenth century. In both cases, men who hoped to become rich ruthlessly exploited the animals until the industry collapsed. The former is deplored by today's conservation-minded populace, while they cheered the latter from the sidelines.

In New Zealand, deer and other big game such as tahr remain controversial. There is a strong fraternity of hunters who value the animals as a national asset and deplore the loss of hunting opportunities caused by the venison industry and the restrictions on access placed by some landowners who permit commercial hunting safaris. Another lobby group considers wild game to be alien species whose modification of indigenous habitats is unacceptable. A middle ground views the introduction of wild ungulates and their modification of native habitats as regrettable but accepts that their eradication is unlikely and not necessarily desirable. Removal of deer will not allow forests to revert to their primeval state, something that by definition became impossible with the extinction of the moa. A discussion document issued by the Department of Conservation in 1997 identifies the issues relating to deer management. The situation is complex, with many groups of stakeholders who have markedly different perceptions of deer. Within the hunting fraternity alone, there are recreational and commercial hunters, plus the game-safari operators, all with quite different aims. Other interest groups include deer farmers, other farmers, the Ministry of Agriculture, Department of Conservation and various conservation groups.

Now that deer numbers are low and forests have partially recovered from the damage done when they were abundant, can deer be kept at a level where conservation threats are alleviated while commercial and recreational values are met? Challies[56] suggested that, on some blocks of land, long-term control of deer could be achieved by allocating commercial operators exclusive rights, while in return the operator would be required to maintain deer numbers at a low and stable level. This would also require that deer be accepted as a permanent feature of the country's biota, and that in some places recreational hunting be prohibited. This management regime, while probably sensible, will no doubt be resisted by both deerstalkers and conservationists.

In 1977, 10 recreational hunting areas were gazetted where recreational hunting was to be the main method of controlling deer, but hunters have had only moderate success in deer control. Game animals have recently acquired yet another set of values to yet another user group: overseas hunters who come to New Zealand to bag a trophy deer or tahr. These hunters usually hire a local guide, often use helicopters for access to good hunting areas and pay a local taxidermist to mount that trophy, thus generating income and employment. A guided tahr-hunting expedition will cost the visiting hunter a minimum of $4000.[57]

Introductions in the twentieth century

By the beginning of the twentieth century the enthusiasm for acclimatisation had waned but the desire to hunt had not. Despite the zeal of the acclimatisation societies when that century began, opportunities to hunt remained limited. Most game species still had restricted distribution and small populations. Pigs and red deer were well established, but the only other 'big game' available were some smaller deer, goats and wallabies. Pheasants, quail, ducks and certain native birds could be hunted, but an increase in both the diversity and abundance of game birds was desired. Since 1910 the only new bird or mammal species introduced to New Zealand have been game birds and farm stock, although possums, hedgehogs, deer, goats and ferrets continued to be spread from one part of the country to another, with official sanction up till 1923, and unofficially and mostly illegally since. During the twentieth century, government agencies were responsible for most new introductions. The heyday of the acclimatisation societies as the major agencies of faunistic change had passed.

In 1901, the government created the Department of Tourism and Health. Thomas Donne, its founding head, was an enthusiastic hunter and a vigorous acclimatiser who believed New Zealand could become a mecca for game hunters. During his time in office, wapiti, rusa deer, white-tailed deer, moose, tahr, chamois and Canada geese were successfully added to the New Zealand fauna and further liberations of the deer species already present took place.[58] Donne also attempted to introduce further species of waterfowl and deer, but one must question the rationale behind his (unsuccessful) attempts to introduce racoons, owls and terrapins. Many of these early-twentieth-century liberations were made in national parks or areas of great scenic beauty later designated as national parks. Hunting, it was thought, would enhance people's enjoyment of the great outdoors.

Today, we are more cautious when introducing new species, but the urge to acclimatise persists: as recently as the 1980s plans were afoot to introduce alligators, mink and beavers. There is now a list of prohibited species, mostly mammals, that have become pests when introduced elsewhere. The Ministry of Agriculture may permit agriculturally useful animals such as fitches (ferrets), llamas, alpacas, chinchilla and Père David's deer to be brought in.[59] These species have all been imported in recent years, but their importation, especially that of fitches and chinchillas, is controversial.

Game-bird hunting

Game birds were among the first and also the most recent birds introduced to New Zealand. Pheasants arrived on some of the first immigrant ships, while red-legged partridge eggs were first brought here in 1980. In between times, over 50 species of game birds had been liberated, but with mixed success. Canada geese are now so numerous they are considered pests in some parts of the country. Mallard ducks are ubiquitous but, like some other now common exotic species, their establishment required persistence and repeated introductions. Mallards were first liberated in 1867, and there were annual releases of British-sourced ducks from 1896 to 1918, though the numbers increased only slowly.[60] Between 1939 and the 1950s, American mallards were introduced, which led to rapid population increases. Mallards now occur in virtually all wetland habitats throughout the country. Most introduced game birds failed to establish, and most that did acclimatise remain uncommon and localised.

Waterfowl hunting remains popular. Dawn on the first Saturday in May sees thousands of shooters in maimais (camouflaged hunting stands) on wetlands throughout the country. Graham Nugent[61] estimated that in 1988 about 56,500 people hunted game birds and killed about 700,000 birds, mostly waterfowl. Seven species of wetland birds may be hunted: the introduced mallard, black swan and Canada goose, and the native paradise shelduck, grey duck, New Zealand shoveler and pukeko. Not all can be hunted in all parts of the country. The acclimatisation societies, which for a century regulated game-bird hunting based on advice from the Wildlife Service, were replaced by fish and game councils in 1990. Each year these regional councils determine which game birds can be hunted in their district and set seasons and bag limits for each species. There are fewer opportunities to hunt upland game. About 20,000 pheasants and 48,500 California quail are shot each year.[62] During the first century of European settlement, other native birds, including godwits, oystercatchers, tui, kaka and kereru, were hunted, but these are now totally protected.

Hunting probably contributed to the extinction of the New Zealand quail. However, it has had little effect on the population size of the four native species that are now legally hunted. The paradise shelduck is one of the few native birds to benefit from agriculture and it is more common now than before human settlement. Grey ducks have declined in number through hybridisation with the closely related and more vigorous mallard, and from the loss of the natural wetlands they prefer.[63] They are less wary of hunters and have a smaller clutch size than do mallards, which are adaptable ducks that have also hybridised with closely related native species in Australia and Hawaii. Even in North America they hybridised with other duck species as they spread beyond their native range.[64] It would be desirable to protect grey duck from hunting, but it is impractical to expect hunters to distinguish a flying grey duck from a hybrid or a female mallard. The shoveler appears to be holding its own despite hunting pressure.

Pukeko utilise wetlands and damp, rough pastureland and are

common in many parts of the country, but they are not a popular game bird. Grey teal are not legal game, but a few are shot each year, usually mistaken for the larger but similarly plumaged grey duck. Despite this their population is increasing. Books by Tom Caithness and Murray Williams provide excellent, though now dated, introductions to game-bird hunting and biology in New Zealand.[65]

Duck hunters have probably made larger contributions to wetland conservation, thus benefiting protected native wetland species, than has the conservation lobby. Three examples illustrate this. The Whangamarino wetlands in the Waikato are home to the largest populations of North Island fernbirds and Australasian bitterns. Large areas of these internationally important wetlands would have been drained had the Auckland Acclimatisation Society not purchased 739 hectares for conservation.[66] Horrie Sinclair, a duck hunter with a fascination for wetlands and the animals that live there, developed the Sinclair Wetlands in Otago. He purchased a run-down and rather damp farm, then set about converting the land back to the wetland it had once been.[67] Under his guardianship wildlife flourished and the Sinclair Wetlands now hold significant populations of totally protected fernbirds, bitterns, crakes and rails. Hunting is permitted, but the wetlands are managed to enhance populations of both hunted and protected species. The endangered brown teal was 'adopted' by Ducks Unlimited. During the last 20 years, members of this hunters' organisation have bred in captivity over a thousand brown teal for release into the wild. No track has been kept of the money the organisation has invested in brown teal conservation, but its president estimated that the sum Ducks Unlimited spent exceeded $30,000, excluding the time and travel expenses incurred by individual members. Ducks Unlimited also supports other wetland restoration projects, mostly undertaken by its members.

How successful were introduced species?

The birds and mammals most often encountered by city and rural New Zealanders are introduced species, giving the misconception that these have run rampant, spreading widely and out-competing native species on their home turf. A few introduced species have been extraordinarily successful, but 80 per cent of the bird species and almost half the mammals liberated in New Zealand failed to establish. Of those that succeeded, few spread far from their point of liberation and for most the odds were stacked heavily in their favour. Of the animal groups covered in this book, three species of frogs, one lizard, 39 birds and 33 mammals currently maintain wild populations in New Zealand. At least eight species of amphibians, three reptiles, about 130 birds and 54 mammals were liberated in this country.

Most exotic species were liberated into highly modified habitats that more closely resembled the homelands of the newcomers than the forests where the native species evolved. Introduced species were liberated into a country from which the major browsers and predators had already been removed, thus making indigenous habitats easier to invade. Even with these advantages possums and red deer needed initial protection and many liberations to establish a foothold. Some introduced birds do occur in indigenous habitats, but few species are common. Blackbirds, chaffinches and, less commonly, song thrushes and redpolls are the only introduced birds that have successfully invaded little-modified native forests. Even among mammals, only possums, ship rats, mice, stoats, cats, goats, pigs, chamois and red deer are widespread and numerous in moderately intact indigenous forests.[68] However, those species are very widespread, exceedingly numerous and alarmingly destructive. Elsewhere in the world the story is similar: naturalised species thrive alongside humans, but few successfully penetrate intact ecosystems.

Island systems with few extant native species are more susceptible to invasion than species-rich continental ecosystems.[69] Over 70 species of birds were introduced to North America, many in large numbers, yet only about six are now widespread. In contrast, in Hawaii 94 species were introduced, of which 36 survive.[70] Many species of birds and mammals were introduced to Australia, yet few successfully established there. In Tasmania, only six species of non-Australian birds and seven species of mammals established wild populations.[71] In Victoria, about 15 bird species and 13 species of mammals now have feral populations, even though the efforts at acclimatisation were comparable in both New Zealand and Australia.[72]

Until recently, little thought had been given to the factors that influenced the success of exotic species in colonising new countries. Clare Veltman and Richard Duncan[73] investigated the importance of ecological and life-history factors in determining establishment success of introduced birds in New Zealand. They found that by far the most important influence was introduction effort. Most successful species were liberated more often, in more places and in larger numbers than species that failed. As always there were exceptions: cirl bunting, mute swan and kookaburra became established (but not widespread) through the release of only a few founding individuals, whereas grey partridge failed despite concerted efforts. They also investigated the influence the same factors had on the birds' geographic range within New Zealand.[74] They found that habitat was the major determinant of range, and that the birds whose preferred habitats were widely distributed had the largest range. Other factors associated with wide distribution were life-history traits that lead to rapid population growth, such as large and frequent clutches, rapid development and small size, and non-sedentary behaviour that enabled the species to disperse quickly.

The trend with mammals is apparently similar. Chamois and, to a lesser degree, sika deer were the only species to spread far from their point of liberation without the aid of subsequent releases. Chamois now occur in mountainous regions of the South Island from Nelson south to Fiordland, a distribution attained through their own powers of dispersal from two releases totalling only nine animals at a single location (Aoraki/Mt Cook).[75] The sika deer herd that now extends over a large part of the central North Island originated from a single liberation of six animals, two of which were apparently killed at the time of release.

A few of the animals introduced to New Zealand are now rare or declining in the countries from where they came. The Parma wallaby was believed extinct until rediscovered on Kawau Island. The brushtailed rock wallaby, Himalayan tahr, green and golden bell frog and green bell frog are listed by IUCN as threatened or vulnerable in their native range. In Britain, at least a dozen species of farmland birds have declined significantly since the 1970s,[76] and of these, song thrush, skylark, blackbird, starling, dunnock and yellowhammer are common on farmland in New Zealand. One other species, the cirl bunting, is now so rare in Britain it is listed in British Red Data Book, though it remains relatively common on the European mainland. The role New Zealand should play in the conservation of these species is discussed in Chapter 10.

Introduced frogs and reptiles

The European settlers were apparently less eager to introduce amphibians and reptiles. Eight species of frogs are known to have been liberated, but only three species, all from Australia, became established here. The whistling frog and the golden bell frog are widely distributed through the country, but the green-and-golden bell frog is restricted to the northern half of the North Island. The rainbow skink is the only exotic reptile that has become established in New Zealand. Rainbow skinks probably arrived in Auckland as stowaways in the 1960s, and from there have spread to the Waikato, Coromandel and Bay of Plenty, and are likely to spread further. They can be found in suburban gardens, grasslands and lightly wooded habitats.

Feral farm mammals

Over the years feral populations of goats, sheep, pigs, cattle and horses have become established in many parts of New Zealand. Some of these populations are descendants of animals that were liberated on islands as food for castaways. The ancestors of other populations were escapees or were abandoned when farming ventures failed. Some of these populations have been feral since the earliest days of European settlement and are descended from breeds that are now rare. Goats have been on Arapawa Island since at least 1839 and are possibly descended from those left there by Captain Cook. Even if they are, there has no doubt been extensive interbreeding with more recently introduced strains. Kune-kune pigs are of special interest because they show high fertility and are good at converting poor-quality food to meat.[77] The kune-kune is an ancient breed but, contrary to some claims, does not pre-date European arrival.

Farm stock has been subject to many decades of rigorous selection to improve productivity and profitability, so contemporary herds have a very narrow genetic base. Some of the old breeds, while not currently favoured, may prove valuable as farming practices and consumer demands change.[78] Feral stock are often better suited to low-level care and poor food quality typical of conditions in less affluent nations than are the highly inbred stock used in the intensive farming systems of the developed world. Some feral populations have been subject to selection under unusually rigorous conditions and have developed adaptations that could play a useful role in agriculture. The feral sheep on subantarctic Campbell Island shed their belly wool, which was a decided advantage on that boggy island and perhaps could be advantageous in rough farmland elsewhere.[79] Feral sheep on some wet islands, including Campbell, developed resistance to foot rot, a trait that could be profitably introduced to flocks in damp climates. The wild Saxony merino sheep on Pitt Island in the Chatham group are descendants of stock brought there in 1841. They are now self-shedding and were considered to be of sufficient potential value that in 1981 a 200-hectare reserve was established to protect them.[80] Several hundred sheep remain in that reserve, though their continued presence in a Department of Conservation reserve is controversial. The possible value of feral stock has been acknowledged, but the actual value of most populations has yet to be assessed.

Many of these feral populations are (or were) found in nature reserves or other areas of conservation significance where they are as much of a threat to native plants as other wild browsing mammals. Most of the island populations have been or are likely to be exterminated to protect indigenous species.[81] Prior to extermination, animals from some islands have been removed and taken into captivity on the mainland, thus preserving them for future use where appropriate.

The Kaimanawa horses

Feral horses of mixed horse and pony stock have been present in the central North Island since the 1870s but, over time, development and changing land-use practices have squeezed them into a restricted area within the Kaimanawa Ranges.[82] Prior to the late 1970s there was little formal interest in or management of the horses, and little concern for the impacts they might have on indigenous vegetation. In 1978, there was increasing interest in the horses, and a count the following year indicated that only 174 remained. Subsequent, more reliable counts suggested that the 1979 estimate was too low, but nevertheless the population had increased to over 1500 by 1994. A protected area was established in 1981 without any research into how that might assist in the preservation of the horses, the impact the horses have on indigenous ecological values, or even whether they were sufficiently unusual to warrant protection. Since then the Kaimanawa horses have proved endlessly controversial, with lobby groups having sometimes diametrically opposed views. The Royal Forest and Bird Protection Society consider the horses introduced pests that threaten endemic plants and natural ecosystems, while the Kaimanawa Wild Horse Preservation Society favours these animals over the plants they may eat. The Royal Society for the Prevention of Cruelty to Animals is more concerned with the humane treatment of the horses, whatever their fate may be.

In 1991, the Department of Conservation released a draft management strategy for the Kaimanawa horses and sought public comment, which was then considered by a working group including representatives of the groups with interests in the horses or their habitats. The department published a Kaimanawa horse management plan in 1995[83] based on the recommendations made by that working party.

The Moawhango Ecological Region, which falls mostly within the horses' 1991 range, is a highly distinctive montane to subalpine area with many outstanding botanical and ecological features. Sixteen threatened or rare plants occur within the horses' range. The only North Island sites for 10 of these plants occur within the area occupied by the horses and, although none is in immediate danger of local extinction, five of those species are vulnerable to disturbance by horses.[84] The rare and vulnerable plants are concentrated in small wetlands that are especially susceptible to horse damage. Disturbance by horses also facilitates the invasion of heather and other introduced weeds into the tussocklands. Horses, of course, are not the only factor that has impacted on this region. Since European colonisation the area has also been subjected to burning, grazing by stock and deer, and military activities.

Ecological studies showed that the area occupied by the horses fell into two more-or-less distinct zones. The northern part of the range contained most of the threatened or rare plants plus large areas of little-modified tussock grasslands and unusual wetland communities. The working party recommended that all horses be removed from this area, which comprised about 70 per cent of the 1991 range. The southern part of the range was more modified, suffering from exotic-weed invasion, so few of the rare plants were present, although there were areas of good-quality tussocklands and wetlands that horses could further degrade. The working group suggested that a herd of about 500 horses be left in the southern region but noted that further research was required to determine what density of horses was compatible with the indigenous ecological values. Ongoing management was needed, including periodic culling or removal of horses. The management plan did not resolve all conflicts, and controversy has re-erupted whenever horse culls or round-ups have taken place.

CHAPTER 7

The forest vertebrate community in the twentieth century

A glance at Tables 4.1 and 4.2 in Chapter 4 will remind you that most of the predators and browsers present in New Zealand 2000 years ago have been replaced by introduced species. When species are removed from an ecosystem, or new species are introduced, the dynamics of that system change. For example, the loss of moa from indigenous forests had already induced changes to the vegetation long before the addition of browsers such as deer or possums. The introduction or removal of predators selects for or against the prey species present. The changes do not go on forever, and eventually a new 'equilibrium' will be reached where the species present adjust to accommodate one another. This generally results in some native species becoming rare and others becoming lost. Most ecologists regard such changes as ecological degradation, but this view is subjective and reflects the way we now value native species and natural ecosystems over their introduced and modified equivalents. To Maori venturing into the forests, the extinction of the fearsome perch-and-pounce-hunting eagle may have been less regrettable than it seems to ecologists today. Deer hunters consider their prey an asset, whereas ecologists regret the modification deer cause to native forest ecosystems.

It is easy to list the species present in modern and pre-human forest communities, but in this chapter I am interested in how those communities function and how the species interact. The most comprehensive study of the dynamics of a New Zealand forest ecosystem was carried out by Ecology Division, DSIR (now incorporated into Landcare Research), in the Orongorongo Valley near Wellington between 1966 and 1990. That research has been synthesised in Bob Brockie's book *A Living New Zealand Forest*. Despite this study, and numerous studies of more restricted scope, huge gaps in our knowledge remain. In this chapter I also briefly discuss the mammals present in the alpine zone and the birds, mammals and lizards in forest remnants and in towns.

The New Zealand forest community

Many forest plants are susceptible to browsing by mammals, and the last century has seen dramatic changes in the forest structure and relative abundance of plant species. These forests were subjected to browsing pressures in the past, but deer, goats, pigs and possums are not the equivalents of moa, takahe, kakapo and kokako. The impact of browsing mammals on the forests has been the subject of extensive research, but suffice to say that in the past hundred years there has been a change in the composition of most indigenous forests as browse-resistant species have increased at the expense of browse-susceptible species.[1] Some forests, in particular the rata/kamahi forests of Westland, have suffered canopy dieback triggered by possum browsing.[2] Deer hinder regeneration, open up the understorey and alter the canopy composition as old trees of preferentially browsed species are replaced by species shunned by deer.[3] The native mistletoes are especially susceptible and have become rare, with one species now extinct owing to possum browsing and the decline in numbers of birds that pollinate their flowers and disperse their fruits.[4]

The relationship between vegetation and mammals is dynamic, and change continues. The most-preferred species, such as tree fuchsia, were first to be eaten out. The least-preferred species, such as pepperwood and rimu, increase in density, while species initially shunned by deer, such as black maire and quintinia, are eaten as their foods of first preference decline. Graham Nugent[5] found that a reduction in the numbers of deer and possums enabled some species, such as pokaka and toro, to recover, but others, such as tree fuchsia, showed limited recovery. Trees are long-lived so it may take centuries before an 'equilibrium' between plants and the new suite of browsers is established. Moa would also have influenced the plants present. The unbrowsed forests described by European scientists prior to the spread of browsing mammals were no more 'natural' than the mammal-browsed forests of today.

Forest predators

Three guilds of predatory vertebrates – rodents, carnivores and birds – are now present in New Zealand forests. The community is complex and space does not enable discussion here of all the intricacies in this mix of native and introduced species. The rodents are omnivores that mostly eat insects and plant foods, but eggs and small vertebrates make up a variable portion of their diet. The carnivores and predatory birds take large insects as well as small vertebrates. These predators all feed on more than one class of food and occur in more than one kind of habitat.

Other introduced mammals further complicate the situation. Hedgehogs are primarily open-country ground insectivores but are common in some indigenous forests and occasionally eat birds, their eggs, reptiles and even small mammals.[6] The impact they have on invertebrate populations, and thus on insectivorous birds, has probably been underestimated. Possums do not just eat vegetation, they also raid birds' nests and eat invertebrates.[7]

Rodents

New Zealand has four species of rodents, ranging in size from the tiny house mouse to the large Norway rat (Table 7.1). The ship rat and house mouse are widespread, while kiore and Norway rat, both once abundant, now have patchy distributions. On the mainland, ship rats and house mice both occur in most forest habitats from sea level to the mountains. They frequently co-exist, although mice are more common than ship rats in the floristically simple beech forests, whereas ship rats are more numerous in mixed forests.[8] In the podocarp/broadleaf forests at Pureora in the King Country, mice and ship rats had reciprocal distributions (Figure 7.1, page 186).[9] Mice were most abundant in disturbed habitats such as cut-over forests or near roads; ship rats dominated in

Table 7.1 Sizes and weights of introduced rodents and carnivores

	Mean weight (g)	Mean head and body length (mm)
House mouse	15–20	115
Kiore	60–80	180
Ship rat	120–160	225
Norway rat	200–300	250
Weasel	F 57 M 126	F 182 M 217
Stoat	F 207 M 324	F 256 M 284
Ferret	F 600 M 1200	F 350 M 417
Feral cat	F 2700 M 3700	F 477 M 514

Data from King 1990

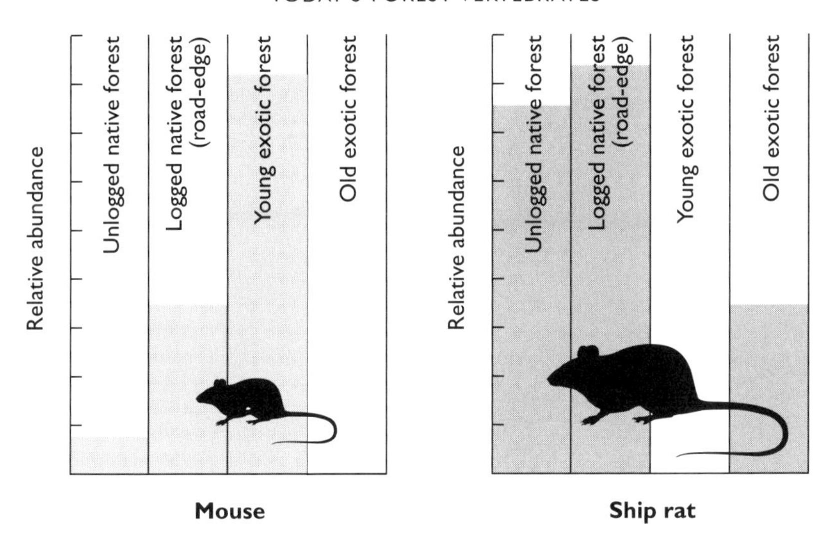

Figure 7.1 Relative abundance of rodents in various habitats, Pureora Forest

the least-disturbed areas and were rare in early-successional habitats. Ship rats were more common in native than in exotic forests and were most often trapped near fruiting understorey plants.

Norway rats generally replace ship rats near human habitation and near water. On islands where ship rats are absent, Norway rats can utilise the full range of available habitats, especially if mustelids are absent.

Prior to the arrival of European rats, kiore occurred throughout the country, but today they are only found in Fiordland, Stewart Island and on some offshore islands.[10] Nowhere in New Zealand do all four rodent species co-exist. Rowley Taylor[11] suggested that kiore and Norway rats are controlled either by competition from the other rodents or predation by mustelids, and that mice are unable to establish where Norway rats are numerous. He suggested that on Stewart Island, the only place where all three rats co-exist, the absence of mustelids enables Norway rats to remain common, and they prevent the establishment of mice, which have reached the island on several occasions. Around Halfmoon Bay village, he trapped Norway rats along the shoreline and under buildings; ship rats in grassland, swamps, scrub and forest; but kiore only in grassy habitats.[12]

All four rodents eat a mixture of plant, invertebrate and vertebrate foods. Most of the relatively few New Zealand studies of rodent diets have been carried out on islands, where their diet probably differs from that in mainland forests.[13] At Orongorongo, ship rats ate mostly plant material in autumn and winter, but insects in spring and summer.[14] Their favoured plant foods were fruits, berries and seeds, especially those of hinau, supplejack and kawakawa. The impact of rodents on forest regeneration is discussed later in this chapter. At Orongorongo, tree weta were the food item ship rats ate most often, while other medium-to-large insects and land snails were eaten less frequently. In the podocarp forests at Okarito, South Westland, ship rats mostly ate invertebrates.[15]

Rodents, including kiore, have had devastating impacts on birds and reptiles (see Chapters 2 and 5), yet today birds, eggs or reptiles are seldom found in rodent guts. Nonetheless, the response of bird populations to rodent control shows that, on the mainland, rats still have an impact on native birds. Kereru breeding success increased markedly after rodent control at Wenderholm, near Auckland, but whether this was due to lowered predation by rats or unintentional kill of possums or stoats is uncertain.[16] Control of ship rats and possums in central North Island forests resulted in an increase in the breeding success of kokako. After eight years of predator control, the Mapara kokako population tripled and numbers of most other native birds also increased.[17] The relative roles of rats, possums and other mammals, predation and competition for food in that equation are still being teased apart. However, predation has been shown to be a more serious threat than competition, and rats and possums are the most critical predators.

Mouse numbers are generally low in New Zealand mainland forests, but they periodically 'explode' in forests where food availability fluctuates. In the mixed-species forest at Orongorongo, mouse numbers changed only four- or fivefold from one year to the next, but in the near-monotypic Fiordland beech forests the population density fluctuated 73-fold, the large populations coinciding with mast years, when seed production was prolific. Today, in mainland forests, mice mostly eat invertebrates and plant material.[18] At Orongorongo, they relied equally on plant material and arthropods, although plant material was eaten more often in winter, and arthropods in summer.[19] The plant material and fungi eaten varied from season to season, though seeds were preferred when available. Of the arthropods, moth larvae were taken most frequently, while spiders, weta, beetles and their larvae were also important food items.

The carnivores

In New Zealand forests, stoats and cats are the carnivores most often present. Stoats are by far the most numerous and pose the greatest threat to forest-dwelling birds. During a survey of mustelids in national parks, 1599 of the 1695 caught were stoats.[20] They also occur in open country where cover is available.

Ferrets prefer open habitats such as farmland, tussock grasslands, wetlands and braided riverbeds, where they pose a threat to birds such as the sandspit-nesting New Zealand dotterel in northern North Island and the birds of the South Island braided rivers.[21] Ferrets and cats are a major threat to white-flippered penguins on Banks Peninsula and yellow-eyed penguins and royal albatross on Otago Peninsula. While ferrets generally shun forests, they are implicated in the decline of kiwi in Northland and are known predators of kiwi at Lake Waikaremoana.

Weasels, the smallest and rarest of the three mustelids, are most common on farmland and in scrub but are probably too scarce to pose a significant threat to native birds, lizards and invertebrates.

All four species of carnivores occur in the mosaic of logged and unlogged podocarp/hardwood forests and exotic plantations at Pureora and in indigenous forest at Mapara, both in the King Country.[22] Stoats were the most numerous carnivores (68 per cent of the mustelids caught at Pureora) and occurred in all habitats, but were most abundant in older exotic forests. Only a few weasels were caught, all in disturbed habitats such as logged native forest and young exotic plantations or fern-dominated areas, where mice were common. At Mapara, ferrets were common only near the reserve margins and along grassy tracks. Their habitats matched the habitat preferences of their preferred prey. In the Orongorongo study, cats were relatively rare, with only two to five trapped in the 275 hectares of valley-floor habitat surveyed.[23] Stoats were more common but their numbers fluctuated in response to changes in the abundance of their food.

Cat and stoat diets vary seasonally, from year to year and place to place, reflecting changes in the abundance of the mammals, birds and invertebrates they prey on. However, mammals have made up the bulk of the diet in all studies of feral cats on the New Zealand mainland.[24] At Orongorongo, rats were their staple food, always comprising at least 20 per cent of the food taken (on basis of weight). Rabbits, possums, mice and stoats were less important, but mice were important prey in the years when they were abundant.[25] Many of the rabbits taken by cats were juveniles, and some of the possum meat was probably scavenged. Orongorongo cats also ate fish, freshwater crayfish and large insects. Bird remains were found in 12 per cent of cat scats examined, but by weight comprised only 4.5 per cent of the cats' diet. Blackbirds, thrushes

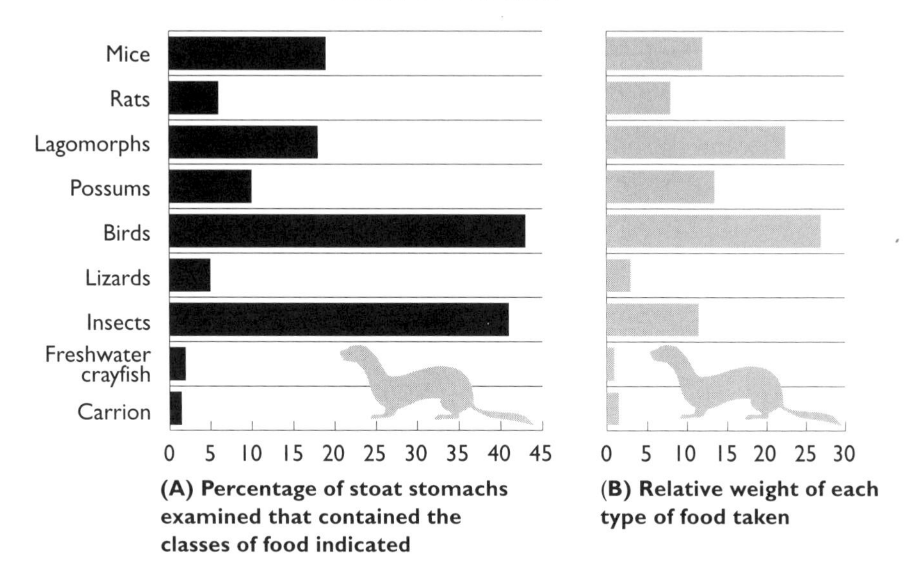

Figure 7.2 The diet of stoats in New Zealand national parks and reserves

Adapted from King & Moody 1982b

and silvereyes were the birds most often eaten. Young cats took mostly small prey such as insects and mice, while adult cats mostly ate possums and rabbits. In the central North Island, cats ate mostly mammals, and birds were seldom recorded as prey items.[26]

Small mammals, birds, large insects and lizards were the main foods found in the stomachs of 1600 stoats killed in 10 national parks and other protected natural areas (Figure 7.2).[27] Female stoats ate more mice and insects than did the larger males. Lagomorphs (rabbits and hares) and rats were taken equally often by both sexes, while males scavenged road-kills more often than did females. The relative importance of these foods varied with location and habitat; for example, rats were more commonly taken in podocarp forests and mice in beech forests, reflecting their relative abundance. There was some seasonal variation in the foods taken: rabbits and hares were mostly eaten in summer, and birds were important prey in autumn when young, naive birds were most common.

Figure 7.2 presents the same data in two ways. Graph A shows how frequently the various foods were found in stoat guts and indicates that birds and insects were the most common prey. Graph B shows the weight of the various foods and shows that small mammals and birds furnished the greatest amount of food. The first method of analysis overemphasises the importance of frequently eaten small prey such as insects but is useful to assess the impact stoats have on their prey species. The second analysis is the better measure of the importance of each kind of prey to the stoats.

Subsequent studies show that there are important differences in stoat diet between different forest types and from place to place. Rats are the preferred food and, if available, dominate the diet.[28] In the King Country podocarp/hardwood forests, stoats mostly ate birds, rats, possums and lagomorhs.[29] While taken frequently, insects made up a small part of the diet. The relative proportions of rats and birds shifted, depending on the abundance of rats. During the six months following rat-control operations, stoats ate more birds.[30] Their dependence on mice or lagomorphs did not change after rat-poisoning operations.

Beech forests generally have a lower diversity of plants and, consequently, fewer animals than podocarp forests. The beech forests of the Eglinton Valley in Fiordland support low densities of rats, cats and possums. Mice are abundant in mast years but scarce at other times. Elaine Murphy and John Dowding[31] found that ferrets and weasels were absent and rabbits and hares restricted to the open grassy flats alongside the road or river. Eglinton Valley stoats had large home ranges (averaging 124 hectares for females and 206 for males), and birds were the main food of both sexes. Murphy and Dowding suggested that pure beech forest is marginal habitat for stoats and that in poor seasons they may not breed.

Other, less intensive studies show that the stoat is an opportunistic predator with a flexible diet.[32] In Westland, lowland podocarp forests insects made up the bulk of stoats' diet, whereas at the Hutton's shearwater colony in the Kaikoura mountains it was shearwaters or their eggs and chicks. At this colony, alternative foods were scarce and the stoats showed an unusually high reliance on carrion. In the rabbit-prone Mackenzie Basin, birds and rabbits were the most common prey, but no rodent remains were found in the 49 guts examined. On the short, grazed grasslands of Birdlings Flat in Canterbury, birds (mostly skylarks), eggs and skinks were eaten more often than they were in less-modified habitats.

The foods eaten by weasels and ferrets are less well known, but because both these species are rare in forest habitats they are not so relevant in this discussion. Most of the 40 weasels Kim King and J. E. Moody caught were trapped in grasslands or at the edge of beech forests. Weasels generally ate smaller prey than did stoats, with birds, mice, lizards, weta and other insects making up the bulk of their diet (Figure 7.3). In the central North Island podocarp/hardwood forests, weasels mostly ate mice;

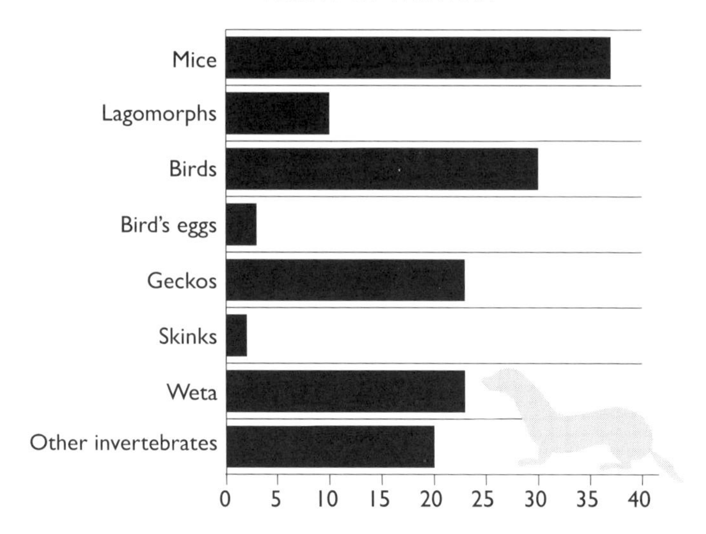

Figure 7.3 Diet of weasels in native forest, showing the percentage of weasel stomachs containing various food categories

Source: King & Moody 1982b

invertebrates were also important, but they took few birds, rats and lizards, while ferrets relied on rat-sized or larger mammals.[33] Ferrets mostly utilise open habitats, including farmland, forest remnants, riverbeds, tussock grasslands and coastal areas. Although there is regional variation in ferret diets, at most locations mammals, in particular rabbits and hares, made up the bulk of ferret prey.[34] Birds were the main prey in two coastal Otago studies, where ferrets ate yellow-eyed penguins, blue penguins and sooty shearwaters.[35]

Impact of introduced predators on invertebrates

Mustelids, cats and rodents all eat weta and other large invertebrates, but their impact on forest invertebrates is virtually unknown. Various species of large invertebrates now have disjunct distributions or occur only on rat-free islands.[36] For example, the large Stephens Island weta is now found on Stephens, Trio and Mana Islands, but not on the rat-infested North and South Islands. The large flightless Stephens Island weevil and the Cook Strait click beetle are now restricted to islands, yet recent fossils have been found in Canterbury. Land snails are eaten by pigs, rats and other introduced mammals, and some endemic species are now in danger of extinction.[37]

One study suggests that rats have greater impact on invertebrate populations than do stoats.[38] In that study, invertebrate numbers were compared on three islands in Fiordland: one with stoats and red deer, one with Norway rats, and one without any mammals. Most invertebrate groups sampled were scarcest on the island with rats. Only centipedes and large weevils were significantly scarcer on the island with stoats.

Effect of stoats and cats on their prey populations

Predators kill animals, but ecologists are not concerned by the loss of individuals: they are more interested in the effect on the populations those prey belong to. In this chapter we are concerned only with the way introduced predatory mammals affect the native species that remain, rather than their role in extinctions. Kim King[39] suggested that on the New Zealand mainland the damage was already done and that, except for the protection of a few threatened birds, predator control on the North and South Islands was not a good use of scarce conservation dollars. However, she warned that 'we must be alert for evidence that might modify this conclusion . . .' Since then, evidence to modify that conclusion has mounted.

Predator/prey dynamics are complex and if our objective is to protect native species, not merely kill predators, we need to understand these dynamics before we once again alter the 'balance'. Killing stoats and feral cats in our precious native forests may seem a good thing to do, but because rodents kill more eggs, chicks and small birds than do mustelids or cats,[40] control of these top-order predators could be counterproductive. Cats and stoats may buffer the native species by preventing rodent populations from reaching very high numbers when food is abundant. In the Orongorongo Valley, cats ate rats more often than birds, and when cats were removed, rats became more common.[41]

Of the carnivores, stoats pose the greatest threat to native forest birds because they are the most common and the most frequent killers of birds. In Fiordland, 54 of 96 stoats killed had bird remains in their guts.[42] Prey that could be identified included a mohua, a kereru, two yellow-crowned parakeets and three robins. The other birds taken were widespread native or introduced species. Stoats are a major cause of the continuing decline in mohua and other hole-nesting birds (see page 191).[43] However, it was not possible to determine whether predation affected the populations of those other birds.

Predators are the main cause of the current decline in mainland kiwi populations because these birds are vulnerable to most introduced predators at various stages in their life cycle.[44] Ferrets and dogs kill adult kiwi; possums and mustelids eat kiwi eggs; and stoats and cats prey on young kiwi. The continued survival of kiwi on the North and South Islands depends on the control of stoats and ferrets, but to date this has met with limited success.[45] The recent decline in kiwi numbers in Northland has occurred at the very time ferrets became common there.

On the mainland and on some islands, kaka exist alongside a variety of introduced predators and competitors. Kaka numbers have been declining for decades, but the extent to which these different threats affected them was not known. Peter Wilson and co-workers[46] compared the population dynamics of kaka on the mainland, where all threats were present, with island situations with various combinations of predators and competitors. Populations were declining and breeding success was low on all islands where stoats were present, yet kaka populations were less affected on islands where rats, possums or wasps were present but stoats absent. They concluded that stoats were the major threat.

All species have some ability to ajust their reproductive effort to compensate for predation, but in many native species this ability is limited. Robins have been more successful at this than many successful species. In Kowhai Bush, near Kaikoura, two-thirds of the 521 robin nests observed produced no fledglings, and most of the losses were attributed to predation by mustelids or ship rats.[47] In comparison, on mammal-free Te Kakaho Island (Chetwode Islands) less than 10 per cent of robin nests were lost to predators, and the main predator was the native morepork. To compensate for predation, Kowhai Bush robins laid larger clutches, re-laid more often and fledged more young than Te Kakaho robins. At the time, most of the nest predation was attributed to mustelids, but subsequent research suggests that the impact of ship rats was underestimated, some rat predation having been wrongly attributed to stoats. Kerry Brown[48] attributed about three-quarters of the predation on North Island robin and tomtit nests to rats.

When rodent populations decline suddenly after poisoning operations, mustelids eat more birds. Elaine Murphy and Philip Bradfield[49] studied the diet of stoats at Mapara in the King Country before and after an aerial drop of 1080 poison. Before poisoning, rats were the main prey and birds a minor component. Poisoning achieved a 90 per cent reduction in rat numbers but did not affect stoats. They switched to birds as their main prey, but the impact, if any, this had on bird populations could not be assessed. The costs and benefits of poisoning operations are yet to be fully evaluated.[50] Some predators are killed by eating poisoned prey and this may under some conditions be an effective way of reducing populations of ferrets and possibly other predators. Native species are also killed during poisoning operations, but increased breeding success owing to the reduction in predators has usually more than compensated for the individuals lost from those native populations (see Chapter 10). Birds that finish breeding by the end of December appear to benefit more from pre-breeding-season poisoning operations than those such as kiwi, kereru and kaka that have longer breeding seasons.

Ferrets and stoats pose a threat to ground-nesting seabirds and shorebirds, including all mainland-breeding penguins and petrels, New Zealand dotterel, black stilt, wrybill, variable oystercatcher and fairy tern.[51] However, on the braided rivers in the Mackenzie Basin cats are apparently the major threat, with 40 per cent of predation events caused by those predators.[52] Ferrets were responsible for 22 per cent, hedgehogs 19 per cent and stoats only 5 per cent.

Hole-nesting birds and predators

A large proportion of New Zealand's threatened or endangered birds nest in tree cavities. Hole-nesting birds are vulnerable to introduced predatory mammals because most nests have a single entrance, so neither chicks nor incubating adults are able to escape hole-investigating mammals. Species such as mohua are unusually vulnerable, as only the females incubate, meaning that nest predation results in a skewed sex ratio. Their incubation period is longer than that of most similar-sized birds, so predators have longer to find the nest; and the chicks are noisy, so they attract the attention of predators.[53] Cavity-nesting may have been an appropriate defence against the native predators, but New Zealand species lack both the predator-defence and breeding strategies exhibited by overseas hole-nesting birds.

Last century, mohua were abundant and widespread in the forests of the South Island and Stewart Island. Today, they remain in only small parts of that former range and most populations surveyed were in decline.

Colin O'Donnell, Graeme Elliott and Peter Dilks[54] made an intensive study of mohua and yellow-crowned parakeets to answer three key questions that concern the impact introduced predators have on these and other hole-nesting birds: Do contemporary predators have a significant impact on the long-term survival of extant native birds? Can we predict when predators will affect forest bird populations? Can we increase the productivity and viability of forest bird populations by controlling predators?

Most of their research was conducted in lowland beech forests in the Eglinton Valley, Fiordland, where both mohua and yellow-crowned parakeets remained common. Additional work was done in higher-altitude forests in the Hawdon Valley, near Arthur's Pass.

Both species were found to be vulnerable to introduced predators and, while other predators took their toll, the prime culprit was the stoat. In most years, predation by stoats was of minor consequence and did not cause a reduction in the populations of either species. However, beech forests have mast years (see below), when huge amounts of seed are produced, mouse and stoat populations increase and many more birds are killed. Elliott[55] concluded that in the lowland Eglinton Valley forests, where masting occured at about five-yearly intervals and mohua usually produced two broods of chicks each year, the birds could probably survive in the presence of stoats. The Hawdon Valley mohua produced only a single brood and their numbers declined from 30–40 birds in the early 1980s to fewer than five after the major masting event of 1990.[56] The Hawdon mohua population was more-or-less stable between mast years, but halved following mast years in 1986 and 1990. Elliott concluded that the decline in mohua populations was due to predation on incubating females. Predation on chicks or eggs was of less consequence to the long-term viability of the population. He recommended that, for two-brooded mohua populations. stoat control was warranted only in mast years, when stoat densities were high. Single-brooded populations are far more vulnerable and annual control of stoat numbers is beneficial. Stoats also prey on yellow-crowned parakeets, but stoat control did not significantly increase breeding success of the parakeets.

Seeds, mice, stoats and birds: the beech mast cycle and predator/prey interactions

New Zealand beech forests are subject to 'mast years' during which trees seed prolifically, with intervening years when there is little or no seed production. In the summer of a mast year, mouse populations increase greatly, then crash with the onset of winter. It was thought that when mice were abundant stoat numbers also increased, and in winter, when the mouse population crashed, the stoats preyed on birds, including rare species such as mohua. In reality, this oft-quoted story is a misleading oversimplification.

Kim King[57] found that mouse and stoat populations did increase during mast years, but there was little difference in the number of birds eaten per stoat during mast and non-mast years. Of course, during mast years there were many more stoats, so the threat they posed to bird populations was greater. Subsequent studies[58] found that the densities of invertebrates, seed-eating birds and insectivorous birds also increased during mast years. Mike Fitzgerald suggested that moth larvae were an important link in the web (Figure 7.4). Moth larvae feed on those beech flowers that drop to the ground before seeding occurs, and the larvae are eaten by mice and insectivorous birds. The subsequent seedfall is followed by an increase in numbers of seed-eating birds, invertebrates and mice, which in turn leads to increases in insectivorous birds and stoats. The stoats eat invertebrates and mice, plus seed-eating and insect-

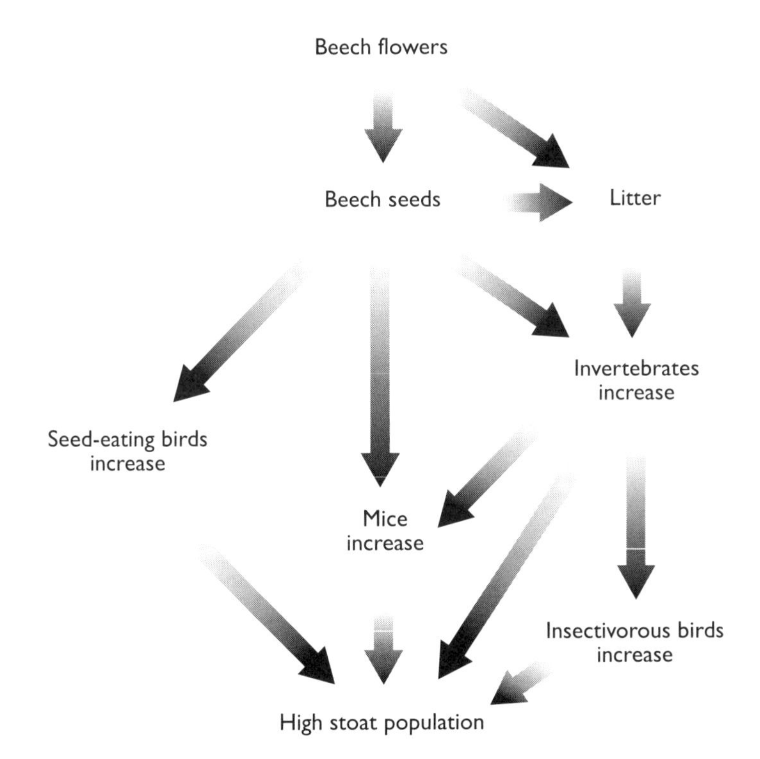

Figure 7.4 The relationship between beech seedfall and the densities of animals in New Zealand beech forests

Sources: Murphy & Dowding 1995, Fitzgerald et al. 1996

eating birds, which are taken during the summer and not just in the following winter, after mice become rare. Thus the increase in mouse numbers is not only linked to the abundance of beech seed, and the increase in stoat numbers is not solely the result of the increase in mice.

The impact these cyclical changes in stoat numbers have on threatened native birds is still not fully understood. Elaine Murphy and John Dowding found that without stoat control bird populations could take two or more years to return to the normal low densities; thus stoats could continue to pose a threat to native birds for two, even three years after masting. Mick Clout and Rod Hay[59] suggested that increased predation by stoats following beech mast years could have contributed to the probable extinction of the beech-forest-dwelling South Island kokako. North Island kokako survived in mixed podocarp broadleaf forests, where the effects of masting and subsequent peaks in stoat numbers were less marked.

Forest bird communities

Only those native birds that can exist alongside the suite of introduced predators remain on the North and South Islands, but even for those species population densities are generally lower on the mainland than they are on Stewart, Kapiti and Little Barrier Islands, where fewer introduced mammals are present. In the Orongorongo study site, there were 18 species of resident forest birds (Table 7.2), although other species were associated with the river, restricted to forest edges or visited the study site from time to time.[60] Silvereyes made up about half of all birds recorded at Orongorongo; grey warblers and fantails were also common, while kingfishers, kereru, cuckoos and whiteheads were the least common of the birds that bred locally.

About 20 species of forest-dwelling birds reside in Kowhai Bush, near Kaikoura (Table 7.2).[61] Whereas the Orongorongo study was conducted in a vast area of tall, contiguous and diverse native forest, Kowhai Bush is a narrow 240-hectare patch of regenerating kanuka forest containing gaps with scrubby vegetation, and is bounded by farmland and riverbed. The patchy nature of Kowhai Bush is reflected in the presence of magpies, redpolls, thrushes and goldfinches, species that are rare in closed-canopy native forest. Kereru sometimes visited Kowhai Bush, but none was resident there.

Most birds at present living in New Zealand forests have generalist diets comprising varying proportions of invertebrates, nectar, fruit, seeds and sap. Of the 18 species Colin O'Donnell and Peter Dilks[62] studied in South Westland, only kereru exclusively ate plant material (foliage and fruits). Native birds with omnivorous diets are best able to adapt to the habitat and faunistic changes that have taken place, and introduced species with broad diets are best able to penetrate the indigenous ecosystems. The birds that have a large component of plant foods in their diets are larger and more mobile than the normally territorial insectivorous

Table 7.2 Resident forest birds in the tall contiguous broadleaf forests of the Orongorongo Valley and the kanuka forest remnant at Kowhai Bush, Kaikoura

	ORONGORONGO	KOWHAI BUSH
Insectivores		
	Rifleman	Rifleman
	Dunnock	Dunnock
	Tomtit	Robin
	Whitehead	Brown creeper
	Grey warbler	Grey warbler
	Fantail	Fantail
Insectivores/predators		
	Kingfisher	Kingfisher
	Morepork	Morepork
	Shining cuckoo	Shining cuckoo
	Long-tailed cuckoo	
		Australian magpie
Predator		
	New Zealand falcon	Australasian harrier
Fruit, seed and honey eaters		
	Kereru	
	Tui	
	Bellbird	Bellbird
	Blackbird	Blackbird
	Song thrush	Song thrush
	Chaffinch	Chaffinch
	Silvereye	Silvereye
		Redpoll
		Goldfinch
		Yellowhammer
		Californian quail

Sources: Brockie 1992, Gill 1980, Hunt & Gill 1979

species. Plant foods such as fruits, nectar and new leaves can be distributed patchily in both space and time. The ecology of the native plant-eating parrots, kereru and honeyeaters was discussed in Chapter 4.

Insectivores and their ilk

Most forest birds eat some insects, but we can recognise two guilds of birds for which insects make up the major component of their diet: a guild of small insectivores and a group of larger birds that take insects as well as small vertebrates. At Orongorongo, the small insectivores included tomtit, fantail, grey warbler, whitehead, dunnock and rifleman. At Kowhai Bush, this guild was made up of robin, fantail, grey warbler, brown creeper and rifleman.

At Orongorongo, all six small insectivores utilised all forest tiers.[63] However, dunnocks, fantails and tomtits were most often seen less than five metres above the ground and whiteheads mostly in the middle tiers, while grey warblers and riflemen used all forest tiers equally. Fantails mostly take flying insects, but the other small insectivores are gleaners. Of these, the grey warbler is the smallest and the only species that feeds while hovering. Warblers make more use of foliage and the outer tips of branches than other insect-eating birds. Foraging behaviour of the insectivorous passerines has also been studied in the rainforests of South Westland,[64] where tomtits showed the strongest tendency to feed in lower tiers of the forest and mohua most often fed in the canopy. Brown creepers avoided the ground and were never seen above the canopy, but made equal use of all tiers in between. In parts of New Zealand where tomtit and robin co-exist, the robins mostly exploit ground insects, while tomtits feed mainly from trunks and branches. At Orongorongo, where robins were absent, tomtits mostly fed on the ground. Although grey warblers and tomtits ate a similar range of insects, they obtained them from different forest tiers. These two species often ate caterpillars, which were seldom taken by fantails, riflemen or dunnocks.[65] At Orongorongo and in South Westland, all these insectivorous birds took seeds and fruits when available. Orongorongo whiteheads ate more plant material than the other insectivores. The Orongorongo study found that the range of insects eaten by the introduced dunnock showed little overlap with native insectivores.

Birds in the genus *Mohoua* use their moderately stout bills to probe under bark. They also take larger insects from the surfaces of trunks, branches and foliage. In the North Island, the whitehead is the sole member of the genus. Two species, mohua and brown creeper, occur in the South Island. Mohua are about twice the weight of brown creeper. They are now almost restricted to beech forests, and even then their distributions are greatly fragmented, but they were once more widespread and occurred in a wider range of forest types. In South Westland, mohua and brown creeper showed extensive overlap in the forest strata used, their feeding methods and food types taken.[66] At Kowhai Bush, brown creepers made approximately equal use of leaves, twigs, branches and trunks of kanuka trees.[67] Mohua sometimes feed by ripping into dead wood, a behaviour not exhibited by the other small insectivores.

The silvereye colonised New Zealand during the nineteenth century, and a century later it is among the most numerous and widespread forest birds, occurring in virtually all terrestrial habitats. Adaptability has been the key to its success.[68] In vegetationally diverse forests such as Orongorongo, where fruit of one type or another is available most of the year, silvereyes are primarily frugivorous. In the less diverse kanuka forest at Kowhai Bush, silvereyes ate fruit in those months when it was available, but for most of the year they gleaned insects from foliage and tree trunks. In South Westland, they ate eight of the 11 food types recorded, utilised 52 of the 58 plant species present, they fed in all forest tiers and their diet varied greatly season to season.

Of the larger, more predatory birds, only kingfishers were studied at Orongorongo. There, insects and spiders made up about half the diet, with cicadas, dragonflies and chafer beetles the most common insect foods. The other half comprised lizards, mice, freshwater crayfish and birds, in descending order of importance.[69] At Orongorongo, kingfishers appeared to spend more time in the forest canopy along the river. Kingfishers use a wide range of terrestrial, coastal and wetland habitats, and their diet simply reflects what small animals are present.

Moreporks mostly eat insects, though they do take other invertebrates plus a few rodents, small birds, frogs and lizards.[70] A study of morepork foods found that 98 per cent of the items found in 75 stomachs examined were invertebrates.[71] Moths, their larvae, beetles and weta were the most common prey items, moths being taken most often in summer and beetles in winter. One mouse and 28 birds were among the 1696 prey items identified.

Both species of cuckoos are seasonal migrants, spending the summer in New Zealand and the winter on tropical Pacific Islands. The shining cuckoo takes many insects, especially caterpillars and species that are unpalatable to other birds. It mostly forages in tree canopies.

The long-tailed cuckoo eats small birds, eggs, lizards and insects, mostly at night, but no detailed study of its food or foraging behaviour has been made.[72]

New Zealand falcons prey mostly on small birds, although up to a third (by weight) of their food may be mammals, and they eat some insects and lizards. Falcons mostly use open habitats or forest edges, so many of the birds eaten are introduced species. The main mammals they prey on are rabbits, rats and stoats.

Birds, bats and lizards, pollinators and seed-dispersers

Plants produce nectar and fruit to entice animals to transfer pollen from one plant to another or to disperse seeds away from the parent plant. Certain plants have highly specialised flowers that allow only closely co-evolved animal species access to that nectar but, fortunately, few such close associations evolved in New Zealand. Most native plants have simple, unspecialised flowers that are visited by both vertebrates and invertebrates, and small fruits that are eaten by several species of birds or other animals. Most of these small flowers are white, scented and have small quantities of concentrated nectar, suggesting that they are primarily insect-pollinated. Bird-pollinated flowers tend to be large, brightly coloured and contain large quantities of dilute, unscented nectar. Only about 15 native plant species have flowers specifically adapted for bird pollination, and none is so specialised that pollination is restricted to a single species.[73] Some examples of specialist bird-pollinated flowers are those of flax, rata, tree fuchsia and kowhai. Many of the apparently insect-adapted flowers occur in tight clusters, flower in winter when insects are inactive, and some (including kamahi, five-finger and lemonwood) are regularly visited by birds, in particular tui, bellbird and hihi. Isabel Castro and Alastair Robertson[74] suggested that these structurally simple flowers evolved to attract a range of both vertebrate and invertebrate pollinators. They suggest this is important in New Zealand's climate, where wind could destroy large, ornate, bird-attracting flowers, or cold spells result in insects being inactive during the plant's short flowering season.

The most closely co-evolved plant/vertebrate systems are those between mistletoes and honeyeaters (see page 205) and three plants – *Tecomanthe speciosa*, kiekie and wood rose – that are adapted for pollination by short-tailed bats.[75] In Paparoa National Park, where short-tailed bats are locally extinct, pollination and seed dispersal of kiekie still occurs, possibly by possums, but probably with reduced efficiency. The wood rose is an endangered ground-dwelling parasitic plant that survives in a few North Island forests. It has dull-coloured flowers that produce a large quantity of sweet, musky-smelling nectar. The endangered short-tailed bat is the natural pollinator of the wood rose, although ship rats can also effect pollination. Kiore and possums are also attracted, but their visits destroy the flowers. The tongue of the lesser short-tailed bat is long with brush-like papillae; they eat nectar and pollen from (but probably also pollinate) rata, pohutakawa, rewarewa, kiekie, perching lilies (*Collospermum*) and, less often, other plants. Short-tailed bats also eat fruits, including those of kiekie and perching lilies.[76]

No New Zealand vertebrates eat only fruit or nectar, but many birds, some lizards and the short-tailed bat eat these nutritious foods when available. At certain times of the year, kereru, kokako and kakapo feed solely on fruits, and these birds apparently breed only when fruit is plentiful.[77] Other native birds that rely heavily on nectar, fruit or seeds include tui, bellbird, hihi, parakeets, kaka, kea and weka. In South Westland, kaka, tui, bellbird and silvereye are important pollinators of forest plants. Kaka and tui have brush-tipped tongues suited for gathering nectar, but kea, with their simple tongues, destroy flowers when extracting nectar.[78]

Nocturnal geckos visit and probably pollinate flowers of various native plants, including cabbage tree, ngaio, flax and manuka, although no plants appear to rely solely on lizards for pollination. Pohutukawa flowers are especially attractive to geckos, and up to 50 Pacific geckos have been recorded in a single flowering pohutukawa tree. Duvaucel's and Pacific geckos with pohutukawa pollen on their chins have been seen up to 50 metres from the nearest tree, illustrating their potential for transferring pollen from one plant to another.[79] Most of the fruits eaten by skinks and geckos are small (3–6 millimetres in diameter), and Tony Whitaker has shown that lizards can deposit viable seeds 10 or more metres from the parent plant. In rocky habitats, lizards are more likely than birds to deposit seeds in favourable crevices. Lizards and bats were once widespread and common, but the impact their modern restricted ranges have on plant regeneration is not known.

About 70 per cent of New Zealand's native trees, 40 per cent of the shrubs, 30 per cent of vines and all parasitic plants have fleshy fruits and rely on animals for their dispersal.[80] Today, kereru, kokako, silvereye, tui, bellbird, blackbird and chaffinch are the major forest frugivores, but in pre-human forests there was a wider range of fruit-eating birds, including moa, raven, Finsch's duck and piopio, some of which were able to eat larger fruits than any extant bird. Parrots and parakeets consume fruit but, in doing so, use their powerful bills to crack and eat seeds. They are seed-predators rather than seed-dispersers, and thus of little service to the plants.

New Zealand has 58 species or subspecies of *Coprosma* that occur in a variety of habitats and exhibit many growth forms.[81] About half have red or orange fruits, but other species have black, blue, purple or white fruits. The red, orange and black fruits are mainly eaten by birds. In the lowlands, most red-fruited *Coprosma* have large, bright green leaves, the contrast making the fruits conspicuous to visually cued birds. At higher altitudes, where birds are the major frugivores, most *Coprosma* species have red fruits. *Coprosma* fruits of other colours are apparently attractive to lizards, and some of these fruits are found in thick tangles of twigs that could be traversed by lizards but not by birds. Some other divaricating plants probably also have lizard-dispersed seeds.[82]

While a number of frugivorous animals are present in New Zealand forests, the regeneration of those forests is not ensured. Only two of the remaining frugivores – kereru and kokako – are large enough to eat fruits more than 12 millimetres in diameter. A few tree species, including tawa, taraire, karaka, kohekohe, mangeao, tawapou and puriri, have large fruits with a single large seed, and in all but the few North Island forests where kokako survive, these species are totally reliant on kereru for dispersal.[83] Fortunately, kereru are good seed-dispersers, they eat fruits from at least 70 species, are still widespread and may fly several kilometres in the time it takes a seed to pass through their gut. The kereru is one of few fruit-eating birds that readily transports seed from one forest remnant to another, which is especially important in areas such as Banks Peninsula where forest remnants are small and some plant species are restricted to few of those remnants. Now that only one large frugivorous bird remains widespread, the regeneration of some native trees may be threatened, particularly on the Chatham and Kermadec Islands, where the native pigeon is rare or extinct.

Peter Williams and Brian Karl[84] investigated the fruit-eating habits of birds in three forest remnants near Nelson and found that the two endemic frugivores, bellbird and tui, mostly ate fruits from native trees. Of the two, the bellbird had the more restricted diet and most fruit eaten was from Hall's totara or karamu. The tui had a more varied diet, including fruits of introduced species. Silvereyes ate a wide range of native and introduced fruits, while blackbirds, thrushes and starlings mostly ate the fruits of introduced plants. Birds, particularly introduced species that move between towns and bush remnants, conveyed weed seeds into those remnants. On the other hand, Williams and Karl found that both introduced and native birds conveyed native seeds from remnants into early successional vegetation, thus aiding the spread of native plants. Silvereyes, with their dispersive behaviour, are especially effective seed-dispersers.

We can only speculate about the role that extinct birds played in seed dispersal. Moa in the genus *Dinornis* are known to have ingested fruits of at least 24 species, including trees such as matai and cabbage tree, shrubs including *Corokia* and *Coprosma*, and also certain herbs and vines.[85] Huia ate hinau, pigeonwood and some *Coprosma* species, and piopio were recorded eating fuchsia fruits, but doubtless this represents only a small sample of the fruits taken by now-extinct birds. The role of moa in seed dispersal is uncertain because the gizzard stones of the larger species would have destroyed at least some of the seeds they ingested.

Today, possums, deer, rats and mice are major consumers of forest fruits, but their role in seed dispersal is debatable. Rats are primarily seed predators, but in a trial with captive animals some very small seeds germinated after passing through ship rat guts, whereas none eaten by kiore or mice germinated.[86] While ship rats can disperse small seeds, it appears the damage they do outweighs any good, and some of the seeds that germinated were of introduced weeds. At Orongorongo, about a fifth of all hinau and miro nuts found on the forest floor had been opened by rats.[87] Elsewhere, rats have been known to take almost the entire crop of certain podocarp and hardwood fruits, including rimu, kahikatea and matai. Rats (initially kiore, then the European species) have the potential to inhibit regeneration of favoured plant species, but their actual impact on forest regeneration has not been quantified.[88] On Breaksea Island in Fiordland, there was an increase in the number of seedlings of 10 out of 17 fleshy-fruited species following the eradication of Norway rats, the only rodent species present on the island.[89] Seedling numbers of two species declined, and it was suggested that in the case of one – broadleaf – the decline in seedling numbers may have been due to an increase in numbers of herbivorous insects resulting from rat eradication. This study also presented evidence to suggest beech seedlings had failed to establish during the century that rats had been present on the island.

Possums appear to eat most available fruits of both native and introduced plants, and a proportion of the seeds from at least some of those species will germinate after passing through possum guts.[90] Possums certainly have the potential to disperse seeds and have been implicated in the spread of introduced weeds into indigenous forest remnants.

The role of frugivores in New Zealand forests has been discussed by a number of authors,[91] all of whom stress how little is known about the ecology of fruit-eating animals and their role in seed dispersal and forest regeneration.

Mistletoes and honeyeaters

Mistletoes are parasitic plants that grow on host trees or shrubs and depend on animals for pollination and seed dispersal. There are six native species, one of which is now extinct and the other five are less common than they were a century ago. Until recently the decline of the mistletoes was attributed to habitat loss and possums, but research now suggests that a reduction in range and abundance of their bird pollinators has also contributed.[92] Two species, *Peraxilla tetrapetala* and *P. colensoi*, have flowers that only open when twisted in a certain way by birds. Tui and bellbirds are the most common and most effective pollinators of these mistletoes, although at Lake Ohau silvereyes and chaffinches have also learned how to open *P. tetrapetala* flowers. Other birds have been observed at *Peraxilla* flowers but have not been shown to effect pollination. Bellbirds and tui seek out mature unopened flowers and seldom visit developing or opened flowers. If birds do not open *Peraxilla* flowers and transfer pollen from one plant to another, self-fertilisation can occur or small native bees can open *P. tetrapetala* but not *P. colensoi* flowers. However, seed set is low after self-pollination or bee pollination.[93]

The dependence of the two *Peraxilla* mistletoes on tui and bellbirds makes them vulnerable. Although both these birds remain widespread, there are parts of the country where only one or neither species is now found. In other areas tui and bellbirds are so uncommon that only a minority of the *Peraxilla* flowers are opened. In parts of the North Island, possums have so greatly reduced *Peraxilla* numbers that, even if the mistletoes are protected, their flowers are ignored: perhaps the local bellbirds no longer know how to open the flowers.[94] The extinct mistletoe species had an even more specialised flower, but its pollinator is unknown. A fourth species, *Alepis flavida*, is primarily bird-pollinated but will readily set seed through self-fertilisation, so it has been less affected by declines in bird numbers. The remaining two species are pollinated by insects.

All native mistletoes depend on native birds, principally bellbirds, tui and silvereyes, for seed dispersal, and seeds that do not pass through a bird gut do not germinate. At sites such as Craigieburn in the Southern Alps, the low density of fruit-eating birds may restrict seed dispersal of mistletoes. Introduced birds seldom consume native mistletoe fruits, even when both the mistletoes and the birds are common.

Honeydew and honeyeaters

Honeydew is an energy-rich, sugary substance that exudes from the anal tube of scale insects that live on beech trees. The scale insects pierce the tree's phloem tissues with their mouthparts and the pressure in the phloem cells forces sap through the insect. It uses what it needs, then unused sap seeps out of the insect in small droplets.

Honeydew is an important element in the beech forest ecosystem. It accumulates on tree trunks, where it supports a sooty mould that may thickly coat the tree. Birds eat insects that live in the mould. Honeydew is an important food for tui, bellbirds, silvereyes, kaka and, to a lesser extent, other forest birds. In the Nelson Lakes beech forests, the density of tui and bellbirds was correlated with the number of honeydew-producing trees.[95] In those same forests, kaka use honeydew as their main energy source and extract wood-boring longhorn beetle larvae for protein. This is a marginal diet and kaka breed only during mast years when beech seed is abundant.[96] Some native lizards, moths and ants also eat honeydew, and bees produce a distinctively flavoured honey from it.

Today, introduced wasps are major consumers of honeydew. The German wasp first invaded beech forests in the 1950s, then in the 1970s the common wasp arrived. Wasps now occur in beech forests throughout central New Zealand, where up to 360 of them have been recorded on one square metre of tree trunk. The wasps are now the greatest consumers of honeydew in Nelson forests and, by sheer weight of numbers, have significantly reduced the availability of honeydew to birds and other native animals.

Browsers

Red deer, goats and pigs are now the most important ground-dwelling browsing animals, and possums the major tree-canopy browsers. In some parts of the country, a second species of deer or even chamois may also be present. Rabbits graze adjacent open habitats but seldom enter the forest proper. Pigs are omnivores that eat plant foods, carrion and invertebrates.

Deer and goats generally prefer subcanopy hardwood species, but the actual plants eaten vary from place to place. At Orongorongo, the diets of goats and red deer overlapped. Both ate many species, but large-leafed plants – including broadleaf, five-finger, pate, kawakawa, large-leafed coprosmas, raukawa, mahoe, rata and putaputaweta – were favoured.[97] On Stewart Island, white-tailed deer prefer broadleaf, putaputaweta, stinkwood, lancewood, tree fuschia, red mapou, kamahi and muttonbird scrub. The smaller, stiffer-leafed plants, such as miro, Hall's totara, manuka, weeping mapou, prickly mingimingi and inaka, are avoided if fleshy-leafed species are available.[98] Deer do not just browse living plants: on Stewart Island white-tailed deer obtain up to 60 per cent of their food by scavenging fallen leaves and fruits.

In Pureora Forest, woody plants make up 70–80 per cent of the weight of food consumed by both red deer and possums. Both mammals ate about a hundred plant species, but about half of those were eaten rarely (Figure 7.5).[99] Seven broadleaf species – broadleaf, lance-wood, pokaka, kamahi, mahoe, putaputaweta and *Coprosma grandifolia* – were particularly important to red deer. Ferns made up 17 per cent and grasses 10 per cent of the Pureora red deer's diet. Most of the leaves from

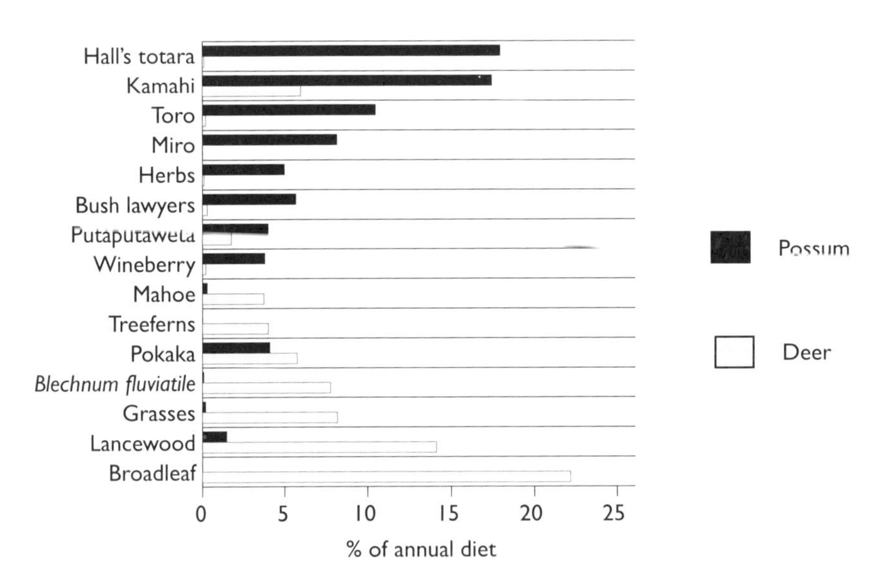

Figure 7.5 The composition of red deer and possum diets at Pureora Forest

Source: Nugent et al. 1997

woody plants consumed by deer were old and yellowed, showing that they were fallen leaves taken from the ground. Whereas deer mostly ate leaves, fruits were of greater importance to possums. Possum diet varied year to year: in 1991, leaves and fruits each made up about 40 per cent, whereas the following year leaves made up nearly 90 per cent of their food. Hall's totara, kamahi, toro, miro, bush lawyer, pokaka, putaputaweta and wineberry were the preferred woody plants. Ferns were seldom eaten. Possums are mostly canopy feeders, but 10–20 per cent of their food is obtained at ground level.

It is easy to list the native plants these introduced herbivores eat, but assessing the damage done to indigenous ecosystems is far more difficult. Two examples will illustrate the types of changes that have taken place. First, browsing herbivores may change the vegetation composition by eating out palatable species. For example, various palatable species of herbs, grasses and shrubs were more common on small goat-free islands in alpine tarns in Paparoa National Park than on the goat-infested land surrounding those tarns.[100] Alternatively, the herbivores may change vegetation composition and, in the long term, change forest to scrub or fernland by impeding regeneration. Glenn Stewart and Larry Burrows[101] compared the regeneration of podocarp/broadleaf forests on deer-infested Stewart Island with that on deer-free Bench Island. On Bench Island, tree ferns and hardwood seedlings quickly colonised tree-fall gaps, leading to the re-establishment of the southern rata and kamahi canopy. On Stewart Island, the establishment of tree ferns and hardwood shrubs was prevented and the few seedlings that did manage to establish were heavily browsed. Eventually, impaired regeneration is likely to result in conversion of those forests where the main canopy species are eaten by deer to shrubland or grassland.

Alpine herbivores

On the eastern side of the Southern Alps, up to five species of introduced herbivorous mammals may co-exist, and these show extensive overlap in the foods eaten and habitats used (Table 7.3).[102] Tahr and chamois are goat-like browsing mammals of similar body size, native to the mountains of Europe (chamois) and the Indian subcontinent (tahr). Only in New Zealand's Southern Alps do these two species co-exist. Chamois and tahr both prefer grassland and shrubland habitats, provided that snow cover allows access to them. Rocky bluffs are used more often by tahr than by chamois. The diets of these two species

Table 7.3 Population density, body size and habitats used by introduced herbivores in the Southern Alps

	Mean body weight (kg)	Maximum density (per km²)	Maximum biomass (kg/km²)	Habitats used
Tahr	45	>30	>1350	g, b, s
Chamois	31	5	155	g, s
Red deer	58	10	580	g, s
Possum	2.8	?	?	s
Hare	3.5	<1	<3.5	g

(g = grassland, s = shrubland, b = bluffs)

Adapted from Forsyth et al. 2000

overlap, one with the other and both with possums, although tahr eat more tussock but fewer shrubs and herbs than do chamois.[103] Their diets in part reflect the habitats used by the mammals: tahr mostly frequent the alpine grasslands, chamois alpine grasslands and subalpine scrub, and possums are mostly found in or below subalpine scrub.

Chamois occupy a greater range of habitats than do tahr and their diet shows greater variation between localities. West of the Main Divide, chamois eat more woody plants, in particular more forest plants, than do eastern chamois. Chamois appear to actively search out preferred plants, whereas the diet of tahr better reflects the relative abundance of species present.

While habitat use and diets may overlap, the two species differ in other ways. Tahr are more sedentary and gregarious, often occurring in large groups, while chamois are solitary or occur in small groups.[104] When first introduced, chamois spread quickly and now occur along almost the entire length of the Southern Alps. Tahr by comparison have a restricted distribution centred on Aoraki/Mt Cook. Chamois move larger distances, sometimes retreating downslope to utilise forest habitats in winter. While chamois may appear to be the more adaptable of the two, they occur at much lower densities and are excluded from places where tahr or red deer are common. Hunting has apparently restricted both the distribution and the range of habitats used by tahr, which, in the absence of hunting, can live year round at altitudes as low as 1000 metres.[105]

These differences in feeding ecology between tahr and chamois mean that these browsers have the potential to damage the alpine vegetation in quite different ways. Chamois seek out their preferred foods, so are more likely to reduce the abundance of such plants as the Mt Cook lily.

Tahr eat whatever is common, but because they are gregarious, occur in higher densities and are sedentary, their impact on the vegetation can be localised but intense. Owing to the overlap in distribution, habitats and foods of the different herbivores, it is impossible to quantify the damage done to the alpine communities by each. Intensive helicopter hunting between 1967 and 1983 resulted in large reductions of tahr in certain catchments, and this was followed by a substantial recovery of snow tussocks.[106] There has been a decline in the condition of tussock in catchments where tahr have returned to high density, but not in those with few tahr.

Possums, the all-round bad guys

There are about 70 million possums in New Zealand, which among them consume about 21,000 tonnes of vegetation every night.[107] Possums are ubiquitous: they occur in native and plantation forests, scrub, tussock grasslands, all rural habitats and even in city parks and gardens. They damage forests and other indigenous vegetation, and cause economic losses to plantation forests, horticultural crops, orchards and gardens. They also damage power and telephone lines.

As if that were not bad enough , these Australian imports also carry bovine tuberculosis and act as a reservoir of the disease, re-infecting cattle after tuberculosis control. They are likely vectors of other animal diseases and possibly spread giardia from one stream to another. Possums threaten native plants, animals and ecosystems in numerous ways. They eat foods favoured by the endangered kokako and other herbivorous birds. They prey on large insects, land snails and native birds, their eggs and nestlings. In the past possums opened up forest canopies, enabling the spread of deer into those forests. The economic losses attributable to possums, including the cost of control, mean that they cost New Zealand more than $35 million each year.[108]

The impacts possums have differ from one forest type to another, usually being least in monotypic beech forests, which support few possums, and greatest in mixed broadleaf forests. In northern New Zealand, possums have caused the deaths of many pohutukawa trees. Possums can cause depletion or even local extinction of mistletoes or the wood rose, but the loss of these naturally uncommon species will not result in dramatic changes to forest structure. On the other hand, dieback of canopy-dominant southern rata in the rata/kamahi forests of the West Coast appears to be triggered by possum browsing, and this may greatly alter the entire forest ecosystem. Both rata and kamihi are highly palatable to possums, but the extent of canopy dieback varies from place to place. Glenn Stewart and Alan Rose[109] identified two main factors that predisposed those forests to canopy dieback. One was the proportion of other palatable species present, dieback being most prevalent where highly palatable tree fuchsia, wineberry, mahoe and pate were abundant. The other predisposing factor was the proportion of large old canopy trees present: these are less able than vigorous young trees to recover from possum browsing.

Dieback is uncommon on stable granite hills where mixed-age forests occur. It is most common on the earthquake-prone schist ranges near the Alpine Fault, where frequent disturbance induces a mosaic of even-aged forest patches. Stewart and Rose thus identified possums as the trigger but geological factors as the predisposing agents to canopy dieback. In these and other indigenous forests, canopies that have been weakened by possum browsing are susceptible to windthrow, pathogens or storm damage.

In the Orongorongo Valley, the most favoured foods such as tree fuchsia and titoki became rare or even absent because possums killed the adult trees and goats or deer the seedlings.[110] Even after possum numbers stabilised, forest composition continued to change as kamahi, northern rata, black tree fern and tawa declined in abundance and less palatable species such as pigeonwood, mahoe, pukatea and other tree ferns increased to compensate. Tawa was not eaten by Orongorongo possums in the 1950s, but as their preferred foods declined, tawa began to be regularly eaten. Recent studies have shown that flowers, fruit, fungi and invertebrates are important components of the possums' diet, thus accentuating their impact on indigenous ecosystems.[111]

The extent to which possums inhibit forest regeneration is unclear because in most forests it is impossible to tell which herbivores eat seedlings. Intuitively, we might expect possums to have less impact on regeneration than deer or goats have, yet on Kapiti Island, where possums were the only browsing mammal present, a large proportion of northern rata, tawa and tree fuchsia seedlings were eaten.

The impact possums have on native animals is equally poorly understood. In 1991, possums were first found to be predators at kokako nests,[112] and since then they have been shown to eat eggs, chicks and even adults of brown kiwi, saddleback, kereru, fantail and harrier. The sign left at nests by possums is similar to that left by rats, thus it has been impossible to accurately assess the impacts possums have on nesting birds. In the only study where all nest visitors were recorded by video, possums caused the failure of four of the 19 kokako nests under surveillance. Of all the nests studied (not all were filmed), possums were probably responsible for 10 out of 33 predation events recorded. Possums have been recorded eating stick insects, cicadas, beetles and land snails, and by eating out preferred tree species, they presumably affect the native insects that rely on those trees. The impact of possums on invertebrate populations is unknown.

Biomass of birds and mammals in New Zealand forests

Bob Brockie[113] calculated that the Orongorongo forest supported an average of 25 kilograms per hectare of possums, 0.6 kg/ha of birds and only 0.25 kg/ha of rodents and carnivores combined. Elsewhere in New Zealand the density of possums can be even higher, up to 76 kg/ha in some Westland forests. This is an unusually high biomass for a single species of herbivore. New Zealand plants evolved in the absence of herbivorous mammals, and most do not have defense chemicals characteristic of plants elsewhere. The biomass of possums could be the result of a susceptible flora being eaten out by a recent invader. As the forest composition alters in response to browsing mammals and browse-tolerant species replace the susceptible, possum biomass may decline. When possums and rats were the only introduced mammals present on Kapiti, the island had a bird biomass about four times that of Orongorongo. Since the eradication of rats and possums, bird biomass on Kapiti has become even higher.

Competition between introduced and native species

The most significant competition between native and introduced animals is not between different species of birds but between native birds and introduced mammals. The overlap in diet between takahe and deer will be discussed in Chapter 9. Many foods eaten by the endangered kokako are also eaten by the possums, goats and deer with which birds now coexist, and circumstantial evidence suggested that competition for food contributed to the decline in kokako numbers.[114] At Pureora and Mapara, the leaves of raurekau, broadleaf, pigeonwood, mahoe, kaikomako and supplejack are all eaten by kokako and at least two of the mammal species present. Possums appear to be the most serious competitor because they feed in the canopy and eat buds and flowers (thus reducing fruit set) as well as the fruits themselves. Both possums and kokako eat tawa, putaputaweta, kohekohe, hinau, pigeonwood, rewarewa, mahoe, kaikomako and supplejack fruits. Possums also eat foods favoured by other fruit- and nectar-eating birds. Subsequent control of possums, rats and goats has been followed by a dramatic increase in the numbers of kokako and most other native birds.[115] Bellbird, kereru and some other native birds increased in abundance after possums were eradicated from Kapiti Island. Scientists are still unravelling the relative roles of predation by possums, competition for food between them and the birds, and improvement in forest health following possum control in triggering increases in bird numbers. However, predation now appears to be the most immediate threat, at least to kokako.

Competition between native and introduced birds is relatively minor because they generally use different habitats. However, it does occur between the native grey duck and the introduced mallard. The mallard, with its higher reproductive rate, has usurped the grey duck in modified areas and is now even replacing it in remote wilderness wetlands.

The introduced Australian magpie is often accused of competing with native birds, including kereru. Magpies are aggressive, conspicuous and noisy; they harass many animals, not just native birds, but whether they cause native bird populations to decline is debatable. Research to investigate competition between magpies and kereru is under way and preliminary results suggest the impact of magpies can be significant.

Birds and mammals in small forest remnants

Scattered throughout New Zealand are numerous small patches of bush that in intensively farmed parts of the country may be all that now remains of the once-extensive forests. Although these remnants are all somewhat modified and few contain nationally rare birds or reptiles, they are important in conserving local biodiversity. Forest remnants are also important because they are the places where New Zealanders can most easily experience native bush and native birds.

Amanda Freeman[116] compared the bird fauna of Kennedys Bush, an 86-hectare reserve on the Port Hills above Christchurch, with the birds present in Kowhai Bush and those in large tracts of indigenous forests near Reefton. She recorded a similarly restricted range of native birds in both Kennedys Bush and Kowhai Bush, and, of those native species present, densities were lower than they were in the forests near Reefton. Conversely, she found that the remnants supported a more diverse range of introduced species than did the larger forest tracts. With the exception of chaffinch and blackbird, exotic birds were at higher densities in the remnants.

Ten species of introduced mammals, three herbivores (possum, rabbit and hare), the insectivorous hedgehog, two rodents (ship rat and house mouse) and all four carnivores (stoat, weasel, ferret and cat) were resident in Kowhai Bush.[117] In addition, pigs and deer were occasional visitors. Stoats, weasels and mice occurred throughout the bush. Ship rats were more common in broadleaf than kanuka forest, while hedgehogs preferred areas with an open canopy, forest clearings or the bush edge. The population size of all mammal species varied seasonally and year to year, but rodent populations showed more marked fluctuations than did other species.

The Kowhai Bush robin population also fluctuated, doubtless influenced by changes in predator density. Fluctuations were least in areas of continuous forest and greatest in areas of patchy forest, where populations periodically declined to local extinction, only to be recolonised by young birds from more favourable sites.

The Kowhai Bush robins illustrate the hazards facing populations in forest remnants. Any small population faces many hazards, and the smaller it is, the more prone to local extinction, whether this be through deterministic factors that drive down populations, or unpredictable stochastic events. Robins seldom cross open spaces, so while they were able to recolonise one part of Kowhai Bush from another, they would be unlikely to recolonise an isolated fragment of forest once it became surrounded by farmland.

The robin is one of many native species that are physically unable or behaviourally unwilling to cross farmland to move between forest remnants. In Australia, the importance of habitat corridors – strips of scrub or forest linking forest remnants – in allowing birds and mammals to move from one remnant to another has been well studied.[118] In New Zealand, it has been assumed that corridors help maintain species diversity in remnants, but few studies have systematically assessed this. On Banks Peninsula, tomtits and riflemen appear to mostly survive in remnants that are connected to other remnants by habitat corridors. Brown creeper survive in connected remnants and in some that have small 'stepping-stones' of scrub or trees in between bush remnants, suggesting that brown creepers will cross small gaps, but riflemen and tomtits seldom cross any gap.[119] The role of habitat corridors in maintaining biodiversity in remnants is controversial. Corridors can encourage the invasion of pests or weeds, or speed up the re-invasion of possums into those reserves after control operations.

In Northland, radio-tagged brown kiwi regularly moved from the forest reserve in which they lived and crossed farmland to feed or roost in small patches of fern thicket, scrub, native forest or even pine forest where there was a dense understorey and a rich soil invertebrate fauna.[120] Kiwi readily crossed hundred-metre gaps between forest or scrub patches, or used bush patches as stepping-stones to reach more distant forest patches. Murray Potter suggested that even very small patches of poor-quality habitat close to forest remnants containing kiwi could be important for their conservation.

The management of native birds in isolated forest fragments can be very different to the management of the same species in larger areas of forest, and if our aim is to enhance native bird populations, we may have to accept or even encourage the presence of introduced fruiting plants. Management of the land surrounding habitat remnants is perhaps as important as management of the reserve itself. Tui, bellbird and kereru will forage outside the forest remnants they reside in, but the extent to which they rely on exotic plants growing inside reserves or trees growing outside the reserve has seldom been assessed.

Birds, lizards and mammals of our towns

Few native forest birds have adapted to city life. Silvereyes, fantails and grey warblers are common city dwellers, and where pockets of native bush remain in cities, bellbirds, tui or kereru may also be present. Introduced species are much more numerous. There has been little interest taken in city-dwelling birds and mammals, and so far little effort has been made to manage the city environment in ways that encourage native animals. Restoration plantings of native trees and allowing some parks and 'wasteland' to revert to unmanicured native scrub would encourage back to town those native forest birds that have the ability to recolonise and are able to cope with introduced predators. Most New Zealand cities contain some coastal, wetland or estuarine habitats where native birds still prevail, but these habitats are beyond the scope of this chapter.

Riccarton Bush in Christchurch is a 7.8-hectare remnant of the kahikatea forest that was characteristic of the wetter parts of the Canterbury Plains. Adjacent to the bush is an area of oak woodland and a tributary of the Avon River. Surrounding the reserve is the densely populated suburb of Riccarton. The natural history of Riccarton Bush has been described in a book edited by Brian Molloy,[121] with chapters by Colin O'Donnell and John Parkes discussing the birds, mammals and reptiles present. The only resident native birds were grey duck, kereru, fantail, grey warbler and silvereye. Only silvereyes were numerous, the bush supporting just one to three breeding pairs of kereru, fantails and warblers. Fantails and grey warblers are widely but thinly distributed through much of Christchurch, but the city supports few bellbirds or kereru, and most of these visit Riccarton Bush from time to time. The only reptile still found there is the common gecko.

Possums, ship rats, Norway rats, mice and cats live in Riccarton Bush, which also appears to function as a refuge for an unusually high density of hedgehogs that mostly forage on neighbouring lawns and gardens. Riccarton Bush cats are a mixture of feral and domestic animals, the latter no doubt hunting to satisfy an instinctual urge or to supplement what their human owners feed them. The density of cats in a mosaic of native bush and residential land on the fringes of Lower Hutt city was also far higher than in non-residential forested land.[122]

The Auckland Domain is mostly parkland with small patches of forest. The forest is mostly exotic trees, with native and exotic shrubs forming the understorey. Brian Gill[123] recorded 22 species of bird in the Domain, 17 of which occurred in the forest. Eight native species used those forest patches: silvereye; fantail and grey warbler were common; tui and kingfisher were regularly recorded; and kereru, morepork and shining cuckoo were rare. One other native species, the welcome swallow, used the open habitats.

Most of the widespread introduced passerines were present. Dunnocks were absent from Auckland Domain – surprisingly, as they are common in parks in other cities. Skylarks were present on rough grassy areas in the 1960s but have disappeared, presumably when these areas were converted to mown lawn. The Malay spotted dove occurs in Auckland but not in other New Zealand cities, and the eastern rosella occurs only in Auckland and Wellington.

Virtually no studies have been done of the ecology of city-dwelling birds, let alone other vertebrates that dwell there. Tim Day[124] assessed the abundance of native and introduced birds in 30 Hamilton gardens and related abundance to the presence and biomass of native plants in those gardens. He recorded 15 species of birds, six of which were native. Nearly three-quarters of the individual birds recorded were introduced, and most of the rest were silvereyes. House sparrows were most common, making up nearly half of the birds seen. More species were seen in gardens with higher plant biomass, which presumably reflected a greater diversity of habitat features and plant species. Day's study showed that silvereyes, fantails and grey warblers were more common, and house sparrows less common, in gardens where native plants predominated.

Lizards are overlooked residents of our cities. Alastair Freeman[125] surveyed the lizards of Christchurch and found that only common skinks remained within the built-up urban area, and even they were restricted to undeveloped sites. Common geckos remained common on the Port Hills. Spotted skinks were found in the areas fringing the city as recently as the 1980s, but none was found in his survey. Freeman suggested that cats and the continuing urbanisation of rural land on the outskirts were the major threats to city-dwelling lizards.

CHAPTER 8

Seabirds and marine mammals

> The waters of Foveaux Strait and to the eastward of the Island are darkened as far as the eye can see by millions of resting birds.
>
> – Basil Howard (1940)

Marine birds and mammals are more abundant and the range of species is more diverse in and around New Zealand than virtually anywhere else in the world. About half the world's species of whales and dolphins and nearly a third of all seabirds occur in the waters surrounding this country. The most abundant birds in our region are not the familiar house sparrows or black-backed gulls, but the less well-known sooty shearwaters and fairy prions. Most marine birds and mammals feed tens, hundreds or even thousands of kilometres offshore, and now breed only on offshore islands; consequently, few New Zealanders are familiar with these fascinating creatures.

Marine birds and mammals are important both ecologically and economically. Before the arrival of people and predatory mammals, seabirds bred at numerous sites on the mainland and there, as on certain islands today, petrels were keystone species playing a vital role in the transfer of nutrients from sea to land. Since human contact there has been a reduction in the ranges and numbers of many species owing to introduced predators and human exploitation. Recently, fisheries interactions have threatened further species.

In this chapter I describe the main groups of marine birds and mammals and some of the ways they exploit marine resources. I briefly describe the marine environment surrounding New Zealand and how this has influenced the diversity and abundance of marine birds and mammals. Conservation of these animals differs from that of their terrestrial counterparts, and this will also be discussed.

Marine mammals are easily defined, as all the New Zealand species feed almost exclusively in the ocean, but seabirds are less clearly distinguished from their wetland counterparts. In this chapter I discuss only those birds that take the majority of their food from the open ocean. Wading birds that feed in the shallows of estuaries and harbours are not included. Gulls, some shags and terns feed in freshwater habitats, harbours, estuaries or inshore waters, but seldom venture more than a few kilometres from land; they get scant attention in this chapter. The black-fronted tern and black-billed gull breed inland but feed in coastal waters for part of the year. These species were discussed in Chapter 3.

Seabird and marine mammal watching in New Zealand

New Zealand presents some excellent opportunities to observe seabirds and marine mammals, and viewing of these animals is important to the economies of a number of our coastal towns. On Otago Peninsula, one can visit colonies of royal albatross, yellow-eyed penguin, spotted shag and Stewart Island shag. Whale watching is a popular tourist attraction at Kaikoura, and boat trips from Stewart Island offer the chance to view albatrosses, shags, penguins and shearwaters, often in astounding abundance. Dolphin-spotting trips operate from several parts of New Zealand, including Banks Peninsula, with its Hector's dolphins, and Kaikoura, where dusky dolphins are common. The Milford Sound scenery may be superb, but visitors should also look out for bottlenosed dolphins and Fiordland crested penguins. Seabirds and dolphins come close inshore in many parts of the country where they may be seen from coastal cliffs or from inter-island ferries. My favourite places for watching marine life are Otago and Kaikoura Peninsulas, but I have seen interesting seabirds from coastal lookouts at many places from Northland to Stewart Island.

Penguins

> Long before the islands were reached the odour of penguin was borne across the water by a breeze off the land. The distance being lessened, the smell became more and more pronounced, and increasingly disagreeable.
>
> – Edgar Waite, in Chilton (1909)

Penguins are specialised diving birds: their torpedo-shaped bodies, reduced wings (which function as hydrofoils just as other bird wings are aerofoils), their stiff, oily plumage and subcutaneous fat deposits are all adaptations to this lifestyle. They are restricted to the southern hemisphere and most occur in cool temperate and subantarctic regions. Six species breed in the New Zealand region: three on the mainland, and three on the subantarctic islands. Five of the 18 species are restricted to the New Zealand region. A further three species breed on Macquarie Island. The smallest, the blue penguin, weighs about a kilogram and is found in New Zealand and Australia; the largest, the emperor penguin of Antarctica, weighs up to 40 kilograms.

Among birds, penguins are the most accomplished divers, but only emperor and king penguins habitually dive deeper than a hundred metres and remain submerged for more than a few minutes. All penguins are pursuit-divers that feed mostly on fish, crustaceans and cephalopods.

Most penguin species breed in dense colonies, but not the New Zealand species. Blue penguins breed in burrows or caves, usually in small groups. Fiordland crested penguins breed in rainforest in loose colonies where nests are usually at least a metre apart. The yellow-eyed penguins of southeastern New Zealand are the most solitary: they like to hear but not see their neighbours. Like other seabirds, penguins are monogamous. They each take turns at incubating while their mate feeds at sea, and it takes two parents to gather enough food for the ever-demanding chicks. Pauline Reilly's *Penguins of the world* provides a good introduction to these birds.

Albatrosses, muttonbirds and other petrels

While penguins are highly specialised diving birds, the albatrosses and other petrels (Procellariiformes) are equally specialised consumers of surface and near-surface marine foods. Petrels have a worldwide distribution but are most common in temperate and subpolar southern hemisphere oceans. They are the most oceanic of all birds, and most species visit land only during the breeding season. None can dive to great depths – sooty shearwaters, which can descend to 60 metres, are probably among the deepest-diving petrels – and some species are confined to the top-most metre of the water column. Many petrels feed at night, when deep-living, often bioluminescent prey migrate to surface waters. They lay a single egg each breeding season, and some albatrosses breed only every second year. Petrels are long-lived, even the smaller species living 20–40 years. One Buller's mollymawk is known to have lived for at least 51 years, and a northern royal albatross for over 60 years.[1]

The Procellariiformes span a greater range of body sizes than any other order of birds. The smallest storm petrels weigh only 19.5 grams; the largest albatross about 8.7 kilograms. Their adaptations to the marine environment revolve around conserving energy while travelling vast distances in search of widely scattered food resources. All have a lower body temperature than most other birds, and egg and chick development occurs slowly, extending the time during which adults have to find food. Petrels convert food into a lightweight energy-rich oil that enables them to provision their chicks efficiently when food is found far from the breeding colony. Most have some variation of a long, narrow wing, adapted for gliding flight.

Albatrosses are consummate gliders, their huge wings being held stiff and almost motionless while the birds utilise differences in wind velocity in the few metres immediately above the sea surface, or gain lift off heaving waves to fly vast distances with minimal expenditure of energy. Albatrosses usually feed by alighting on the water and seizing fish or squid while floating. They can manage only the briefest and shallowest dives. At the other end of the scale are the delicate storm petrels. Their wings are broader and more rounded than those of other petrels and are used in conjunction with their long legs to patter over the water surface, from which they peck small crustaceans.

Diving petrels are small, dumpy birds with short wings, distinguished from other petrels by their rapid, whirring wingbeats. Their flight may be cumbersome, but they are adept divers. Diving petrels occur only in the southern hemisphere, though in the northern hemisphere there are small auks of similar appearance, size and ecology. The remaining petrels include the small prions, some of which filter-feed; the pelagic soft-plumaged highly aerial gadfly petrels; the sea-skimming, compact-plumaged shearwaters; and the generalist fulmars. The shearwaters fly just above the water with stiffly held wings, plunging into the water then diving in search of prey. The gadfly petrels are buoyant in flight and seize most of their prey at the surface in a manner similar to that adopted by terns.

Of the world's 110 species of Procellariiformes, 41 breed in the New Zealand region, 10 others regularly occur in New Zealand waters and a further 13 have been recorded as vagrants. John Warham's *The petrels: their ecology and breeding systems* and *The behaviour, population biology and physiology of the petrels* provide superb accounts of petrel biology.

Shags, gannets and their kin

The 62 species of Pelecaniformes include the pelicans, shags, gannets, frigatebirds, tropicbirds and boobies. Most Pelecaniformes are tropical or subtropical in distribution, so in New Zealand only the cooler-water birds – shags (cormorants) and gannets – are regularly present around the mainland. They are less pelagic than the Procellariiformes, and only the tropicbirds, some boobies and frigate birds commonly venture far from land. On mainland New Zealand, there are seven species of shag, with five further species on the Chathams and subantarctic islands. The Australasian gannet occurs around most of New Zealand, and the red-tailed tropicbird and masked booby breed on the Kermadec Islands.

The shags are long-necked, fish-eating birds that use their feet for propulsion under water. They are inshore birds, mostly found in shallow waters, typically making short dives to depths of less than 10 metres. All shags nest in colonies that, for the New Zealand species, usually comprise from a dozen to a few hundred pairs. New Zealand's shags can be divided into three groups. There are four species that utilise estuarine and freshwater habitats as well as sheltered marine waters. They also occur in other countries, but the remaining species are all endemic. The spotted shag and its sister-species, the Pitt Island shag, are widespread around the mainland and Chatham Islands respectively. The New Zealand king shag group contains six closely related species, all with restricted distributions in Cook Strait, southern New Zealand, Chatham, Bounty, Auckland and Campbell Islands.

The Australasian gannet breeds at 22 locations around New Zealand. Gannets and boobies catch fish by plunging, usually from heights of 10–15 metres. They rely on the momentum of the plunge to carry them under water, but once beneath the surface they have limited ability to swim.

Skuas, gulls and terns

The most familiar seabirds are gulls, which are generalists, feeding in inshore marine, estuarine, freshwater and terrestrial habitats and utilising a wide variety of foods. In New Zealand, there are three species, one of which, the black-billed gull, makes little use of marine habitats. Gulls exploit offal, carrion and rubbish, and with the much-increased availability of these foods gulls are probably now more common and less marine than ever before. Skuas are often associated with seabird colonies, where they prey on eggs, chicks and sometimes adult birds and carrion, but they also exploit a variety of marine foods. One species breeds on

Stewart, Chatham and the subantarctic islands, but four other species visit the seas around New Zealand. The Arctic skua is a common but often overlooked summer migrant.

Terns are lightweight, buoyant in flight and mostly feed by hovering several metres above the water, then dropping to the surface to catch fish or other small creatures. They rarely become completely submerged. Eleven species breed in the New Zealand region, including five wide-ranging tropical species on the Kermadec Islands. The black-fronted tern breeds on the South Island's braided rivers, obtains most of its food from rivers and farmland, and many move to the coast after breeding.

Whales, dolphins and porpoises

The cetaceans include some of the most wonderful of all mammals, whose special qualities have inspired novelists, poets, eco-groovers and scientists alike. The large whales impress through their immense size; the smaller dolphins by their playful bow-riding antics. These are some of the most highly specialised of all mammals. The teeth of baleen whales have been replaced by filter-feeding plates. Some toothed whales echo-locate with precision equal to that of bats, while the low-frequency calls of the large baleen whales enable them to communicate over vast distances. Several species are known only from a few dead animals stranded on beaches, and the ecology of even the common species is poorly understood.

The cetaceans range in size from New Zealand's Hector's dolphin, at 1.2 metres one of the world's smallest, to the 30-metre blue whale, which can weigh up to a hundred tonnes. The Mysticeti, or baleen whales, and the Odontoceti, toothed whales, are very different from one another and only distantly related.

The smallest of the Mysticeti is the five- to six-metre pygmy right whale, which weighs a mere 4.5 tonnes. The other baleen whales are much larger. These whales filter small crustaceans and fish using sieve-like plates of baleen that hang from the roof of the mouth. Most species of baleen whale are nearly globally distributed and most are migratory, feeding in polar and subpolar seas in summer and breeding in tropical or subtropical seas in winter. Eight of the 10 species occur in New Zealand waters.

The Odontoceti are a more varied group. Worldwide there are about 66 species, including the giant sperm whale, the killer whale, dolphins, porpoises and the little-known beaked whales. All have teeth, but the

The South Island's braided rivers, such as the Rakaia, support a unique community of wading and wetland birds.
Photograph by Kerry-Jayne Wilson

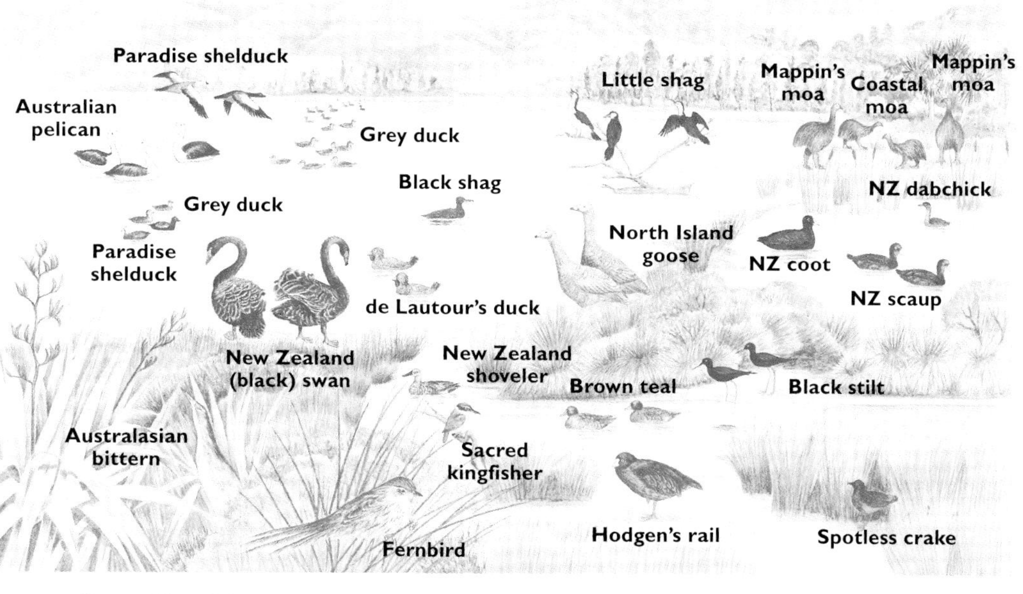

Key to Plate 3 (overleaf): Pauatahanui Inlet, near Wellington, during the Holocene, 2,000–10,000 years ago. This was one of the rare occasions when the Australian pelican, once considered part of New Zealand's wetland fauna but now thought to be a vagrant, was present.
Painting by Pauline Morse

Three seabirds: wandering albatross (above), black-bellied storm petrel (below) and Fiordland crested penguin (right). Note the huge wingspan of the albatross, which enables it to forage over vast distances, and the long legs and big feet of the storm petrel, which patters over the water surface searching for plankton.

Photographs by Kerry-Jayne Wilson (albatross and penguin) and John Marris (petrel)

number and form of these is enormously variable: some beaked whales have only two teeth, while certain dolphins have over 200. The many-toothed dolphins and porpoises generally eat fish, while the larger Odontoceti have fewer teeth and mostly eat squid. The killer whale is an exception: it preys on marine birds and mammals. The sperm whale is the largest of the toothed whales and can dive to over a thousand metres and remain under water for an hour and a half. Dolphins and baleen whales seldom dive below a hundred metres, although a bottlenose dolphin has been known to descend to 274 metres during a four-minute dive. Alan Baker's *Whales and dolphins of New Zealand and Australia* and Steve Dawson's *New Zealand whale and dolphin digest* offer useful introductions to our cetaceans.

The beaked whales

The family Ziphiidae is one of the least-known groups of mammals. There are 18 species of beaked whales, half of which have been recorded in New Zealand seas. They are seldom seen close inshore and most species are probably rare. Several species have never been seen alive, including Shepherd's beaked whale, which is only known from 13 stranded specimens, eight of which were found on New Zealand beaches. This species has two large teeth on the tip of the lower jaw and 20 or more simple peg-like teeth on each side of both the upper and lower jaws. Other beaked whales typically have only two or four teeth, and only on the lower jaw.

In some species the teeth grow to full size only after sexual maturity; or only males have teeth, or their teeth are larger than those of females. The size, shape and position of the teeth, together with the tooth-shaped scars common on adult males, suggest that in most beaked whales teeth are used in mating rituals or harem defence, not food gathering.

The most unusual teeth belong to males in the genus *Mesoplodon*. Their teeth protrude from the side of the mouth, the most extreme case being the strap-toothed whale, whose teeth may be up to 35 centimetres long and curve out and over the mouth to almost meet above the upper jaw.

Beaked whales mostly feed on squid, and some species are capable of diving to great depths and remaining under water for 30 minutes. Recent evidence suggests that some beaked whales use the expandable floor of their mouth to suck in prey like giant vacuum cleaners.[2]

Seals and sea lions

> The male, or sea-lion, is a large bulky animal of dark-brown colour and the old bulls have a thick mass of long hair, disposed mane-like over the neck and shoulders. The female, or sea-bear, is smaller and sleeker, and grey in colour. She is not so ferocious as the male, unless she has a young one by her side, when she defends it with true maternal instinct.
>
> – Edgar Waite, in Chilton (1909)

The seals belong to two groups: the true seals (phocids), which are most closely related to the mustelids; and the eared seals (otarids), which are more closely related to bears than to phocids. Phocids have a torpedo-shaped body with small foreflippers and hindflippers that are directed astern. The otarids have larger flippers, on land can turn their hind flippers under the body and are surprisingly agile. The elephant seal (a phocid), the New Zealand fur seal and Hooker's sea lion (otarids) breed in the New Zealand region. The most common of the three is the fur seal, which occurs around much of the mainland coast and on the Chatham and subantarctic islands. Currently the population is increasing and fur seals are recolonising many parts of the country from which they were exterminated,[3] first by Maori from northern and eastern regions, and then early in the nineteenth century by European sealers in southern New Zealand. This species also occurs in southern Australia.

Hooker's sea lions are endemic to the New Zealand region. About 90 per cent of them occur on the Auckland Islands, with smaller numbers on Campbell Island, the Snares Islands and in southern New Zealand. Elephant seals are scarce in New Zealand but more numerous on Macquarie Island.

The breeding biology of all three is similar. They breed in dense colonies on beaches, where males defend territories in which the females aggregate. Females mate about a week after giving birth to a single pup. Elephant seals suckle their pups for little over a month, but in that time pups grow from their birth weight of about 40 kilograms to 140–180. Fur seals may suckle their pups for up to a year; sea lions even longer.[4]

Ecology of marine birds and mammals

The oceans are difficult places for air-breathing birds and mammals to inhabit. One of the factors limiting their ability to catch food is the time they can remain under water and the depth to which they can dive. Only a few mammals can dive to great depths and remain under water for long periods. Sperm whales, elephant seals and some beaked whales make long and deep dives, but for other mammals and all birds dive times and depths are less impressive and they make their living in the topmost hundred metres of the sea. Many seabirds do not dive at all and can exploit only the topmost metre of the ocean.

For flighted birds, the adaptations required for diving and flight are so different that no bird can excel at both. There is a trade-off between the ability to dive and exploit food at depth, and ability to cover long distances in search of patchy resources. Penguins are far more proficient divers than flighted birds but cannot travel long distances as quickly. For this reason they are probably more influenced by local food supply. For instance, blue penguins at Motuara Island, in the Marlborough Sounds, made longer-duration foraging trips, and left their eggs unattended more often, than blue penguins at Oamaru, which suggested that food supply was more limiting for the Motuara birds.[5] Shags dive well but their flight is laboured, so they are restricted to inshore waters. Spotted shags and Stewart Island shags forage up to 15 kilometres offshore, but other species are found even closer inshore. Of the shags, king and Stewart Island shags are the most proficient divers, but even they dive to only 30 metres for, at most, two or three minutes.[6] Albatrosses, on the other hand, are capable of flying vast distances with minimal expenditure of energy. Wandering albatrosses breeding on the Auckland Islands can feed in mid-Tasman, a round trip of 5000 kilometres, returning at about 14-day intervals to feed their chicks.[7] Their long wings and light build are advantageous in flight, but on the water they bob like corks, so they can exploit only the uppermost few metres of the ocean.

Shearwaters are the consummate seabirds. Their long, thin, albatross-like wings enable them to travel fast and forage over huge distances, but their smaller size and dense, oily plumage enables them to dive. Sooty shearwaters can dive to depths of 60 metres[8] and are sufficiently manoeuvrable under water to pursue and catch fish or cephalopods.

About 70 per cent of the planet's surface is ocean, yet over most of this vast area productivity is low and food available to birds and mammals is patchy. Seabirds and seals must breed ashore, and suitable breeding sites may be far from places where food is abundant. The numbers of birds or seals breeding on islands or coastal cliffs strategically located near zones of high marine productivity can be astounding. For instance, close to three million pairs of sooty shearwaters, plus hundreds of thousands of other seabirds, nest on the Snares Islands (328 hectares).[9] The number of shearwaters alone is roughly equivalent to the total number of all seabirds that nest in the entire British Isles. More than 1.3 million pairs of seabirds nest on Rangatira Island (218 hectares) at the Chatham Islands.[10] The ground there is so honeycombed that people have to wear 'petrel-boards' – plywood squares – strapped to their boots to avoid collapsing the burrows used by breeding birds. On the Snares, Rangatira and North Brother Island in Cook Strait, I have found two petrel burrows per square metre in prime habitats.

The Bounty Islands probably have the greatest density of marine birds and mammals. This cluster of storm-swept rock stacks supports more than 250,000 pairs of seabirds plus about 16,000 fur seals.[11]

Whales and dolphins give birth to their young at sea, so they can breed wherever their food is found. The baleen whales adopt a different strategy, feeding in polar regions during summer then calving in the tropics during winter. Most do not feed during the winter: the costs of migration and lactation are met from fat laid down during the polar summer.

Primary productivity depends on two factors: light and nutrients. Nutrients that algae could utilise for growth are mostly contained in dead plants and animals, which, being heavier than water, accumulate on the sea bottom. Only where nutrients are brought close to the sea surface can significant primary production take place. Vertical mixing of water, important in bringing nutrients into the photic zone, is influenced by wind, boundaries between ocean currents, and eddies on the lee sides of islands or over submerged banks. Light penetrates most deeply in clear blue tropical seas because they are not murky with suspended organic material.

Warm surface waters float on top of cooler deeper waters and prevent the vertical mixing of waters that could bring nutrients into the photic zone. For these reasons marine birds and mammals are generally scarce in tropical seas. Nutrients are more likely to be returned to the photic zone in cool subantarctic waters, which are thus more productive and support far more birds and mammals. As most of the sea's primary production is by microscopic algae, marine animals large enough for mammals and birds to eat are themselves several steps up the food chain, so little of that limited marine productivity is available to birds and mammals.

The seas around New Zealand

The unusual diversity of marine birds and mammals is a result of both the latitudinal range that New Zealand spans and the complex bathymetry and hydrology of the surrounding seas. The subtropical Kermadec Islands are home to some tropical species, while subantarctic seas support not only species characteristic of that zone but are also visited by some truly Antarctic birds. The New Zealand land mass is the above-water portion of larger submerged plateaus that sit astride the Subtropical Convergence (Figures 8.1, 8.2, pages 226, 227).[12]

The at-sea distribution of marine birds and mammals is presumably influenced more by the distribution of their prey than by oceanographic features, but these do directly affect the occurrence of prey. John Warham[13] suggested that the main factors that influence the distribution of petrels (or their prey) were latitude, sea-surface temperature, currents, upwellings, salinity, nutrients, tidal streams over underwater banks, sea mounts and reefs, water depth and wind. Presumably these same factors influence other seabirds and mammals.

The continental shelf edge is an important feeding zone for a number of seabirds, and off southwestern New Zealand this is very close to land. The importance of this feature has not been well studied in the New Zealand region, but a few examples illustrate its significance. Westland petrels mostly forage over the shelf edge or the inner continental slope in waters 250–780 metres in depth, seldom feeding close inshore or in truly oceanic waters.[14] Buller's mollymawks mostly utilise continental slope waters when feeding off the coast of the South Island, but during incubation they use deep mid-Tasman waters.[15] Reasons for the switch were not studied. Several New Zealand breeding petrels visit the New South Wales coast, where grey-faced petrels occupy continental slope waters while black petrels and 'Cookileria' petrels (including black-wing and Cook's petrels) occur in waters deeper than 1400 metres.[16]

To the southeast, east and northwest of the South Island are extensive continental shelf plateaux. Marine birds and mammals are especially numerous over two of these, the Campbell Plateau and the Chatham Rise. The West Wind Drift pushes nutrients up on to these relatively shallow plateaux, where the strong westerly winds characteristic of these latitudes ensure thorough mixing, bringing nutrients into surface waters.

Two ocean currents, the Trade Wind Drift and Tasman Drift, bring warm subtropical waters to the northern part of the country.[17] By contrast, the cool West Wind Drift bathes the south and the subantarctic islands. The hydrology around New Zealand is complex, with many smaller currents originating where these main three are deflected around the country. The general pattern of surface-water circulation around the country is shown in Figure 8.2 (page 230).[18]

New Zealand sits astride the Subtropical Convergence, where warm subtropical seas and cool, less saline subantarctic seas meet. Many marine organisms are sensitive to changes in temperature or salinity, so

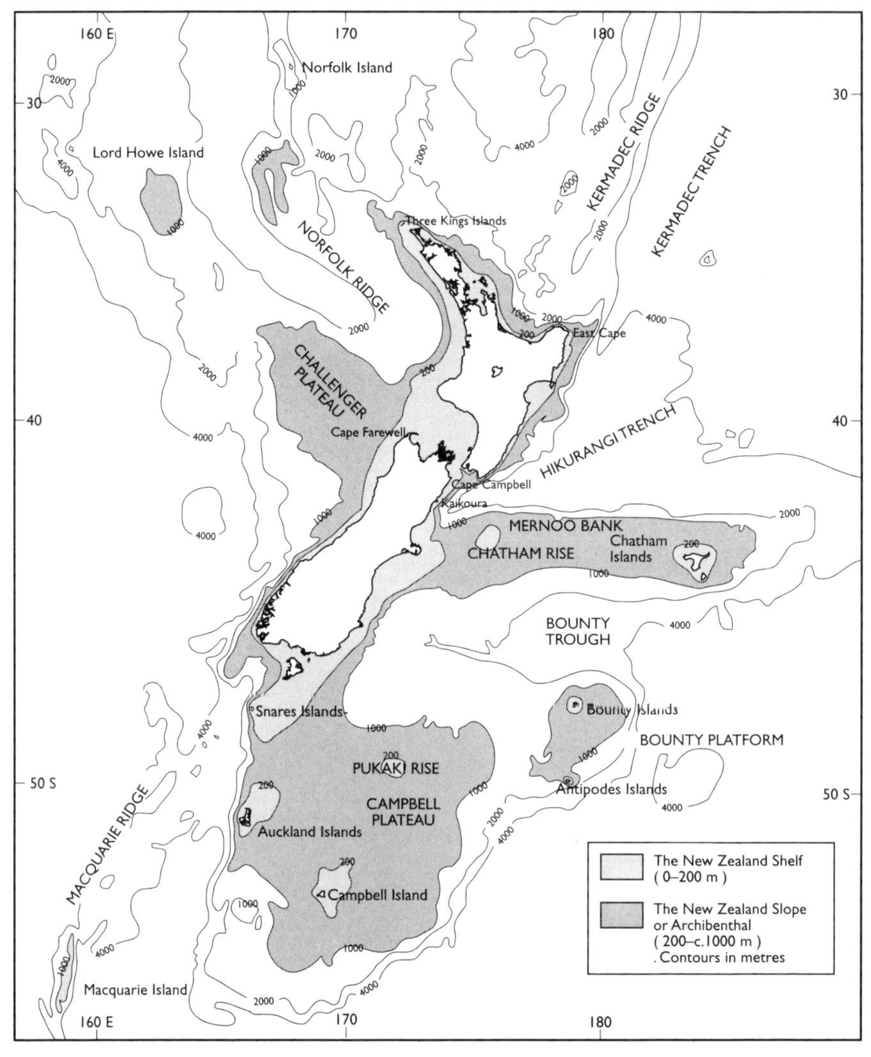

Figure 8.1 Bathymetry of the New Zealand region

Adapted from Knox 1975

the marine life present on either side of such a zone differs markedly. This in turn is reflected by the distribution of birds and mammals. Convergence zones, where water masses both meet and mingle, can be highly productive, and birds and mammals are sometimes abundant there. The Subtropical Convergence extends right around the world between latitudes 40–50° S, and New Zealand is the only land mass that intersects it. The latitude of this convergence varies from winter to summer and

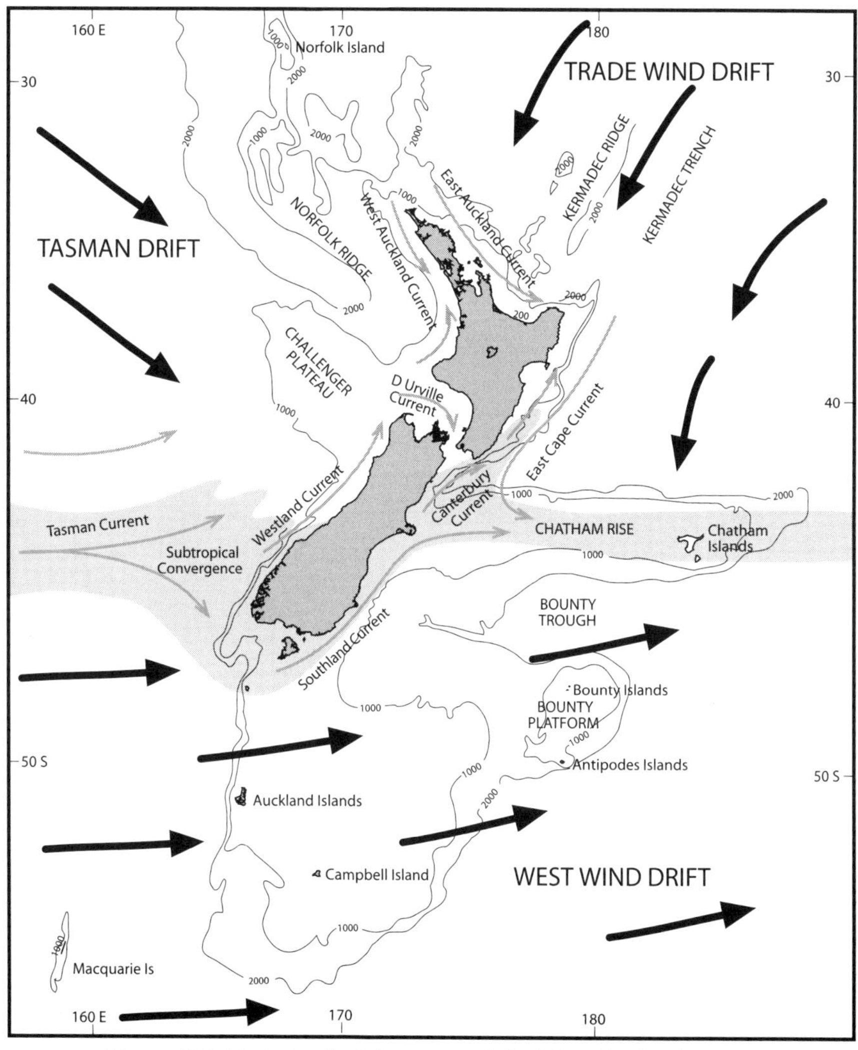

Figure 8.2 Hydrological features of the New Zealand region

Adapted from Knox 1975

year to year, but generally extends across the Tasman Sea at about 45° S. A zone of convergence waters extends around Fiordland and Stewart Island, up the South Island east coast and eastwards from Canterbury or Kaikoura toward the Chatham Islands.

The funnel-like Hikurangi Trench originates northeast of New Zealand. Water flows down this ever-narrowing trench, gathering nutrients from a vast area of ocean, then is deflected to the surface where the arms of the trench end near Cook Strait and Kaikoura.[19] Near Kaikoura, deep-water species such as sperm whales occur close inshore, along with a rich mixture of birds representing both inshore and deep-water species, as well as species from both subtropical and subantarctic waters.

It is impossible to relate the distribution and abundance of marine birds and mammals to oceanographic factors in any but the most general way. Insufficient is known about the distribution of the animals to know where exactly they are most common. Likewise, insufficient is known about oceanic productivity to know where food could be most abundant. Productivity in the New Zealand region is highly variable and patchy.[20] The subantarctic waters south of the convergence are more productive than the subtropical waters to the north. To the north, surface waters warm up and 'float' on top of the cooler, denser subsurface layers, preventing mixing and thus resulting in nutrient depletion of the upper layers where photosynthesis could occur. In general terms, primary productivity and zooplankton abundance increases from north to south, is higher to the east than to the west of New Zealand, and is generally higher inshore than offshore. There are areas with high primary productivity near the Challenger Plateau, off East Cape, around the Chatham Islands, in the Mernoo Gap, over Pukaki and Campbell Rises and directly north of New Zealand.

Upwellings occur at various places around the country, such as off Westland, Northland, Kaikoura, Cape Campbell, Cape Farewell, Three Kings Islands, East Cape, over the Mernoo Gap and Wanganella Bank, and in Cook Strait.[21] Plankton production and zooplankton abundance is enhanced in some of these upwellings, and Janet Grieve (née Bradford) and co-workers suggested that upwellings, in particular those north of the subtropical convergence, appreciably increase productivity around New Zealand. Upwelling is less important in well-mixed subantarctic waters. While all trophic levels probably benefit from upwellings, both nutrients and plankton drift down-current and by the time the nutrients made available through upwelling are incorporated into the trophic levels exploited by birds or mammals, they may have travelled far from the point of upwelling. For example, a tongue of nutrient-rich water extends down-current from the upwelling off Cape Farewell.[22] Zooplankton abundance is high immediately north of New Zealand, near the Three Kings, southeast of East Cape, over the Mernoo Gap, the Bounty Trough, southwest of Fiordland, west of the Challenger Plateau and in Cook Strait. In winter, grey petrels frequent the shelf waters off East Cape, where upwelling and localised zones of high primary productivity and zooplankton have been reported.[23]

The distribution of certain dolphins appears to be correlated with sea-surface temperatures. David Gaskin[24] found that common dolphins occur in waters north of the Subtropical Convergence; dusky dolphins mostly near it and southern right whale dolphins south of it. Hourglass dolphins are found in cold waters on either side of the Antarctic Convergence.

Sea-surface temperature is an important factor in determining the distribution of petrels. For instance, in the North Pacific during the northern hemisphere summer, New Zealand-breeding mottled petrels and sooty shearwaters are found in cold waters; flesh-footed and Buller's shearwaters in cool waters; and Cook's, black-wing and Pycroft's petrels in warmer waters.[25] This only roughly corresponds with their distribution around New Zealand; for example, in the North Pacific, flesh-footed shearwaters are found in waters as cold as 7°C, whereas around New Zealand they are found north of the subtropical convergence, where the temperature is at least 14C° during the months the birds are in the region.

The distribution of fur seals around mainland New Zealand appears to be influenced by cooler, more productive waters. Most of the seals are found south of Otago and South Westland in areas influenced by the Subtropical Convergence. Further north there are breeding colonies at Banks Peninsula, Kaikoura, Cape Foulwind, Cape Farewell and Cook Strait, all close to upwellings.[26] After the breeding season most males migrate to overwinter in areas such as Kaikoura, Cook Strait, Cape Farewell, Cape Egmont and the Three Kings Islands, all close to zones of upwelling. During the breeding season females forage over the continental shelf close to their breeding colony, but once their pups are larger the females feed in deeper waters up to 220 kilometres from their colony.[27]

There is little information on how currents, fronts, upwellings, salinity and water depth influence the distribution of New Zealand seabirds. John Warham[28] noted that in seas off Chile wandering and royal albatrosses and sooty shearwaters prefer more saline waters, black-browed albatrosses occur over a broad range of salinities, and the distribution of white-chinned petrels is influenced by temperature rather than salinity. Whether this reflects the distributions of the same species around New Zealand is unknown.

Migratory seabirds

Many seabirds, especially petrels, disperse widely, often travelling thousands of kilometres from their breeding base, but in this section I am concerned only with truly migratory species, where all or most of the individuals travel from that part of the world where they breed to distant geographically discrete non-breeding areas. At least eight species of petrels breed in New Zealand and winter in the North Pacific. Three of them – wedge-tailed shearwater, black-winged petrel and Kermadec petrel – breed in northern New Zealand and winter in the tropical North Pacific. Cook's petrels winter in the temperate North Pacific, but the remaining four species – sooty, Buller's and flesh-footed shearwaters and mottled petrel – spend the northern summer in Arctic or subarctic waters.[29] Sooty shearwaters and mottled petrels probably range further toward both poles than the Arctic tern, which is often said to be the bird that migrates furthest. Both these petrels range south into Antarctic waters and north into the Bering Sea, with mottled petrels extending further south and further north than the shearwaters. It is assumed that the mottled petrels seen in Antarctic waters include birds nesting on the Snares Islands and southern New Zealand. Incubation spells for this species are about two weeks in duration, ample time for the off-duty bird to visit Antarctic waters.[30]

Buller's, flesh-footed and sooty shearwaters and mottled petrels all finish breeding in April or May, then rapidly fly north to reach the North Pacific about a month later. Non-breeding birds lead the way and fledglings depart weeks after their parents (Figure 8.3, pages 231–32). The route followed is not known with any precision, but sooty shearwaters are first seen off Japan in May, then in Alaskan waters by June. Numbers off California peak in August and September, as birds begin the return journey to New Zealand. Their Pacific circuit is not this simple. Some sooty shearwaters overwinter near Japan, and others, possibly South American-breeding birds, are seen off California in May and June.[31]

Figure 8.3 ▸ The inferred migration routes taken by sooty shearwater, Hutton's shearwater and white-faced storm petrel (page 232)

Based on information in *HANZAB*

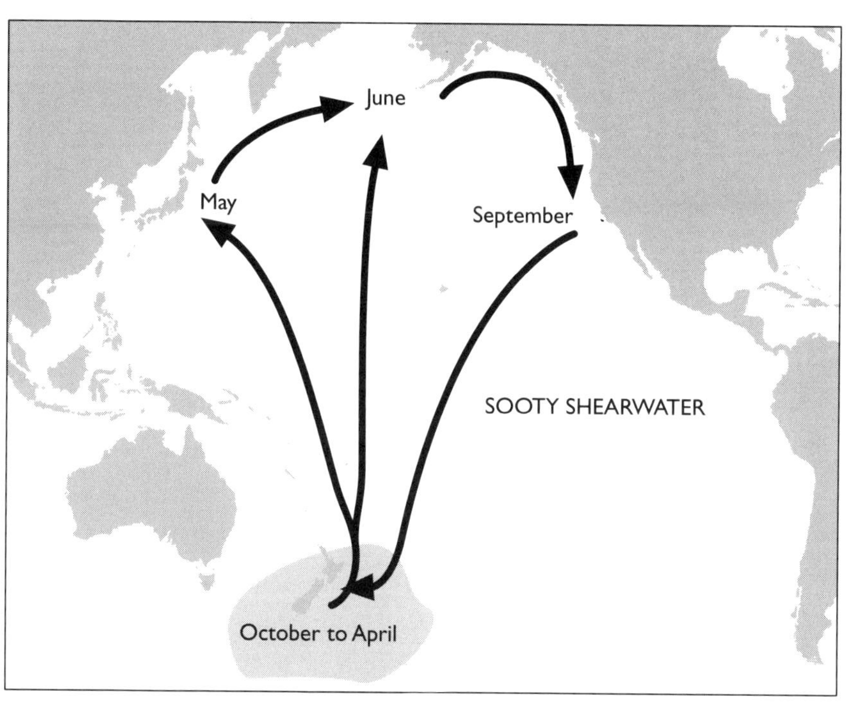
June
May
September
SOOTY SHEARWATER
October to April

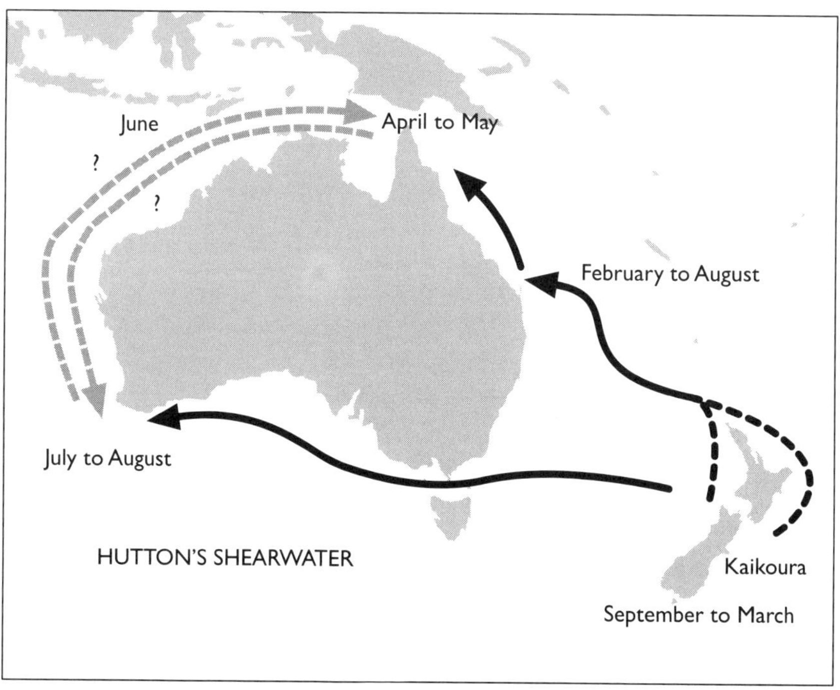
June
April to May
?
?
February to August
July to August
HUTTON'S SHEARWATER
Kaikoura
September to March

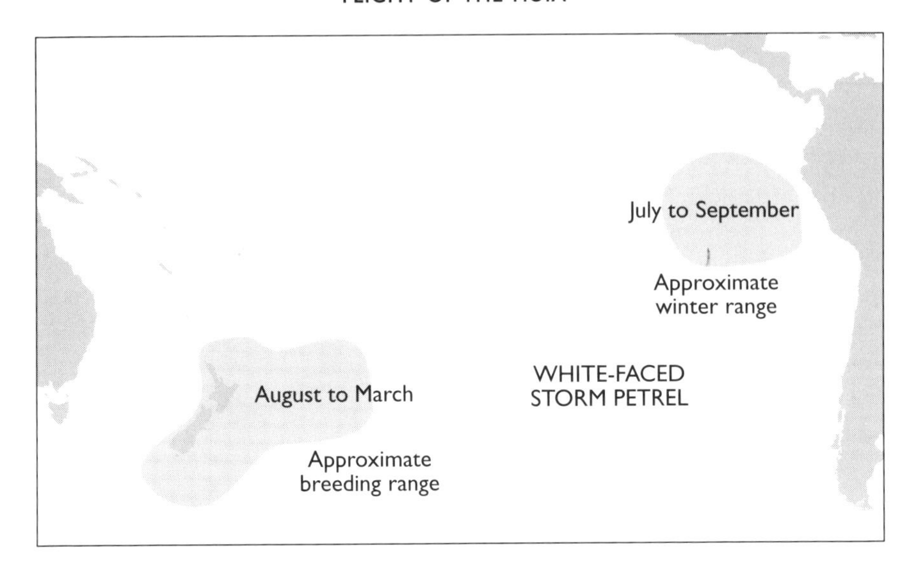

New Zealand-breeding black petrels and white-faced storm-petrels overwinter in the eastern tropical Pacific, some as far east as the Galapagos Islands.[32] Most Hutton's shearwaters spend the non-breeding months in Australian seas, perhaps circumnavigating that continent. After breeding has finished, many New Zealand-breeding gannets migrate to eastern Australia while others apparently remain in New Zealand waters.

Foraging ecology of albatrosses

During incubation albatrosses and mollymawks (and also other petrels) remain on the nest for days or even weeks while their partner is foraging at sea. While rearing their chick, both parents may travel huge distances on foraging trips that last a week or more. Until miniature transmitters were developed that could transmit position and other data to satellites for downloading to biologists' computers, we knew virtually nothing about the movements of these birds. Ship-based observations suggested that the various species had different foraging habits, but because individuals could not be followed, it was impossible to interpret the little information available. During the last decade, however, several New Zealand-breeding species have been tracked, both during and after the breeding season.

Three species of great albatrosses breed exclusively in the New Zealand region. Both species of royal albatross mostly feed in waters less than 500 metres deep, although birds breeding at Taiaroa Head also frequent waters 500–1000 metres in depth.[33] Wandering albatrosses generally feed in deeper waters.[34]

During incubation, two female wandering albatross breeding at the Auckland Islands rapidly flew north along the continental slope west of the Snares Islands and Fiordland, then flew westwards into the Tasman Sea to feed in deep mid-Tasman waters (Figure 8.4, page 234). Each trip ended with a rapid flight home.[35] These trips were of 11–13 days duration and covered 3950–5600 kilometres. A bird tracked in April while it was raising a chick, fed along the continental slope west of the Auckland Islands and off New Zealand in waters at least 1000 metres deep. Another bird from this colony was captured northeast of Sydney (3300 kilometres away) while it had a dependent chick at home.[36]

Buller's mollymawk is the best studied of New Zealand's smaller albatrosses.[37] While incubating their egg, both sexes make trips of 9–13 days feeding either in the Tasman Sea, some reaching Tasmanian waters or in continental slope waters off the South Island east coast. During the three weeks immediately following hatching, when one or other parent was at the nest brooding the chick, feeding trips were short and the adults foraged within a few hundred kilometres of the breeding islands. Once the guard stage was over, they alternated between long trips along either coast of the South Island, and short local trips. Late in chick-rearing, males made more short trips while females made trips of longer duration, foraging further from the breeding islands.

Several species of albatross leave New Zealand after breeding and spend the time between breeding seasons off South America. Northern royal albatrosses, which breed in alternate years, move quickly between New Zealand and South America, traversing about 10 degrees of longitude each day. Both legs of their circumpolar migration are from west to east, so the birds circle the southern hemisphere between breeding years.[38]

Chatham mollymawks also migrate to South America between breeding seasons, but unlike the royal albatrosses, which utilise shallow waters to the east of Patagonia, they forage in deeper waters off Chile and Peru[39] and take a more northerly trans-Pacific route when returning to the Chathams. One pair of Buller's mollymawks breeding on the Solander Islands departed on their migration early owing to the death of their chick, and were tracked as they crossed the Pacific to South America.[40] Both birds followed the Chatham Rise but, once east of the Chatham Islands, increased speed and were off the coast of Chile nine and 10 days after leaving Chatham waters. Most of the distance was achieved during short periods when the birds were caught up in low-pressure weather systems.

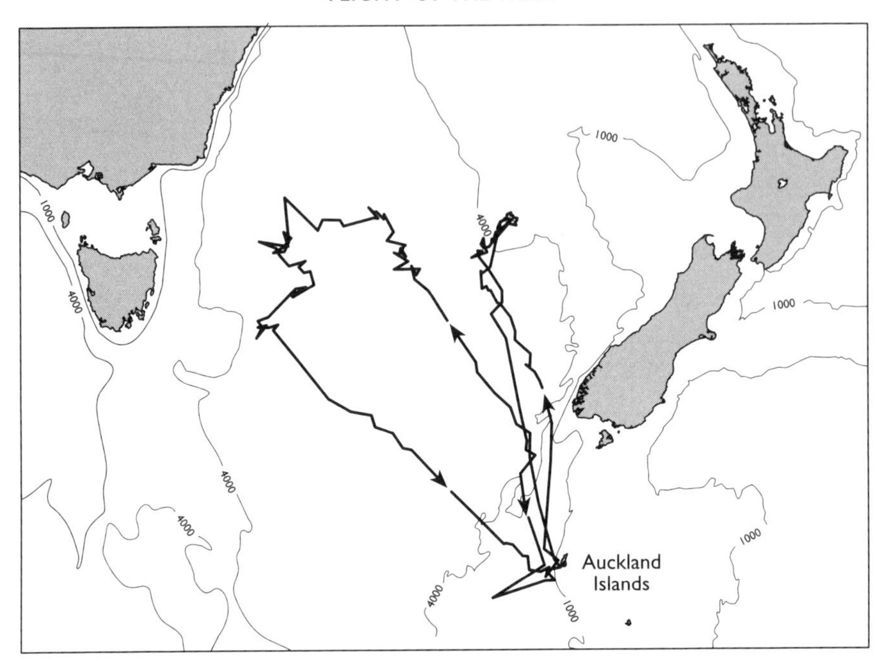

Figure 8.4 Foraging trips made by an Auckland Island-breeding wandering albatross (above) and Buller's mollymawks (below)

Adapted from Walker et al. 1995, Sagar and Weimerskirch 1996

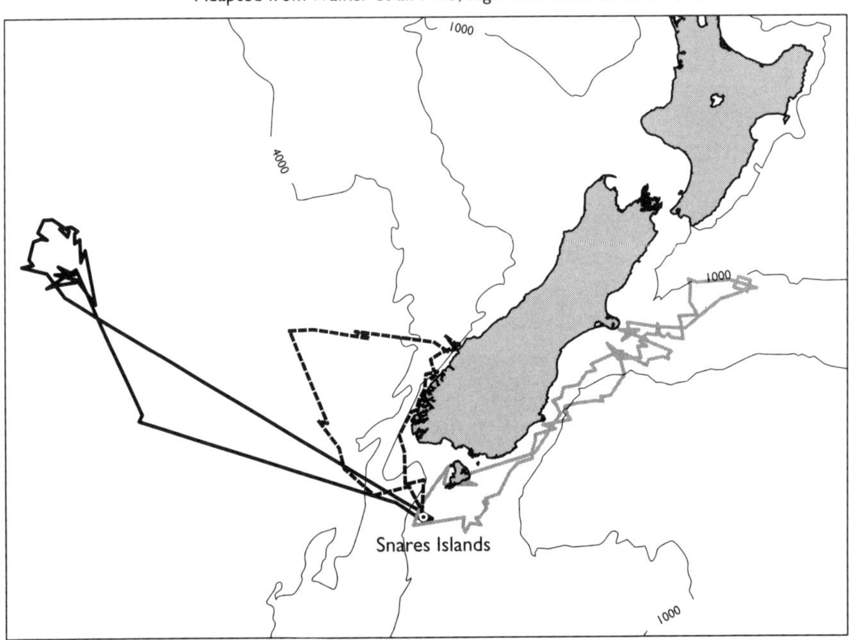

A community of seabirds

There have been few studies of the foods and foraging ecology of marine birds and mammals in New Zealand waters, so we know little about how they divide the food resources among them. In this section I will list the seabirds commonly found in the eastern Cook Strait region, then describe the foods they eat and their feeding methods.[41]

Inshore, the cool Southland Current – a mixture of subantarctic and subtropical waters[42] – influences the eastern Cook Strait region (between Kaikoura and the Wairarapa coast). To the north and east the region is bounded by the warm East Cape Current, and to the west the turbulent, mixed waters of the Cook Strait narrow. In the eastern entrance to Cook Strait and near Kaikoura, cold, nutrient-rich water upwells from deep-sea trenches. Because this is within the subtropical convergence zone, birds characteristic of both northern (for example, Buller's shearwater) and southern New Zealand (for example, some alba-trosses) are present in this region. Tables 8.1 and 8.2 (pages 236–39) list the bird species most commonly present, the feeding methods they use and the foods they eat.

Black-backed and red-billed gulls are common along the coast but seldom venture more than a few kilometres offshore. Gulls are opportunists that take offal and refuse from boats. At sea, black-backed gulls take small fish and invertebrates, while red-bills mostly feed close inshore on krill and other small animals. Gulls are the only New Zealand seabirds that also feed ashore.

The most aerial of the seabirds are white-fronted terns and Arctic skuas. The usually gregarious terns are rarely seen more than a few kilometres offshore. Their flight is graceful and buoyant; they hover several metres above the water, dropping down to catch small surface-schooling fish. During summer they are often harried by Arctic skuas. These strongly flighted, agile birds force the terns to drop their food, which is then eaten by the piratical skuas. Two other species also feed mostly in flight. White-faced storm petrels peck individual planktonic crustaceans from the water surface while using their wings and feet to patter across the water. Fairy prions can take crustaceans while flying close to the sea surface but, unlike the storm petrels, their feet seldom make contact with the water. They also seize prey while floating on the water surface.

Many petrels scavenge around fishing vessels. In the Cook Strait region, Cape pigeons, giant petrels, six types of albatross, Westland petrels and flesh-footed shearwaters commonly follow fishing vessels and

	DIVING		PLUNGING			
	Pursuit	Surface	Surface	Shallow	Deep	Pursuit plunge
Arctic skua						
White-fronted tern			✓✓✓	✓✓		
White-faced storm petrel						
Fairy prion			✓			
Red-billed gull		✓	✓✓	✓✓		
Black-backed gull			✓	✓✓		
Cape pigeon		✓✓	✓✓			✓
Giant petrel		✓				✓
Wandering albatross				✓		✓
Royal albatross				✓		
Black-browed mollymawk		✓	✓			✓
Shy mollymawk			✓			✓
Salvin's mollymawk			✓			✓
Buller's mollymawk		✓	✓			
Westland petrel		✓✓✓				✓
Sooty shearwater	✓✓	✓		✓✓		✓✓
Flesh-footed shearwater	✓		✓✓			✓✓✓
Buller's shearwater		✓	✓			✓
Fluttering shearwater	✓✓	✓✓				✓
Hutton's shearwater	✓✓	✓✓				✓✓
Gannet			✓	✓	✓✓✓	
Diving petrel	✓✓✓	✓		✓		✓✓✓
Blue penguin	✓✓✓					
Spotted shag	✓✓✓					

Table 8.1 Feeding methods used by seabirds of the eastern Cook Strait region

Adapted from Bartle 1974, Harper 1983, Harper et al. 1985, M. Imber unpublished observations and *HANZAB*

SURFACE FEEDING				FLIGHT FEEDING		
Seize	Filter	Foot paddling	Scavenge	Dipping	Pattering	Piracy
✓			✓			✓✓✓
				✓		
✓				✓✓	✓✓✓	
✓✓✓				✓✓✓	✓	
✓✓		✓	✓✓	✓✓		✓
✓✓✓			✓✓✓			
✓✓	✓✓	✓✓	✓✓	✓	✓✓	
✓✓✓			✓✓✓			
✓✓✓			✓✓			
✓✓✓			✓✓✓			
✓✓✓			✓✓			
✓✓✓			✓✓✓			
✓✓✓			✓✓✓			
✓✓✓			✓			
✓✓✓			✓✓			
✓						
✓			✓✓			
✓✓✓	✓			✓✓		
✓						
✓						

✓ Feeding method occasionally used
✓✓ Feeding method often used
✓✓✓ Primary feeding method/s

	FOODS				
	Cephalopods and other molluscs	Fish	Crustaceans	Fish waste	Carrion seabirds
Arctic skua	✓	✓✓✓	✓		✓✓
White-fronted tern		✓✓✓	✓		
White-faced storm petrel		✓	✓✓✓		
Fairy prion		✓	✓✓✓		
Red-billed gull	✓✓	✓	✓✓✓	✓✓	✓✓
Black-backed gull	✓✓	✓✓✓	✓✓	✓✓	✓✓✓
Cape pigeon	✓✓	✓	✓✓✓	✓✓✓	✓
Giant petrel	✓✓	✓✓	✓✓	✓✓	✓✓✓
Wandering albatross	✓✓✓	✓✓	✓	✓✓	
Royal albatross	✓✓✓	✓✓	✓	✓✓	
Black-browed mollymawk	✓✓	✓✓✓	✓✓	✓✓✓	
Shy mollymawk	✓✓✓	✓✓	✓	✓✓✓	
Salvin's mollymawk	✓✓✓	✓✓	✓	✓✓✓	
Buller's mollymawk	✓✓✓	✓✓	✓✓	✓	
Westland petrel	✓✓✓	✓✓✓	✓	✓✓	
Sooty shearwater	✓✓	✓	✓✓✓		
Flesh-footed shearwater	✓✓✓	✓✓	✓	✓	
Buller's shearwater	✓	✓✓	✓✓✓		
Fluttering shearwater		✓✓✓	✓✓		
Hutton's shearwater		✓✓✓	✓✓		
Gannet	✓	✓✓✓			
Diving petrel			✓✓✓		
Blue penguin	✓✓	✓✓✓	✓		
Spotted shag		✓✓	✓✓✓		

Table 8.2 Foods, distribution, abundance and body size of seabirds of the eastern Cook Strait region

Adapted from Bartle 1974, Harper 1983, M. ImberM. Imber unpublished observations and *HANZAB*

Offshore Inshore	Seasonality	Body size (kg)	Numbers
I	M	0.4	●
I	S	0.13	●●●
O	M	0.05	●
I	R	0.13	●●●
I	R	0.3	●●●
I	R	1.0	●●●
OI	S	0.45	●●●
O	S	4.5	●
O	S	6-11	●
O	S	8-10	●●
O	S	3-5	●
OI	S	3-4	●●●
OI	S	3-4	●●
O	S	3	●
O	S	1.2	●●
O	M	0.8	●●●
O	M	0.6	●●●
IO	M	0.4	●●●
I	S	0.3	●●●
IO	M	0.35	●●
I	S	2.3	●●
I	R	0.13	●●●
I	R	1.1	●●
I	R	1.1	●●

✓ Food type occasionally eaten
✓✓ Food type often eaten
✓✓✓ Primary food type/s

R A resident species present in the eastern Cook Strait region year-round
S Present in the New Zealand region year-round but seasonal changes in their abundance in the eastern Cook Strait area
M Migratory

scavenge waste thrown overboard. The most abundant of the albatrosses, and the one most frequently associated with fishing vessels, is the inappropriately named shy mollymawk. The natural foods of albatrosses consist mostly of fish and cephalopods. Their primary feeding method is surface-seizing (catching prey while floating on the water surface), although they occasionally drop on to the water or even briefly submerge to catch prey observed while in flight. Several species of albatross can be seen virtually anywhere around New Zealand, but we know little about how the different species co-exist. Royal and wandering albatrosses are

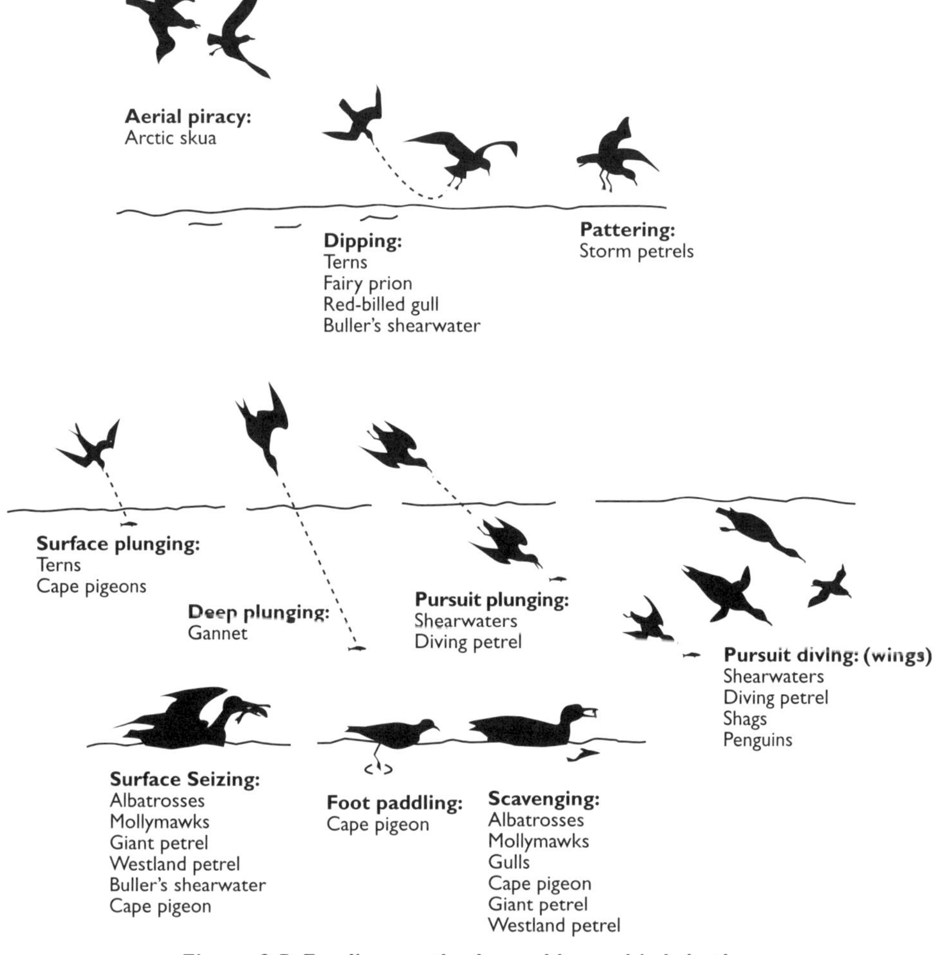

Figure 8.5 Feeding methods used by seabirds in the eastern Cook Strait region

Adapted from Ashmole 1971 and the feeding methods listed in Table 8.1

much larger than the medium-sized shy and Salvin's mollymawks, which in turn are larger than the black-browed and Buller's mollymawks. Larger birds usurp their smaller kin when scavenging around fishing vessels, but how size influences their natural prey is unknown.

The giant petrels are similar in size to mollymawks and among the most opportunistic of the petrels. They and the smaller Cape pigeon scavenge offal and refuse thrown overboard from vessels, and in the past fed on waste from abattoirs, whaling stations and even sewer outfalls. Male giant petrels prey on seabirds and scavenge dead birds or mammals. Females more often feed on crustaceans, squid and fish taken while floating on the sea surface. Of the smaller petrels, Cape pigeons are the most flexible in the ways they catch food and the foods they eat. They mostly feed from the water surface, seizing larger items or filtering small prey, but from time to time they use most other feeding methods.

All of the petrels use several feeding methods, while none dives to great depths; many species feed at night when prey species that live at depth by day migrate to surface waters.[43] The largest albatrosses rely mostly on surface-seizing; the shearwaters mostly on pursuit-plunging, while the tiny diving petrel is the most agile beneath the water. Just how the various species partition the food resources is unknown. There appears to be a great deal of overlap in the diets of seabirds, and mixed-species feeding flocks often appear to be exploiting the same foods. Competition is lessened by differences in the feeding methods used, and more detailed studies may show dietary overlap to be more apparent than real.

Tables 8.1 and 8.2 suggest that the six species of shearwater common in the eastern Cook Strait region exhibit extensive overlap in both diet and the methods used to catch those foods. In size they vary from the 0.3-kilogram fluttering shearwater to the 1.2-kilogram Westland petrel. Buller's shearwaters are mostly seen over the continental shelf.[44] This species is very different from the other shearwaters: it eats more crustaceans, it seldom dives and is only half the weight of the similar-sized diving sooty shearwater. The Westland petrel also catches much of its prey by surface-seizing and, while these birds dive more often than Buller's shearwater, their dives are mostly made from the water surface and are shallower than those of other shearwaters. The remaining four are typical shearwaters, relying mostly on pursuit-plunging. How deep they dive is not well known. Sooty shearwaters regularly dive to 30 metres, and the deepest dive recorded was 67 metres.[45] Fish is the main prey for most species; the larger species appear to take more cephalopods, and the smaller species more crustaceans. During autumn in southern New Zealand, sooty shearwaters took crustaceans more often than they took squid, and fish made up a relatively minor component of their diet, but the foods consumed varied from year to year.[46] Some Westland petrels with dependent chicks in burrows at Punakaiki forage in Cook Strait or south to Kaikoura and the Mernoo Bank, 800 kilometres as the petrel flies from its breeding colony.[47]

Hutton's and fluttering shearwaters are so similar that it is difficult to tell them apart. They are both common in this region, so how do they co-exist? The two species overlap in size, but on average the migratory Hutton's shearwater has longer wings than the non-migratory fluttering shearwater.[48] Both are commonly seen inshore, but fluttering shearwaters often enter harbours and bays whereas, when close to land, Hutton's are usually found on exposed coasts. They generally feed further offshore and range further afield than fluttering shearwaters. Fluttering shearwaters apparently dive to greater depths than do Hutton's.

In addition to the diving petrel and shearwaters, three other species also find their food beneath the water surface. Gannets plunge from heights of up to 20 metres, the momentum of their dive taking them beneath the water. Unlike petrels, they have little manoeuvrability under water and rely on the height and angle of their plunge to catch prey spotted while in flight. In northern New Zealand, gannets feed mostly in inshore waters, eating fish 10–20 centimetres long.[49] Pilchards, anchovies and jack mackerel were the main species taken.

The blue penguin and spotted shag both dive from the water surface in search of fish or cephalopods, the penguin using its wings, the shag its feet, for propulsion.

Seabird islands

> . . . the hours of darkness are filled with an indescribable medley of sound poured forth from the throats of all the petrels. Peace returns at dawn, when the birds leave the island and fly out to sea.
>
> – Basil Howard (1940)

Seabird islands are strange places. The vegetation tends to be simple, with just a few hardy plant species present. The ground may be so bare that it appears as though the leaf litter has been neatly swept away. Trees have often toppled over, their roots undermined by numerous burrows. By day the islands are strangely silent, with the birds either at sea or dozing in their burrows. After dusk the island comes alive. The returning birds plummet through the canopy and their weird cacophony of wailings continues through the night until silence returns with the coming of dawn.

Such dense aggregations of birds breeding cheek by jowl greatly influence the nature of the soils and vegetation. Only a few hardy plants can withstand the trampling of thousands of little feet and the constant cropping of seedlings for lining burrows. Where the island is forested, the canopy is usually dominated by just a few species. On the southern islands, muttonbird scrub or tree daisy forests are characteristic. Tree daisies are shallowly rooted and during gales on densely burrowed islands the ground can heave and whole trees sway.

Further north, rata or pohutukawa are common. The gnarled trunks of these trees, many of which are partially toppled over, have regrown upwards and are used each morning by the smaller, more agile petrels as take-off stages. The larger shearwaters must nest close to clifftops or hillside clearings, or walk sometimes hundreds of metres to the nearest take-off rock. Other petrel colonies are in scrub or grassland. On some subantarctic islands, tussocks form tall pedestals under which birds burrow and between which are highways worn bare by their comings and goings. On these islands, the tall tussocks can hide the birds from the predatory skuas that inhabit many southern islands.[50]

Seabirds can inhibit plant regeneration, induce erosion, kill or bury plants, and damage roots. Their burrows can allow friable soils to dry out, or peat soils to become sodden. The birds transfer nutrients from sea to shore, thus increasing soil fertility. Vegetation growth may be lush, but nitrogen and phosphorus levels can become so high that just a few coprophilous (dung-loving) plants predominate. Seabirds previously nested at many locations on the New Zealand mainland, mostly in coastal locations, although some colonies were as far inland as the Canterbury foothills and Lake Waikaremoana. Their role in transferring nutrients from sea to land and in maintaining soil fertility was very important: even now, some centuries after petrels became locally extinct, soil nitrogen may still be derived predominantly from seabirds and phosphorus and cadmium levels may still be high.[51] Cadmium levels in sites where seabirds once nested may match those achieved through fertiliser application.

Gannets and shags can be even more destructive to vegetation. When over 3000 petrels nested on tiny (0.2-hectare) Whero Island, an outlier of Stewart Island, it was covered by tree daisy scrub and grass. Whero was later colonised by Stewart Island shags, which usurped the petrels and virtually denuded the island.[52] Other shags often kill the trees they nest in. Closely packed gannet colonies can be denuded of vegetation during the breeding season, and the few nitrogen-loving annual plants that establish while the birds are absent die when the gannets return the following summer.

Large invertebrates and reptiles are unusually abundant on seabird islands, their populations a product of nutrient enrichment. Certain seabird islands are important refuges for some threatened reptiles and invertebrates (see Chapter 2), and Fiordland skinks are most common on islands with fur seal colonies.[53]

Motunau Island, a story of rabbits, weeds and seabirds

Rabbits were introduced to Motunau Island, off Canterbury, in the 1850s. Until 1935, the vegetation was periodically burned and grass and clover sown to encourage the rabbits, yet the island still remained an important nesting station for thousands of seabirds. Between 1958 and 1962, the rabbits were exterminated and the distribution and abundance of both seabirds and vegetation was mapped.[54]

It was assumed that once the rabbits were removed, native plants would gradually re-establish, but they did not. Today, rank grass and introduced weeds are edging out the few native plants that remain. Fairy prions become impaled on boxthorn plants, and petrels do not nest beneath boxthorn thickets, but otherwise weed invasion has had minimal impact on birds. The distribution of the penguins and prions has changed, but their total numbers seem little affected.[55] Skinks and sooty shearwaters slightly appear more common, and white-faced storm petrels possibly less numerous, than before. Boxthorn has been removed, but is control of other weeds warranted? Restoration of Motunau would be difficult and expensive. It is a refuge to protect seabirds and lizards, so if they are little affected by weeds, why spend scarce conservation resources on weed control? Condoning weeds in a nature reserve does seem anathema, but it may be the most pragmatic management option.

Muttonbirding

> . . . the descendants of those nominated in the Deed of Sale as sole proprietors of the mutton-birding rights, betake themselves to the islets with their wives and families, provisions, appliances and camping gear, and settle down to six weeks of busy labour.
>
> – Basil Howard (1940)

Muttonbirding, the traditional Maori harvest of fledgling shearwaters and other petrels, was once practised in most parts of the country and most species with accessible nesting grounds were probably harvested. Buller's shearwaters on the Poor Knights, Hutton's shearwaters near Kaikoura, Westland petrels at Punakaiki and the now-endangered Chatham Island taiko were all subject to past harvest. Muttonbirds were taken from islands in the Marlborough Sounds up until 1960.[56] Muttonbirding was not restricted to coastal locations: early last century, petrels breeding inland at Lake Waikaremoana were taken.[57] Today, only sooty shearwaters and grey-faced petrels may be legally taken, and only from few specified islands. The harvest appears to be sustainable and it is possible that some other common species could also withstand exploitation.

The best known example of muttonbirding is that of titi, or sooty shearwaters, from islands around Stewart Island.[58] Muttonbirding is an important tradition to the southern Ngai Tahu people. Athol Anderson suggested that it has a long history and that the socio-economic structure of the southern tribes was based on harvest and exchange of titi. He calculated that in the 1840s the annual take of 250,000 birds was probably worth more to Ngai Tahu than the £5000 they realised for the sale of Otago and Canterbury to Pakeha settlers. Others have suggested that the large-scale titi harvest is a relatively recent development.[59] Titi were an excellent food source: the shearwaters nested in huge numbers and produced a seasonally reliable, readily preserved, fat-rich food just as the cold southern winter set in. Titi remained an important economic commodity through most of last century, but, while the titi harvest is still a commercial venture, its cultural importance now outweighs its economic value. Although people now use helicopters and diesel-powered boats to reach the islands, once they are ashore the harvest is conducted in a largely traditional way.[60]

When Stewart Island and its adjacent islands were purchased by the Crown in 1864, the Deed of Cession guaranteed the right to take titi to those beneficiaries with existing rights plus their descendants, male or female. The right to take birds is jealously guarded and the islands are visited only by those with muttonbirding rights or their spouses. Only fledglings are taken, and are removed from their burrows as they complete development (nanao) or caught in the open as they prepare to leave the island (rama). Titi were traditionally preserved in their own fat, then stored in bags made from bull kelp. Today, most are salted and packed in plastic or metal containers. No adults are killed and care is taken to protect the burrows in which the birds breed.

Sooty shearwater chicks fledge about a week after their parents have departed on their migration to the North Pacific (Figure 8.6, opposite).[61] When the chicks leave their natal burrows and first go to sea, it is early winter and they must fend for themselves in oceans they have never before seen. Mortality of any petrel during those first weeks after fledgling is high, so muttonbirding in effect replaces high natural mortality with human harvest. Sustainable harvest is usually possible if animals are hunted just prior to a stage in their life history when mortality is naturally high. There are no limits on the number of chicks that can be taken. The topography of the islands and the number of burrows that are too long or inaccessible seems to ensure that enough chicks fledge to maintain the population.

The annual catch is not reliably documented but may still be up to 300,000 birds. Poutama is the only one of over 30 islands for which harvest figures are available. There, during the 1990s, about 20,000 chicks, a fifth of the total raised, were harvested each year.[62] The daily catch on Poutama has declined since 1990, but because sooty shearwater numbers have also declined elsewhere, overharvest is unlikely to be the cause.[63]

Conservation

Marine birds and mammals pose very different conservation challenges than do their land-dwelling counterparts. The well-proven methods used to save terrestrial species such as black robins, kokako or takahe (see Chapter 9) are largely inappropriate to the conservation of taiko, sea lions or Hector's dolphins. Indeed, when it comes to conservation, seals and penguins have more in common than penguins and passerines. Table 8.3 (page 248) highlights some important differences between marine and terrestrial species that are pertinent to conservation. While a large proportion of the terrestrial and wetland birds are now extinct, only five seabirds have become extinct during the last 2000 years. However, about 14 species of petrels and the brown skua have become extinct on the New Zealand mainland, while others have gone from the larger islands of the outlying island groups. Many species – for example, the yellow-eyed penguin – are now found in only a small part of their former range, having become extinct over much of the country through hunting and other threats in both Maori and European eras.[64]

Today, a variety of conservation issues face marine mammals and birds and about half our seabird species are endangered, threatened or vulnerable. Sea lions, fur seals, Hector's dolphins, most species of albatross,

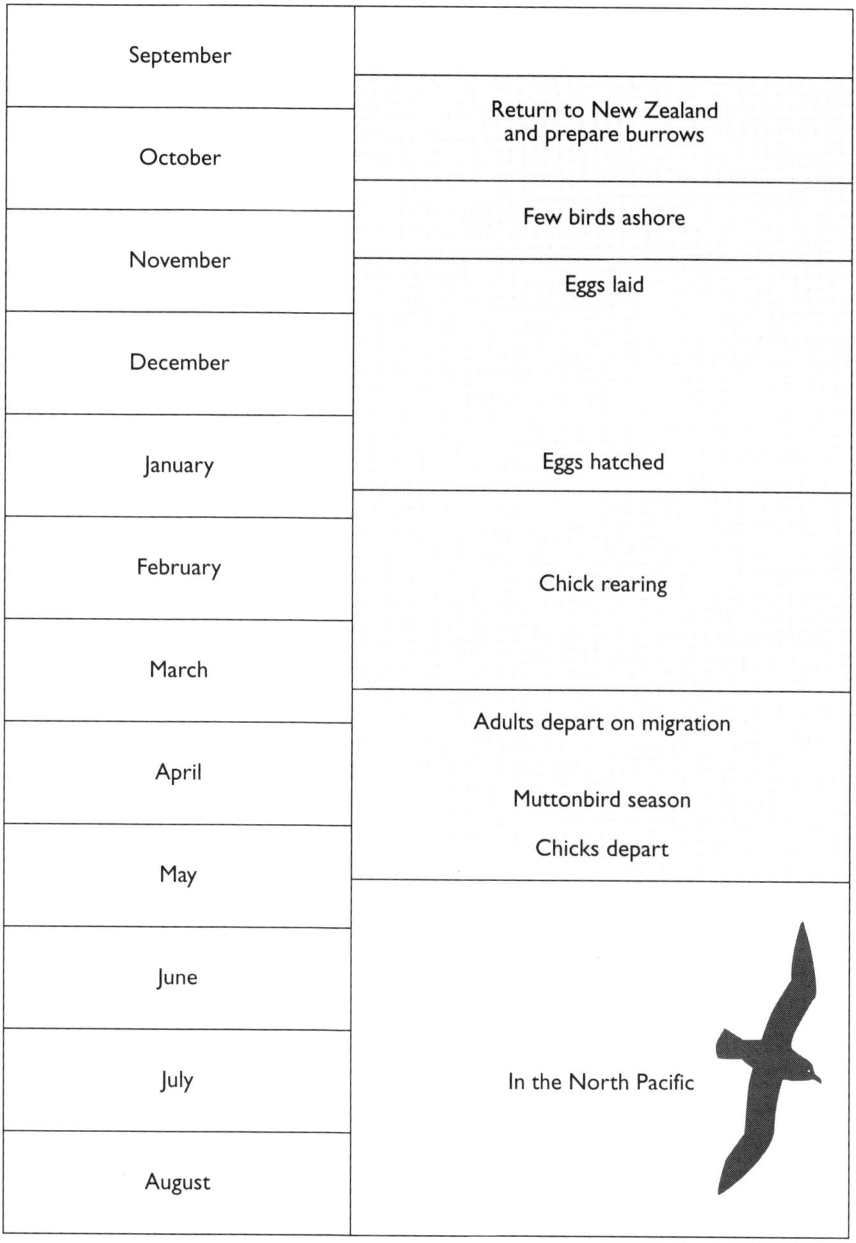

September	
October	Return to New Zealand and prepare burrows
November	Few birds ashore Eggs laid
December	
January	Eggs hatched
February	Chick rearing
March	
April	Adults depart on migration Muttonbird season
May	Chicks depart
June	
July	In the North Pacific
August	

Figure 8.6 The annual cycle of the sooty shearwater

Adapted from Warham et al. 1982

Table 8.3 A comparison of conservation problems facing New Zealand's marine and terrestrial mammals and birds

	Terrestrial species	Marine species
Movements	Localised	Distant
Major threats	Habitat loss or modification, Introduced mammals	Human-induced mortality, Loss of breeding habitat, Predation
Scale	Local to provincial	National to international
Endemism	High	Medium

Source: Wilson 1992

some petrels, penguins and shags are killed as fisheries by-catch.[65] Fur seals are perceived as competing with fishers, who from time to time seek permission to cull them.[66] A claim has been made to harvest royal albatrosses at the Chatham Islands,[67] and further claims under the Treaty of Waitangi to take species including fur seals and some shearwaters are likely. The main threats faced by each species of seabird breeding in New Zealand and steps required to mitigate these have been identified by Graeme Taylor.[68] Introduced predators and interactions with fisheries are major threats. Disease, loss of nesting habitat, nest-site competition, marine debris, oil spills and other pollutants are important at a local level for some species. For example, nesting-burrow competition with the locally abundant broad-billed prion is the most serious threat to the endangered Chatham petrel. Without human intervention, more than half of all breeding failures are caused by prions interfering with Chatham petrel chicks.[69]

Most threatened terrestrial vertebrates have relatively localised distributions and are restricted to well-defined habitats. While some species such as kereru move between different habitats seasonally, these movements are over small distances. In contrast, marine species range over much wider areas, with many of our albatrosses, petrels and whales migrating beyond the exclusive economic zone (EEZ). For instance, sooty shearwaters that breed in New Zealand spend the southern winter in Alaskan waters; New Zealand-breeding royal albatrosses feed off the coast of South America; and most baleen whales that pass through New Zealand waters feed in Antarctic seas in summer, and in winter breed in the tropics. The international repercussions can be challenging:

albatrosses killed by foreign-owned vessels in New Zealand's EEZ may have bred on islands under the jurisdiction of a third nation.

Saving endangered seabirds such as the taiko and Chatham petrel is especially challenging. Not only are their nests exceedingly difficult to find, but these birds spend most of their lives at sea and we know virtually nothing about what they do away from their breeding burrows. Translocations to predator-free islands have saved land birds, but moving seabirds is much more difficult. Adults are mobile and imprinted on the colony at which they have bred, so the chicks must be translocated before they imprint on their natal colony; this then involves feeding the chicks until they fledge. Fluttering shearwater, diving petrel, fairy prion, Pycroft's petrel and Chatham petrel chicks have now been transferred from one island to another.[70] The techniques are still experimental and, because these birds are several years of age before they return to land to breed, the success of these operations is uncertain. However, early results suggest there is reason to be optimistic. Overseas, petrels have been induced to colonise new locations by broadcasting calls and placing decoy birds in locations suited to the species.[71]

Exploitation of New Zealand fur seals

> It is recorded that the industry was carried on so assiduously, and the south-west portions of the coast were hunted so industriously by sealers, who killed females and their young for food, that there was a fear of the seal in these parts becoming extinct.
>
> – Edgar Waite, in Chilton (1909)

Most species of marine mammals and birds are vulnerable to exploitation and predation because their low reproductive rates mean limited ability to recover from drastic reductions in numbers. Other factors also make them vulnerable. Many breed in colonies, so each year the entire breeding stock is assembled in just a few locations. Young seabirds and seals grow slowly, so both young and adults may be found on these colonies for many months. Hunters or predators can quickly reduce numbers of breeding animals and offspring. I will use the history of exploitation of fur seals as an illustration.

New Zealand fur seals were subject to intense exploitation by both the early Polynesians and the first Europeans. Seals appear to have been a mainstay in the diet of Archaic Maori in many places, even in the northern North Island hundreds of kilometres north of places where seals now occur.[72] Seal bones are common in the oldest layers in these archaeological sites, but are replaced by smaller game in later strata, suggesting that fur seals were hunted to local extinction in northern New Zealand long before the European sealers arrived. When European sealing began, fur seals were virtually confined to areas south of Otago Peninsula and the Open Bay Islands (South Westland), with an isolated colony at Cape Foulwind.[73] Archaeological evidence shows that they were present at Kaikoura and in the Cook Strait region during the eighteenth century. New Zealand sea lions also once bred as far north as Northland and became extinct from the New Zealand mainland because of overexploitation by Maori.[74]

Fur seals were New Zealand's first export commodity. European sealing began in 1792, but by 1830 the boom was finished. The seals had been hunted to extinction in most parts of their 1790s range and were rare in the few areas where they survived. The number of seals killed is unknown, but the numbers taken from the Bounty and Antipodes Islands alone[75] probably exceed the total number of fur seals now present in the entire New Zealand region, despite more than a century of protection and spectacular increases in numbers in some parts of the country.[76]

Fur seals are popular scapegoats whenever fish stocks decline, and claims under the Treaty of Waitangi for customary harvest have been discussed. Hardly a year goes by without pressure from some quarter or another to cull seals. How feasible would sustainable harvest be? The last open season was in 1946, but it was very badly managed. That year, the government bowed to pressure from fishermen who claimed that the fur seal population was increasing and the seals were adversely affecting fish stocks. Certain fishermen in Otago, Southland and Fiordland were permitted to kill seals and market the skins.[77] This open season met with considerable criticism, both before and after the event. The rationale for the cull was based on misinformation about the quantity and types of fish taken by seals and the numbers of seals eating (or not eating) those fish. A total of 6187 fur seals were killed, but their skins were generally in poor condition and provided little profit to the sealers. Killing was indiscriminate and did nothing to improve the fishery. The season, the pressures leading to it and the findings of the scientists have been described in detail by J. H. Sorensen.[78] The 1946 debacle remains a salutary lesson in bad wildlife management and both fishers and conservationists should refer to Sorensen's report whenever the controversy re-erupts.

Sustainable harvest of fur seals may be feasible, although the numbers taken would be small. Female fur seals give birth to their first pup when about five years of age, and each year can produce only a single pup, which is dependent on its mother for at least another eight months.[79] Thus, harvest of females would result in population decline. However, fur seals are polygynous, so a male-only harvest could perhaps be sustained.

Sealing in New Zealand was uncontrolled and seals of all ages and both sexes were taken indiscriminately. In Alaska and southern Africa, closely related fur seals have been harvested sustainably. In Alaska, young males congregate on the periphery of the colonies. Each year about 30,000 three- or four-year-old males (the age at which their furs are best) were killed. In southern Africa, about 60,000 pups were taken each year.[80] In southern Africa and Alaska, the colonies are few in number but huge, and each year the number of pups born was carefully monitored. In New Zealand, fur seals live in numerous small colonies scattered over a larger geographic range. Males do haul out in colonies away from the breeding locations, so a sustainable harvest would be more difficult than in Alaska or southern Africa, but not impossible. Whether harvest would be economically feasible and whether the public would find a cull of these large mammals acceptable, are quite different matters.

Diet of the New Zealand fur seal

It is not easy to find out what fur seals eat because it is not possible to induce them to regurgitate their most recent meal, a technique that works well with penguins. Seals do regurgitate squid beaks and other indigestible bits, but such samples are not properly representative of the diet. Shooting seals to collect their stomach contents is wasteful and these days unacceptable. Their faeces are easily collected and inspected, but the picture obtained of their diet is highly biased: bones of big fish are more likely to be present than are bones of little fish, while squid, crustaceans and soft-bodied animals are unlikely to be found at all.

Studies of faeces and regurgitations indicate that, in New Zealand waters, fur seals eat squid (mostly arrow squid), octopus and fish. The main fish prey appear to be hoki, jack mackerel and lanternfish, with a variety of other species taken in smaller numbers.[81] An earlier study, where seals were shot, found barracouta to be common prey. There are regional differences in prey, with fish the main prey in Cook Strait and squid in Otago. Hoki and arrow squid were the main commercially valuable species targeted by fur seals.

Although there is no definitive study of the diet of New Zealand fur seals, all three studies that have been done conclude that they do not take commercially valuable fish in quantities that would deplete the stocks. Nor is there any reason to suspect that a seal cull would result in increased fish stocks.

Seabirds and predators

In the past, seabirds nested in many places around the mainland, but only the most robust and aggressive species still breed on the North and South Islands. The two larger penguins and Westland petrels appear to be able to co-exist with the full suite of introduced predators now present on the South Island, but even these pugnacious birds lose some eggs and chicks to predators. Some Banks Peninsula colonies of white-flippered penguins have declined by two-thirds through predation of adults, mostly by ferrets.[82] The only small petrel still breeding on the mainland is Hutton's shearwater, which nests near the snowline, where predators are less common. Sooty shearwaters and grey-faced petrels still occur on a few mainland headlands, but their breeding success is low and the colonies seem doomed unless protected from cats and mustelids.[83] Small colonies appear especially vulnerable. I have monitored very small sooty shearwater colonies on Banks Peninsula and in Westland, and found that eggs, chicks and even some adult birds were killed. No chicks survived more than a few days after hatching, and the Westland colony declined from 11–12 pairs in 1995–96 to one pair in 1999–2000.[84] The colony on Banks Peninsula has been predator-fenced and numbers are increasing. At one mainland sooty shearwater colony in Otago, stoats probably killed adults and chicks, and Norway rats got the eggs and young chicks.[85]

Petrels are particularly susceptible. They lay only one egg, and if it is lost they do not re-lay. Once hatched, the chick may be left unguarded for days at a time. Mike Imber[86] suggested that petrels are threatened by any species of rat whose maximum body weight equals, or exceeds, the petrel's mean adult weight. Cooks and Pycroft's petrels (mean adult weights 190 and 160 grams) co-exist with kiore (up to 100 grams), but not with the larger Norway rat (up to 450 grams). The small storm petrels and diving petrels had gone from all the rat-infested islands Imber visited, and fairy prions (130 grams) were less common on islands with kiore than on neighbouring rat-free islands. The larger grey-faced petrels and sooty shearwaters persist on Whale Island (Bay of Plenty) in the presence of Norway rats despite the failure of up to a third of their breeding attempts owing to rat predation. Kiore were responsible for the loss of almost half of little shearwater eggs in a study on Lady Alice Island (Northland), the rats most probably eating the eggs when they were left unattended.[87]

Cats can take larger petrels, and Imber suggested that the loss of grey-faced petrels and the decline in black petrels on Little Barrier Island was due to cat predation. The cats have since been exterminated. Kermadec and black-wing petrels and wedge-tailed shearwaters no longer breed on cat-infested Raoul Island, but they remain on the nearby predator-free Meyer Islets. Diving petrels and broad-billed prions became locally extinct on Herekopere Island (off Stewart Island) after cats were introduced, yet fairy prions and sooty shearwaters survived.[88] The researchers suggested that diving petrels were susceptible to cats because they were the smallest of the four species and also because the adults returned nightly to the island to relieve their incubating mate or feed chicks. Other petrels visit their breeding colonies less often. The expatriated diving petrels and broad-billed prions breed earlier than the other two species, yet it seems surprising that the small fairy prions survived when even the large (787 grams) sooty shearwaters were killed by cats. Today, Hutton's shearwaters breed only high up in the Seaward Kaikoura Range, but in the past they bred at lower altitude in North Canterbury.[89] Shearwaters or their eggs or chicks are the main prey for the local stoats, and Richard Cuthbert suggested that the survival of the shearwaters in the face of stoat predation depends on two factors. First, that the shearwater chicks hatch in December or January, coinciding with the period when the demand for food by stoats is greatest. These long-lived, slow-breeding birds can withstand a high level of chick predation but only a very low level of predation on adults. Second, the birds are absent for five months in winter and the stoat population is limited by the scarcity of alternative winter foods.

Weka were introduced to a number of offshore islands by sealers and by Maori, and on these islands they prey on seabird eggs and chicks. They have been blamed for the scarcity of prions, diving petrels, mottled petrels and possibly the local extinction of little shearwaters on Solander Island. On the Open Bay Islands, where weka were introduced early in the twentieth century, they are the greatest cause of egg mortality and an important contributor to chick mortality of Fiordland crested penguins.[90]

Fisheries interactions

The accidental killing of marine birds and mammals has doubtless occurred since people first went to sea in search of fish, but as the size of the vessels and the scale of the industry has increased, so too has the number of animals killed. A few species have benefited from fishing. Gulls are quick to take advantage of any new food supply, and most albatrosses and some other petrels will converge on fishing vessels when offal is discarded. During the hoki-fishing season, more than half of the food fed to chicks by Westland petrels comprised discards from that fishery,[91] and the availability of offal has perhaps enabled their population to increase.

For other species, interactions with fisheries are far from beneficial. In the New Zealand EEZ, dolphins, albatrosses, petrels, sea lions, fur seals and penguins have been killed by commercial fishing gear.[92] The incidental capture of non-target species (by-catch) has led to serious declines in the populations of some of these species. Because all have low natural adult mortality, and most have only one young each year (or even every second year), even low kill rates can cause their populations to decline. Until 1992, when the offending gear was banned from the New Zealand EEZ, albatrosses were killed through collision with net-sonde trawl monitoring cables used by Soviet squid trawlers. In 1990, about 2300 shy mollymawks, many royal albatrosses and other seabirds were killed by these vessels in New Zealand's subantarctic waters.[93]

Numbers of sooty shearwaters appear to be declining both on their New Zealand breeding colonies and at their overwintering areas in the North Pacific.[94] The reasons are unclear, but by-catch in North Pacific drift nets and overfishing of important prey species such as anchovies or Antarctic krill may have contributed. Males and females of some petrel and albatross species forage in different areas, and if one sex suffers more from fishing-induced mortality than the other, then the effect on productivity is potentially greater than if both sexes were equally impacted. Grey petrels are caught by bluefin tuna long-liners off East Cape, and 15 of the 16 birds autopsied during the 1989 fishing season were females that had probably recently laid.[95] The annual survival of female Auckland Island wandering albatrosses is lower than for the males, and this is likely to be due to long-line by-catch.[96]

Fisheries interactions may be a very local or an international problem. In New Zealand inshore waters, recreational and commercial fishers catch dolphins, penguins, shearwaters and shags.[97] Often the numbers are relatively small, but they may come from small localised populations. These problems can often be solved by local restrictions on the use of offending fishing methods; for example, set-nets were banned from use around Moeraki Peninsula (Otago) to protect yellow-eyed penguins.[98]

One of the few well-researched by-catch issues is the entanglement of Hector's dolphin in set-nets around Banks Peninsula. Assessment of the impact that net entanglement had on the local dolphin population led to the creation of a marine mammal sanctuary surrounding the peninsula. Commercial set-netting is now banned year-round, and during summer amateur set-netting is banned within four nautical miles of land, the zone where dolphins are most susceptible to entanglement. Other fishing methods may be used at any time. The sanctuary was controversial becauses local fishers resented restrictions on the use of a popular fishing method. The biology of Hector's dolphins and the research that led to gazetting of the sanctuary is described in *Downunder dolphins: The story of Hector's dolphin*, by Steve Dawson and Liz Slooten, and in the numerous scientific papers listed therein.

Albatrosses and the bluefin tuna fishery

While the Hector's dolphin by-catch is a strictly local issue, with New Zealand fishers catching dolphins close to home, albatross by-catch is in every way an international problem that occurs throughout the Pacific Ocean and in the southern Indian and Atlantic Oceans. Albatrosses range so widely that birds breeding on islands in the southern Pacific may be killed in the Atlantic or Indian Oceans or vice-versa. Albatrosses breed on islands under the jurisdiction of 12 nations and are caught on the high seas and in the territorial waters of these and other nations by vessels registered in yet different countries.

One of the most serious threats to albatrosses comes from Southern Ocean long-line fisheries. Long-lines average about 130 kilometres in length and are suspended below buoylines typically spaced at 250-metre intervals. Between the buoylines are 'baskets' of branchlines, or snoods, which end in baited hooks. A long-line vessel sets up to 3000 hooks per day, each of which, during the short time before it sinks beyond reach, may be swallowed by an albatross that is then dragged under water and drowned. The rate at which albatrosses are caught is low – only 0.4 per 1000 hooks set – but because 50–100 million hooks are set each year, this is high enough to exceed the sustainable take for certain albatross populations.[99]

Table 8.4 Threats and population trends in New Zealand albatross populations

	Fisheries-related	Plastic ingestion, mortality	Human disturbance pollutants	Introduced predators or harvest	Natural events	IUCN status	Population trends
Antipodes Is wandering albatross	X					Vulnerable	Unknown
Auckland Is wandering albatross	XX	?		x		Vulnerable	Declining ?
Northern royal albatross	X	x	x	x	XX	Endangered	Declining
Southern royal albatross	X	?	x			Vulnerable	Increasing
Black-browed mollymawk	XX	?				Near-threatened	Declining
Campbell mollymawk	XX					Vulnerable	Recovering
Buller's mollymawk	X					Vulnerable	Increasing
Pacific mollymawk	?	?	?		X	Vulnerable	Unknown
Shy mollymawk	XX			x		Vulnerable	Increasing?
Salvin's mollymawk	?	?				Vulnerable	Unknown
Chatham mollymawk	X	?	x		X	Critically endangered	Unknown
Grey-headed mollymawk	XX		x			Endangered	Declining
Light-mantled sooty albatross	X			x		Not determined	Unknown

XX = major threat; X = significant threat; x = minor threat to existing populations but may have been responsible for past declines

Adapted from Gales 1997, Croxall & Gales 1997, Robertson 1997, G.A. Taylor 2000

The number of albatrosses caught cannot be estimated with accuracy, but Nigel Brothers[100] estimated that about 44,000 were killed by Japanese long-liners in the Southern Ocean each year. The most commonly killed species were then thought to be wandering albatross (9625 per year), black-browed mollymawk (19,250), shy mollymawk (10,000) and light-mantled sooty albatross (4125). In New Zealand waters, grey petrel, wandering albatross, black-browed mollymawk and Buller's mollymawk were the species most often caught.[101] However, those estimates are no longer very informative. Recent taxonomic research has increased the 14 species of albatross recognised in Brother's study to at least 20 apparently discrete taxa, most of which are known to be caught on long-line hooks.[102] Twelve of the currently recognised taxa breed in the New Zealand region, eight exclusively on New Zealand islands.* The shy, black-browed and Buller's mollymawks have each been split into two or more species, and we do not know whether the animals killed are spread evenly across all the new taxa or some species are more susceptible than others.

The Japanese fishing fleet, particularly bluefin tuna vessels, is among the few whose impact on seabirds is moderately well documented and one of the few fleets that have attempted to reduce their by-catch. Boats from some other nations, including Indonesia, Korea and Taiwan, provide absolutely no information on their seabird by-catch. New long-line fisheries off South America and southern Africa pose further threats to albatrosses, including New Zealand endemic species such as the Chatham mollymawk.[103] No one knows how many albatrosses are killed by long-liners from those fleets, and long-lining is certainly not the only fishing method that kills seabirds.

There are no counts of any of the New Zealand populations threatened by by-catch that are accurate enough to enable the impact of by-catch to be quantified. For most species, counts are too few in number or insufficiently accurate to even know if populations are in decline (Table 8.4, opposite). The difficulties of interpreting the limited information available are illustrated by changes in numbers of Buller's mollymawks. This species is killed by long-line, trawl and set-net fisheries, yet numbers breeding at the Snares Islands, their main breeding location, have increased since they were first counted in 1969.[104] Numbers breeding at the Solander Islands, their only other breeding station, appear to

* About four pairs of Salvin's mollymawks breed on the Crozet Islands; all others breed in the New Zealand region. I have considered the species to be endemic to New Zealand.

be stable or even in decline. Why population trends should differ between islands only 160 kilometres apart is puzzling. Analysis of population parameters for the Snares birds shows how complex the relationship with fisheries must be. Annual survival of adults was lower between 1969 and 1991 than in the years before or after that period, and this lowered survival rate may be attributable to fisheries by-catch.[105] However, despite lowered adult survival, the numbers breeding increased, suggesting perhaps that fisheries discards provided additional food that enhanced post-fledgling survival.

For most New Zealand albatross populations that have declined, fisheries by-catch is a likely cause, but this cannot be proven to the satisfaction of some industry advocates. Numbers of Campbell and grey-headed mollymawks breeding on Campbell Island have perhaps halved since the 1940s, and for grey-headed mollymawks breeding success is too low to sustain the population. Both species are caught on long-lines.[106] The threats faced by the Pacific and Chatham mollymawks are impossible to assess. Both species forage outside the New Zealand EEZ in waters where long-line fishing takes place, but there is no observer coverage so by-catch cannot be assessed.[107] The Chatham mollymawk is solitary when at sea and its distribution at sea is poorly known. It is the least common and least studied of the New Zealand endemic species, and no reliable estimate of its numbers has been made.

On some islands in the Indian and Atlantic Oceans, the declining number of breeding albatrosses is better documented. On the Crozet Islands, wandering albatross populations halved between 1965 and 1985. The dynamics of these populations are finely balanced. At South Georgia, this species has declined by 1 per cent a year since 1961. Even though 94 per cent of adults survive from one year to the next, only 1–2 per cent below the optimal survival rate, the birds have been unable to compensate for this seemingly low level of adult mortality.[108] Population trends for the New Zealand taxa are not well documented. The first accurate census of Auckland Island wandering albatross was made in 1991, and during the 1990s the numbers breeding varied each year.[109] Comparing these counts with much less reliable earlier counts is problematic, yet evidence suggests that their numbers have declined and, because the birds are known to be caught on tuna long-lines, this seems a likely cause.

While research may be difficult and expensive, prevention is cheaper and simpler. Albatross by-catch can be reduced, perhaps by as much as 90 per cent, if the vessels set hooks at night and tow streamer lines (tori poles), which occupy the airspace behind the ship, thus deterring albatrosses.[110] Night setting may reduce albatross mortality, but not that of the smaller petrels, many of which feed at night. Mechanical bait-throwers, which heave lines clear of the ship's wake, or underwater setting of lines could greatly reduce the numbers of albatrosses killed. Machines to do this are currently being developed.

Other threats to albatrosses

On a global scale, albatrosses are the most threatened group of seabirds and one of the most threatened of all bird groups. Of the about 20 species, only two are not classified as threatened or endangered under IUCN criteria. While fisheries interactions pose the greatest threat to most species, there are other significant threats to some.[111] Of the species breeding in the New Zealand region, only the black-browed mollymawk (less than 1 per cent of which breed in New Zealand) and the light-mantled sooty albatross (about which too little is known to assess its status) are not vulnerable, threatened or endangered.

All three species that breed at the Chatham Islands are subjected to a wide variety of threats. All probably suffer fisheries-related mortality, but the major threat is from seemingly natural events. Since 1970, the climate has been drier and warmer than normal, causing stress to birds, killing chicks and adults; and storm-force winds have resulted in loss of soil and vegetation on the small rocky islets where they breed.[112] The breeding success of Chatham mollymawks and northern royal albatrosses has diminished and a decline in their populations seems inevitable. Historically, royal albatross and Chatham mollymawk chicks were harvested, and although this is now illegal, small numbers may still be taken.

The only other location where the northern royal albatross breeds is Taiaroa Head, where the small colony on the outskirts of Dunedin is a popular tourist attraction. Thanks to stringent management and predator control, the number of Taiaroa albatrosses has increased, although the numbers nesting in sight of the viewing structure has not, indicating that these magnificent birds preferentially breed away from the gaze of their admirers.[113] Tinting the glass in the observatory windows appears to have solved this problem.

John Croxall[114] has reviewed the state of knowledge and research needs for albatrosses. He noted the importance of the New Zealand region, with eight species breeding only there. There is an urgent need for research on population dynamics and foraging ecology, and much more extensive studies of movements at sea. Long-term monitoring studies are required, otherwise populations could slip away unnoticed.

Hooker's sea lions and the Auckland Island squid fishery

Trawlers fishing for arrow squid off the Auckland Islands catch Hooker's sea lions, a species endemic to New Zealand's southern islands and about 90 per cent of which breed at the Auckland Islands. Their numbers are not accurately known, but recent estimates suggest that the total population is about 12,000.[115] Nor is the number of sea lions killed each year known, and whatever figures are produced, they are disputed by either fishermen or conservationists. The Ministry of Fisheries estimated that between 1988 and 1998 the annual kill varied between 17 and 141 sea lions. About 65 per cent of the sea lions killed were females, many of which had a dependent pup ashore that would also have died.

So far it has proven impossible to calculate the sustainable by-catch with any certainty because population parameters such as pup production and survival rates are not accurately known. Most models suggest a knife-edge situation, with by-catch close to, and sometimes exceeding, the maximum sustainable yield.[116]

Hooker's is the world's rarest sea lion, yet there is tremendous opposition to measures that would lower the by-catch. The squid fishery is worth about $40 million a year and there are no obvious solutions that would lower by-catch without adversely affecting the economics of the industry. Commercial fishing is prohibited within 12 nautical miles of the Auckland Islands, and extending this exclusion zone to cover a larger part of the islands' continental shelf might reduce the by-catch. In recent years, a quota of 60–80 sea lions has been set and the fishery closed if and when this number was exceeded. The quota was exceeded every year from 1995 to 1998.[117] To many people, the idea of setting a quota for a threatened species seems abhorrent, yet it may be a pragmatic way to resolve the conflict.

Other threats to marine birds and mammals

Marine pollution, plastic flotsam, old fishing nets and plastic strapping all pose threats to marine birds and mammals, but the severity of these threats to New Zealand species has not been assessed. Plastic flotsam is ingested by many seabirds, particularly albatrosses and petrels. These objects may remain in the gut for a year or more, but whether they actually kill the birds is not known.[118] Mammals appear to be more prone to entanglement than to ingestion of plastic and other non-biodegradable debris.[119] Fur seals are regularly seen with old fishing net or plastic strapping around their necks and, as the animals grow, this cuts into their flesh.

Elsewhere in the world major oil spills have caused the deaths of many marine birds and mammals, but although small spills are common in New Zealand, so far no large-scale spill has occurred. A wide range of chemicals end up in the ocean and accumulate in marine animal tissues. Chemical contaminants do occur in New Zealand species, but their effect on the animals is not well studied.[120]

Impediments to resolving conservation issues

One of the biggest obstacles to resolving the threats discussed in the previous pages is our inability to prove that populations are in decline, let alone to quantify that decline. The population size of most seabirds and marine mammals is not known, and accurate counts have been made of only a few conspicuous species such as gannets.[121] Reliable counts for other species are difficult, perhaps impossible. Fur seals, for instance, breed in many scattered colonies, they are similar in colour to the rocky beaches they inhabit, and they spend much of their time out of sight beneath boulders or at sea. For most parts of New Zealand, counts are of low accuracy and out of date.[122] Burrow-breeding petrels present even greater challenges: few attempts have been made to estimate the population size of these species, and those counts are rough.[123]

Without knowing population size, birth rates and mortality rates we are unable to assess the impact of by-catch, harvest or disturbance. Given that counts of most marine mammals and birds are intrinsically so inaccurate and their reproductive rate is so low, population changes are likely to be obscured merely by counting error. Much of the dispute over the impact set-netting has on Hector's dolphin, or that by-catch has on sea lions and albatrosses, comes down to how poorly we are able to calculate population size and how little we know about other population parameters. In the case of Hector's dolphin, however careful the scientists, cetaceans are notoriously difficult to count and error is intrinsically high. Wise managers recognise the limitations of the data and act conservatively. For some species, counting error is so great that their populations could decline by a third or even a half before we would notice. It could take decades for some species to recover from such losses. In the case of the colony-breeding sea lions and albatrosses, accurate counts are possible and, given adequate resources, we could determine the impact by-catch has on their breeding populations at least. An unknown proportion of the total population of most seabirds is made up of pre-breeding birds that do not visit the breeding colonies. Colony counts cannot include these birds. Baseline counts prior to by-catch or harvest are needed if we are to determine the real impact. For species currently under threat, these data are lacking.

Marine birds and mammals are an important part of New Zealand's heritage and, with the great diversity and abundance of these creatures in our region, New Zealand has an international obligation to ensure their conservation. The fisheries that impact on marine animals are lucrative and by-catch mitigation would reduce profit. The battles between the fishing industry and conservation lobbyists have been acrimonious, and this does not augur well for rational resolution of the disputes. In the absence of sound information on the numbers and population dynamics of the species, responsible biologists must err on the conservative side and recommend that catches be kept low.

CHAPTER 9

Conservation

> In conclusion let a few words be recorded for the preservation of our native fauna. It is a work of difficulty, except with a few, to get folks interested in this subject; amidst the busy swarm of men pressing onward in the struggle for wealth or position, how few out of the entire mass would think of turning aside, and thus lose a fraction of the time devoted to the toilsome climb of the social ladder.
>
> – T. H. Potts (1882)

New Zealanders are among the world's best when it comes to the management of critically endangered species. Don Merton is best known for his role in saving the black robin, at the time the rarest bird in the world, but this is only one of several species he has helped save. The expertise of Merton and his colleagues is recognised internationally and they have been called on to assist with the conservation of critically endangered species overseas. Until Dick Veitch and his team killed the last cat on Little Barrier Island, and Rowley Taylor and Bruce Thomas devised a radically new strategy that exterminated rats from rugged Breaksea Island, the eradication of predators from all but very small islands was considered impossible. Rat eradication is now an almost routine procedure, each year freeing up new islands for conservation.

For most of the last 50 years, endangered species management has mostly been conducted on small islands beyond reach of predators and people. However, the last 15 years have seen the development of innovative management strategies that enable rare species to thrive on the mainland in the presence of the threats that made them rare. For example, during the 1990s John Innes and his team developed a strategy that made it possible for kokako populations to increase despite ongoing threats from introduced predators. The ecological restoration techniques currently being developed in 'mainland island' projects have already seen dramatic increases in most of the native birds present. Until the 1990s, the majority of rare endemic species were restricted to small islands from which the general public was excluded. The restoration of Tiritiri Matangi Island in the Hauraki Gulf developed a new approach to repair of damaged habitats that combined the expertise of scientists with the volunteer labour of hundreds of lay people. This island is now an open sanctuary where the public can view endangered species such as saddlebacks and takahe previously accessible only to conservation workers.

I regard these as seminal achievements, yet they are but a few of numerous efforts made to save our endemic species from extinction. New Zealanders have shown they can save species and heal some of the ecological wounds inflicted in the years since first human contact. In this chapter I will document the history of vertebrate conservation, using case studies that highlight some of the issues conservation managers face when saving rare species. With so many threatened species, it is impossible to report on all of them, or even mention the full range of management actions taken. This chapter is essentially historical in approach and deals with the pragmatic problems faced by conservation managers. It also provides information that will underpin the broader issues discussed in Chapter 10. Some of the problems that are peculiar to the conservation of frogs and reptiles or of marine birds and mammals were also raised in Chapters 2 and 8.

It is generally believed that, before European contact, Maori had developed an understanding of the principles of resource conservation and had taken measures to conserve wildlife. I have found it impossible to obtain specific information about these or to judge their success, thus I have felt unable to include a discussion of wildlife conservation during the Maori era.

The nineteenth century

> In a newly colonised country, where the old fauna and flora are being invaded by a host of foreign immigrants, various natural agencies are bought into play to check the progress of the indigenous species, and to supplant them by new and more enduring forms.
>
> – Walter Buller (1872–73)

Even during the nineteenth century it was evident that many native birds were becoming scarcer. This was regrettable and some of the naturalists of the time even suggested that measures be taken to save these interesting and beautiful species.[1] However, their concerns went largely unheeded in a colony whose development depended on altering the native species' habitats. The conventional wisdom at the time was that things colonial were inferior to things European, whether they were plants, animals or people. It was generally assumed that the indigenous forests and birds would eventually be replaced by more competitive European species. What good, then, was the conservation of native species? It was best that good collections be made quickly while they were still numerous, so they could be preserved as curiosities in the world's museums, if not the New Zealand bush.

During the late 1800s, collectors such as Andreas Reischek and Sir Walter Buller obtained hundreds of specimens of New Zealand's unique birds. Not all ended up in museums: collecting natural history specimens was a popular pastime and some wealthy individuals amassed vast collections of both living and dead animals. Collecting birds was a lucrative trade and rare specimens provided an additional source of income for lighthouse keepers, surveyors and others whose work took them to remote places. Buller, the son of Lord Rothschild of Tring, was one of the most avid collectors, and he had men visit many parts of the world, including New Zealand, to procure specimens. In 1889, he sold his bird collection to Lord Rothschild and for years afterwards continued to collect on his behalf. The last pairs of laughing owls and huia that Buller collected were sent alive to Rothschild.[2] The huia had been collected for release on Little Barrier Island, but by sending then to Tring, Buller hoped that 'my captive birds will breed in their new home, and in this way the race will be perpetrated'.

Rare and unique species such as the huia, kiwi, hihi and kakapo were highly sought after and subjected to intense hunting pressure. Reischek is best remembered for the 150 hihi he shot on Little Barrier Island in the 1880s. The island was their last refuge and, on Reischek's first visit, hihi were so rare he saw none.[3]

Sir Walter Buller was the pre-eminent ornithologist in nineteenth-century New Zealand. He is best remembered for his *History of the Birds of New Zealand,* which remains one of the classic books on New Zealand ornithology. Ross Galbreath[4] described Buller as a 'reluctant conservationist' who, in 1895, was still convinced that the native fauna was doomed, so that his attempts to conserve birds in the wild remained half-hearted (although he did support proposed measures to conserve native species when it was politically astute).

The several editions of Buller's *Birds of New Zealand* reflect the changing attitudes to native species. In the 1888 edition he accepted the demise of the native birds and welcomed the arrival of English species; but in his 1904 supplement he considered the wholesale introduction of exotic birds a mistake and in nostalgic tones lamented the loss of native birds. The way New Zealanders valued their native birds had begun to change.

T. H. Potts was one of the first advocates of protecting this country's native species. He believed that the endemic animals and plants could survive and argued that native species should have the same legal protection that certain introduced animals then enjoyed.[5] Potts argued that the disappearance of native species was a direct result of human intervention, and even suggested that New Zealanders had a moral responsibility to preserve their endemic species. A small band of naturalists supported him, but their ideas held little credence in the late nineteenth century.

In 1878, Potts proposed that certain islands be set aside as sanctuaries for native plants and animals, and eventually Resolution Island in Fiordland was given this status in 1891.[6] After further years of procrastination, Little Barrier Island was declared a sanctuary in 1894, and Kapiti Island in 1897. The battle to achieve so little had been long and frustrating, yet the achievements were notable because at the end of nineteenth century few other nations had reserves to protect their native biota.

Richard Henry was the first person in New Zealand to actively intervene in an attempt to save threatened species. In 1894, the Department of Lands and Survey appointed Henry the curator of Resolution Island in Dusky Sound and he lived there, often alone, for 14 years.[7] During this time he shifted kakapo, brown kiwi, little spotted kiwi and weka from the mainland to Resolution and other islands where he hoped they would be safe from mustelids, which had not yet spread to Fiordland's western shores. By the time he left Dusky in 1908, he had moved almost 750 flightless birds to islands in Dusky Sound. His good work was undone after stoats invaded Resolution Island in 1900 and eventually reached all the islands Henry had used as refuges.

The sad story of Stephens Island encapsulates the ecological tragedy that unfolded in New Zealand during the nineteenth century. The island had seldom been visited until a lighthouse was built in 1892. The lighthouse made it possible for collectors to visit the island, where they discovered a previously unknown species, the Stephens Island wren. Within 12 months the wren succumbed to the onslaughts of one cat and several collectors. While the loss of the wren was lamented, particularly as local museums had not obtained specimens, the island was home to other animals of interest. Piopio and saddlebacks, then rare on the mainland, were abundant, and tuatara so numerous that at night it was nearly impossible not to tread on them. By 1894, sheep, cattle and cats had been introduced and virtually all the forest cleared. Of the native vertebrates, tuatara and Hamilton's frog were about all that survived (rare lizards and invertebrates were discovered later), but a few decades later they were deemed of such great interest that they alone warranted the island's being made a sanctuary. If sanctuary status had been declared in 1892, the extinction of piopio might well have been averted.

The early twentieth century

In the early decades of the twentieth century, conservation continued to take a back seat to development. Laws were enacted to protect native wildlife, but few people yet lamented the loss of native species or recognised the follies of acclimatisation, and habitat destruction continued unabated. In the introductory chapter of one of his early books, naturalist Herbert Guthrie-Smith noted with regret the decline of many native birds, the introduction of exotics and the holus-bolus clearance of indigenous forests. In his last book, written in 1936, he expressed these sentiments much more forcefully and described the impact of European settlers on native birds, forests and landscapes in depressing detail. He was pessimistic about the future of the native fauna and laid the blame firmly with the wholesale clearance of native habitats and the thoughtless introduction of exotic predators.[8]

Initially, protection was passive and came in the form of legislation and protected areas. National parks and scenic reserves conserved places of great scenic beauty but did not always protect the areas that would most effectively conserve native fauna and flora. Lowland forests, where biodiversity was greatest, are less well represented in New Zealand's reserves portfolio than are the less biologically diverse mountains. In 1915, Gouland Downs in Northwest Nelson, an area then well stocked with kiwi, weka and kakapo, was gazetted as a sanctuary in the hope that this action alone would save the native birds.[9] Active management of those species was not then part of the conservationists' repertoire.

In 1907, the Animals Protection Act conferred total protection on some non-game native birds, and in 1921–22 this was extended to most species of native bird. Graeme Caughley[10] considers this to have been a revolutionary law: in 1907 no other country gave total protection to native, non-game species, and it was not just a reaction to scarcity, as many of the protected birds were still common. Even today, total protection of native species is unusual. Not until 1940 did the United States confer total protected status on a native bird, the bald eagle; and as late as 1982 the United Kingdom still gave no species total protection. Modern-day New Zealanders grow up with the mind-set that native birds should be totally protected, and Caughley suggested this has been an important motivating force for conservation, acknowledging that indigenous animals (or at least birds) are part of New Zealand's spiritual wealth. Lizards apparently were not considered part of the nation's spiritual wealth until the 1980s, and even today few invertebrates qualify.

Lance Richdale was one of the first New Zealanders to adopt a policy of direct intervention and active management to tip the balance in favour of native birds. When royal albatrosses began nesting at Taiaroa Head, near Dunedin, in the 1930s, they were molested and no eggs hatched until Richdale intervened. In 1937, and in following years, he camped at the colony during the breeding season, and owing to his persistence the area was eventually protected. Predators were trapped and the area was fenced, but disturbance by people was controlled only by Richdale's presence until, in 1947, a ranger was appointed.[11] When, in the 1940s, cats were killing large numbers of birds on Herekopare Island (off Stewart Island) and government agencies were slow to act, Richdale went to the island and began the job of cat extermination himself. He mailed cat-killed corpses of native birds to Wellington-based officials, intending the maggoty remains to remind the office staff of the need for action. While active management is now considered normal practice, such intervention was then controversial. Richdale's approach was even criticised by E. V. Sanderson of the Royal Forest and Bird Protection Society, who believed that foreign invaders such as deer and stoats should be controlled but the native birds should be left alone to 'flourish' naturally.

Conservation of the takahe

In November 1948, Dr Geoffrey Orbell discovered takahe in the Murchison Mountains, Fiordland. These were the first live takahe seen since 1898 and it had generally been assumed that the species was extinct. Their rediscovery sparked a great deal of public interest, and half a century later this story remains a perennial favourite.[12] Interest was not just confined to New Zealand. In Britain *The Times* reported the rediscovery, and over the next decade the *Illustrated London News* ran a series of articles on takahe. In 1950, the Ornithological Society of New Zealand renamed its journal *Notornis*, which at the time was the generic name for takahe, and adopted the bird as the society's emblem. For the first time, at least in New Zealand, an endangered bird provoked so much public interest that the need for action to save the species was acknowledged.

If takahe should be saved, then surely so too should other species. Orbell's rediscovery of the takahe marked the dawn of a new era in conservation. Takahe have now been a focus for conservation for over half a century, making this one of the longest-running endangered-species conservation programmes anywhere in the world. The way takahe management has changed over this period reflects the growing sophistication of endangered-species management. A recent book[13] celebrates 50 years of takahe conservation, management and research, and discusses the changes in the bird's management in far greater detail than is possible here. Nonetheless, takahe have played such a pivotal role in conservation management and so many lessons have been leaned from takahe management[14] that a brief account of the story is essential.

Soon after the rediscovery, access to that part of Fiordland National Park inhabited by takahe was restricted, and even today the area can be entered only by permit. However, left alone in their protected area as initially proposed by government and conservation groups, the species was likely to become extinct. R. A. (later Sir Robert) Falla counselled against hasty action and recommended scientific studies to determine distribution, abundance and the factors that influenced the takahe's survival. Management of this endangered species fell to the newly formed Wildlife Branch (later renamed Wildlife Service) of the Department of Internal Affairs. Initially, the Wildlife Branch culled deer, controlled possums and, in association with the acclimatisation societies, managed game birds and freshwater fish. When takahe were rediscovered, a new role unfolded for the Wildlife Branch: the conservation of protected native species.

Fifty years ago, saving an endangered species was a step into the unknown. Over the next four decades the Wildlife Service maintained an ongoing programme on takahe. It also saved other species from extinction, protected habitats and did much to heighten public awareness of the plight of native wildlife.[15] The Wildlife Service was merged into the newly fledged Department of Conservation (DOC) in 1987.

Takahe were once widespread in both the North and South Islands,* but by the time of European settlement they were confined to Fiordland. Their range shrank further until, by 1948, they were confined to the area between South and Middle Fiords of Lake Te Anau. After rediscovery their numbers and range diminished further, so that by 1981 all 120 surviving birds were confined to the Murchison Mountains. Since 1981, the Murchison population has fluctuated and a few birds have been released in the nearby Stuart Mountains. In 1998, there were 141 adult takahe in Fiordland, 59 on four offshore islands (Tiritiri Matangi, Mana, Kapiti and Maud) and 14 in captivity.[16]

Captive rearing

One of the first strategies adopted to increase takahe numbers was captive breeding. Beginning in 1957, eggs, chicks and adults were taken from Fiordland to aviaries at Mt Bruce that were later to become the Wildlife Service's native bird reserve.[17] Success was limited: many birds did survive, but it was not until 1972 that the first fertile eggs were laid, and several more years elapsed before chicks were successfully hatched. By 1983, only four chicks had fledged from more than 72 eggs produced. By then, however, the emphasis of management had shifted from captive breeding to management in the natural environment.

Since 1985, captive rearing has again been an important component in takahe management.[18] In the wild, takahe usually lay two eggs but raise only one chick. Soon after egg-laying, the second egg is removed from nests of wild pairs with more than one fertile egg, and given to pairs with infertile eggs. The aim is to ensure that each pair of wild birds will have one chick to raise. Any surplus eggs are taken to the Burwood Bush rearing unit, near Te Anau, where they are placed in incubators. To avoid chicks 'imprinting' on people, contact is minimised. The chicks are fed and brooded using takahe puppets, and tape recordings of adult takahe are played at hatching, feeding and other stages during the chicks' development.

When three months old, the chicks are placed in small pens in the tussocklands surrounding the buildings, then transferred to larger enclosures for winter where, in the company of captive adults, they learn the foraging skills they will need when released into the wild. Takahe raised in the absence of skilled adults never learn to recognise some

* All surviving takahe are of the South Island form. North and South Island takahe are now considered to have been distinct species (Trewick & Worthgy 2001, Holdaway et al. 2001.

important foods such as *Hypolepis* fern rhizomes. Initially, captive-reared birds were released into the Stuart Mountains, where takahe had persisted until the 1960s, but survival there was low.[19] Since 1993, they have been released into the Murchison Mountains population, where it is more feasible to monitor their survival, or they are used to seed the island populations. Captive rearing, despite poor success in the early years, has resulted in a dramatic boost to their overall numbers. All the island takahe and a quarter of the Fiordland birds were reared in captivity or are the progeny of birds reared in captivity.[20]

Takahe and deer

Fiordland takahe have a very restricted diet. For much of the year they feed mostly on three species of snow tussock and the mountain daisy

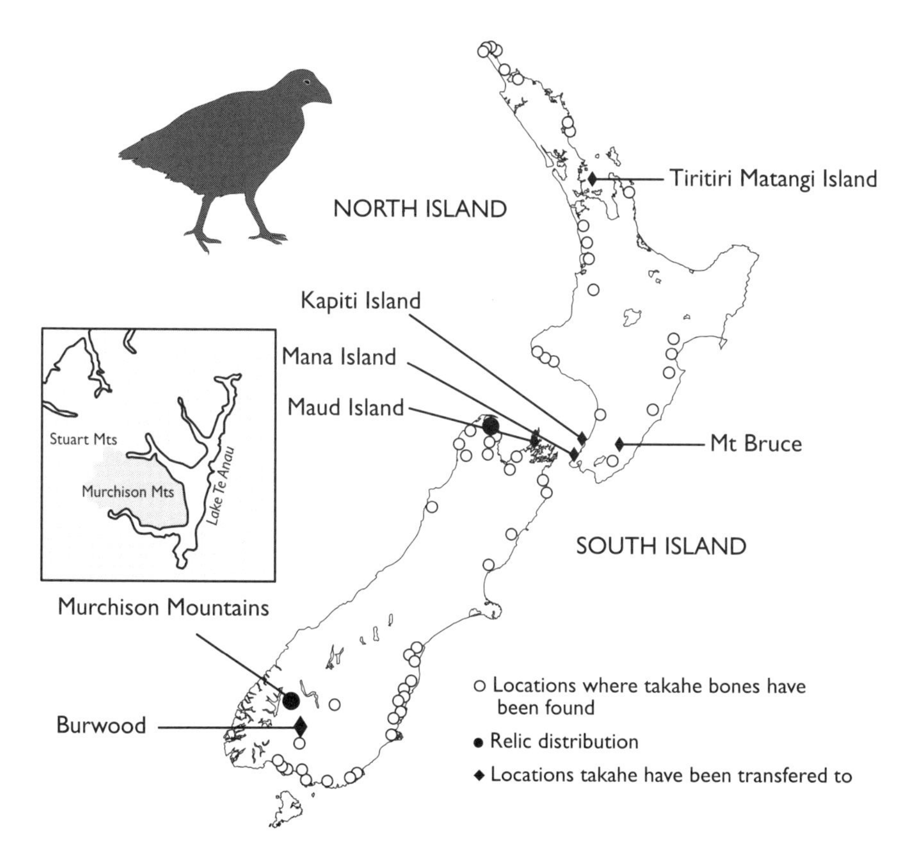

Figure 9.1 Past and present distribution of takahe

Adapted from Lee & Jamieson 2001

Celmisia petriei. When eating tussocks they select only certain individual plants, and then eat only the basal few centimetres of each tiller. During winter, takahe move into the adjoining beech forests, where they eat the rhizomes of the fern *Hypolepis millefolium* and a few species of herbs and grasses.[21]

Wildlife Service scientists studied the diets of takahe and red deer to assess the extent to which these two browsing animals competed for food. They found that takahe would typically sample many tussock plants, then feed intensively on one particular plant before searching out another. In spring and early summer, takahe were selecting plants with high levels of phosphorus and nitrogen, then in late summer and autumn they selected tussocks rich in carbohydrates. Red deer were found to show a similar preference for both the species and individual plants selected by takahe.[22] However, unlike takahe, deer also ate other plant species. While the tussocks could cope with grazing by takahe, deer grazing retarded their growth and vigour for several decades.[23] Both direct competition for favoured foods and degradation of the habitat by deer contributed to the decline of takahe, which by then were confined to the least-modified grasslands.[24] Superficially there appeared to be an abundance of food for takahe, as their favoured plant species were common; however, the Wildlife team found that there does not have to be widespread destruction of the habitat to affect takahe – just the elimination or weakening of the most nutritious plants.[25]

These studies showed that control of deer would greatly assist in the recovery of takahe, but the implementation of this was controversial. The largest area of suitable habitat outside the Murchison Mountains was in the Stuart Mountains, where wapiti were present. These North American deer, larger than but closely related to the European red deer, had been introduced to Fiordland in 1905. They are New Zealand's prime big-game trophy animal and the Deer Stalkers' Association had already been actively saving them from helicopter-borne venison hunters and from a park board that wished to eradicate them. The plan to reduce wapiti numbers and then re-introduce takahe drew flak from the hunters. The Deer Stalkers' Association asked Parliament to reserve the area for wapiti, recommending management practices that would have conflicted with takahe conservation. Some people even suggested that takahe be removed from Fiordland National Park and that wapiti be treated preferentially. The Wildlife Service defended the endangered, endemic takahe against the introduced wapiti and, after a heated battle, the needs of the takahe were mostly recognised. Articles by Jim Mills and John Bamford present the opposing views on this issue.[26]

An alternative method of improving nutrient levels in takahe territories was tried in 1979 and 1981, when fertiliser was applied to five territories where breeding success had been low. This did enable takahe to raise more chicks, but deer control was equally effective and more in line with the ideals of a national park. When the proposal to establish takahe in the wapiti block was abandoned in favour of intensive management in the Murchison Mountains, deer control became a little less controversial.

Despite half a century of management and decades of deer control, the Fiordland takahe have continued to decline in range and abundance.[27] However, the overlap in the diet of deer and takahe, and the reduced vigour of deer-grazed tussock, do not prove that deer limit takahe numbers or breeding success in Fiordland. The relationship between takahe, deer and tussock is complex and the debate remains unresolved.[28]

The role of predators in the decline of takahe is even less clear. Stoats do occasionally kill takahe, usually in the year after mast seeding of beech (see Chapter 7), but the evidence suggesting that stoat predation affects takahe populations is equivocal.[29]

Is the takahe an alpine species?

The Wildlife Service people overseeing takahe conservation assumed that the birds were specially adapted to the alpine and subalpine habitats to which they were then confined.[30] Other scientists disagreed. Takahe bones have been found in many parts of New Zealand, including areas that were under lowland forest at the time the bones were deposited.[31] In addition, fossil bone deposits show that the major decline of takahe occurred during the 800 years they were subjected to hunting by Maori, and not during the post-glacial climate warming.

This has implications for the management strategies we might adopt in saving the species.[32] If the takahe is adapted to the alpine environment, then this is where we should seek to conserve it. If, however, the bird once also occurred in other habitats, then its survival in the Fiordland grasslands does not necessarily mean this is the preferred habitat. It could equally show that this is where the factors that caused the bird's decline were least severe, or were prevalent most recently. Takahe are large, meaty birds that would have been easy and attractive prey for Maori hunters,[33] and they appear to prefer edge habitats where they would have been easy to kill. However, hunting pressure by Maori would have been less intense in Fiordland than in most other parts of the country, so perhaps, like kakapo, the takahe's ability to survive in habitats that were marginal for birds, hunters and competitors alike enabled it to survive into the twentieth century.

It is important when conserving a threatened species that the reasons why it remains where it does are determined accurately. Implicit in many threatened-species management strategies is the assumption that the habitat they survived longest in is the habitat they prefer, even when it is known the animals formerly occurred elsewhere.[34] If takahe also lived in lowland habitats, then maybe they can best be managed closer to sea level. However, before they are transferred to lowland habitats, it is essential to eliminate the agents that made them extinct in those habitats.

Island populations

In New Zealand, the favoured method for saving threatened species has been marooning them on predator-free offshore islands, but for takahe this was assumed not to be an option. To those scientists who assumed takahe were alpine birds that required tussocklands, it seemed that the subantarctic islands were the only suitable places free from competitors and predators. However, these islands were outside the takahe's natural range and of high conservation value in their own right. Other biologists disagreed and argued that the birds would do better on a pasture-grass diet on predator-free islands.[35] After a protracted debate, nine takahe were released on Maud Island in 1984–85 and, because they survived, further birds were taken to Kapiti, Mana and Tiritiri Matangi Islands, where they now thrive in pasture, fernland and forest habitats. The management plan is to keep the Fiordland and island populations discrete, but to periodically transfer birds from one island to another, so the island birds are in effect managed as a single population.

Because self-sustaining populations have established on all four islands, the releases can be considered successful, yet the rate at which the island populations increased was well below expectations, proving the sceptics and the island-release proponents both right and wrong.[36] While adult survival has been high and island pairs lay more eggs per year, only about half of the island-laid eggs have hatched and survival after hatching is even lower than in Fiordland. The reason for this is unknown, but the frequent movement of birds between islands, thus disrupting established pair bonds, may be a factor.[37]

Saddlebacks and rats

> The relative abundance of many very rare species is a position which we should not allow ourselves to accept with complacence. In many cases this remains their last stronghold and active steps should be taken to find more about their habits and life histories and consideration given to transferring some to safer islands with comparable conditions. The accidental introduction of rats is an ever-present threat.
>
> – Brian Bell (unpublished Wildlife Service report, 1961)

By 1960, each species of saddleback was confined to a single location: the North Island birds on Hen (Taranga) Island in the Hauraki Gulf, and the South Island birds on the Big South Cape Islands off Stewart Island. The risk of rats or other predators reaching those islands had long been appreciated and there had been attempts to transfer North Island saddlebacks to other islands, but the birds transferred had not survived long. In 1963, renewed attempts were made to establish a second population of North Island saddlebacks, and in January 1964 the first of many successful transfers took place. This was the first time in New Zealand that a species was returned to part of its former range. The techniques developed with North Island saddlebacks[38] soon proved invaluable in saving the South Island species and were subsequently adapted for the translocation of other threatened species.

The Big South Cape Islands consist of Big South Cape Island (400 hectares) and two small adjacent islands (see Chapter 5). They had been little modified and, until ship rats were found there in March 1964, were free of introduced mammals. After various delays, the Wildlife Service sent Brian Bell[39] to investigate, and he found that rats had reached plague numbers in the north of the large island and were spreading south. Wildlife considered the best course of action was an urgent transfer of saddlebacks, together with three other species that survived only on these islands, to predator-free islands. In August 1964, an expedition was dispatched to accomplish this task. The experience gained with the North Island saddlebacks enabled the capture and transfer of 36 saddlebacks to two nearby rat-free islands, and the species was saved.

The islands were also the last refuge for the greater short-tailed bat, bush wren and the Stewart Island snipe. The wrens and snipe, already scarce, proved difficult to catch and hold pending transfer. No snipe lived long enough for transfer to be attempted, and only by heroic efforts was Bell able to transfer six wrens to Kaimohu Island,[40] but they failed to establish. Nothing could be done to save the bats.

Over the years, as saddleback populations have built up, birds have been moved to ever more refuges so that each species now occurs on nine islands.[41] Saddlebacks have been successfully established on all suitable rat-free islands and on some islands where kiore, mice or weka are present.

Ever since the Big South Cape episode, the Wildlife Service (and subsequently DOC), has endeavoured to prevent a species becoming restricted to a single island. However, not until 1983 was it possible to establish a second population of black robins (see below), and in the 1980s it was discovered that little spotted kiwi only survived on Kapiti Island. Until the late 1990s, Gunther's tuatara occurred only on North Brothers Island, and the Chatham petrel still breeds only on Rangatira Island.

With the success of the saddleback programme, marooning poorly flighted birds on predator-free islands became an important strategy for endangered species conservation, and since then at least 50 species (not just birds) have been moved between islands. Black robin, kakapo and little spotted kiwi are among the species that survive today thanks to island transfers. Translocation to a predator-free island is a convenient, low-cost solution and was the only option for the preservation of many rare species until the 1990s, when effective predator-control strategies were developed.

Black robin

> . . . not only did she capture the hearts and conscience of a nation and make a massive contribution to the conservation cause but she also saved her species from extinction. Old Blue's story must surely be unique.
>
> – Don Merton (1992)

Black robins became extinct on the larger islands of the Chatham group after the arrival of rats and cats in the nineteenth century, and for over 90 years they were confined to seven hectares of forest on top of cliff-bound Little Mangere Island. Even this fortress did not provide a secure refuge. The robin population had declined from at least 30 in 1968 to 18 birds by 1973, and in 1976, when most were transferred to Mangere Island (by then cleared of stock and other mammals), only seven survived. The problems were not over, and in 1979 the population reached an all-time low of five birds, with a single breeding pair.

More than a decade of intensive, close-order management followed, during which all manner of new techniques were tried. No longer was the hands-off approach of the early takahe years acceptable; now all was fair when saving an endangered bird. By 1992, the black robin population had increased to about 138 birds, and there are now more than 200.

This is one of the best known of all conservation stories and it has been documented in detail in a book by David Butler and Don Merton, a retrospective appraisal by Merton[42] (the recovery team leader) and two television documentaries. Yet so many innovative techniques were developed that no account of wildlife conservation in New Zealand would be complete without a brief account of the black robin story and a tribute to 'Old Blue'.

The transfer to Mangere Island was the first high-risk manoeuvre carried out. In 1976, with only two mated pairs remaining in rapidly deteriorating habitat, Merton judged that the risk of leaving the ageing birds where they were was greater than the risks involved in transferring them to sparse but regenerating habitat on Mangere. That season he transferred five birds, and the two males left behind were moved the following year. To avoid competition with the robins, the few Chatham tomtits on Mangere Island were shot.

The transfer was no easy matter. The bush on Little Mangere Island is on top of 200-metre cliffs, and the birds, once they had been caught, were placed in backpack carry-crates and taken down to a waiting dinghy for the trip to Mangere. When two birds were caught simultaneously, half the potential breeding population was transferred in a single operation. A fall or an overturned dinghy at this stage would have jeopardised the entire operation. Permits had been issued for only one breeding pair to be moved, but Merton and the team decided, in view of the fast-deteriorating habitat and poor breeding success on Little Mangere, that both pairs should be moved. All birds survived the transfer and the move was vital to the subsequent management that saved the species. In these days of greater bureaucratic control and the need to have decisions made at head office with the inevitable delays this entails, Merton would have had less freedom to react to the crisis in hand.

On Mangere Island, breeding success remained low and, with only two elderly females left, time was running out. The situation was in even more critical than Merton and the team thought, as only one of those females, 'Old Blue', ever bred successfully; and then only after 1979 when she changed mates.

In 1980, Merton intervened again, this time removing the robin's eggs and fostering them to Chatham Island warblers. This induced the robins to re-lay and raise their second clutch. The warblers incubated the eggs but did not successfully raise the robin chicks. The following year, eggs were fostered to Chatham Island tomtits, which could successfully rear them to independence. The nearest tomtit population was on Rangatira Island, about 15 kilometres away. The eggs were placed in a Thermos flask modified for the job by Merton, and transported by fishing boat. This was the first time cross-fostering had been attempted with any wild endangered passerine, let alone by a species resident on a different island.

Tomtits had not initially been used as foster parents because they are related to the black robins and it was feared that robin chicks would imprint on their tomtit foster parents or learn inappropriate tomtit behaviours. Some robins did imprint on tomtits, but the problem was alleviated by fostering tomtit-raised chicks back to robin parents for their final week of rearing. To accomplish this, eggs had to be transported from Mangere Island to Rangatira Island, and chicks back to Mangere. All of the eggs and chicks survived translocation. The cross-fostering was so successful that sometimes there was a shortage of robin parents to fledge the chicks, and some robin pairs had up to six chicks to look after. Supplementary feeding was necessary. Mealworms were imported from the New Zealand mainland, and the recovery team also collected naturally occurring insects for the birds.

In 1983, some robins were moved to Rangatira Island to establish a second population. The intensive management was successful and by the summer of 1989–90 there were 66 robins on Rangatira Island and 12 on Mangere Island. After a decade of close-order management the efforts could be scaled back. Each pair was carefully monitored and full records were kept of eggs laid and chicks produced. All nestlings were banded but, providing the population kept increasing, the time for close-order management was over.[43] Intensive monitoring of the entire population continued until 1998–99, when conservation priorities had shifted to other species.

All black robins alive today are descended from a single pair, 'Old Blue' and 'Old Yellow'. The genetic variation in the population is extremely low, even lower than in South Island robin populations descended from the translocation of just a few individuals.[44] This suggests that the low genetic variability resulted primarily from the many generations during which only about 20 pairs survived on Little Mangere Island, rather than the extreme genetic bottleneck. The rapid recovery of the population following the genetic bottleneck no doubt minimised its effect. So far no problems due to severe inbreeding are apparent, but the ability of the species to adapt to future ecological changes cannot be predicted. The deleterious effects of inbreeding that commonly concern conservation biologists are probably overemphasised. John Craig[45] has shown that many species of New Zealand bird breed with kin, showing degrees of inbreeding that some conservation geneticists consider deleterious. During the robin recovery, all nestlings were banded and the genetic parents, as well as the foster parents, of all chicks were recorded. This is the only species that has recovered from such low numbers, and up to 1998–99 the genealogy of all black robins was known. This invaluable information has never been analysed.

Most island species are probably descended from the few colonists that originally reached the island, and thus have already gone through a genetic bottleneck. Species such as the black robin have probably had several genetic bottlenecks: the first when New Zealand's ancestral robin arrived from Australia; the second when New Zealand robins crossed to the Chathams; and, for the black robin, when Little Mangere Island was colonised from the larger islands in the group. Some scientists have suggested that bottlenecks purge the species of deleterious genes and those individuals that survive tend not to carry bad genes. These are questions conservation biologists need to answer, and in the long term an important outcome of the black robin project could be the answers it provides to such questions.

A large proprotion of the limited money and manpower available for conservation was allocated to saving black robins, and we must ask if they could have been more profitably invested elsewhere. The black robin is in the genus *Petroica*, species from which also occur in Australia, New Guinea and islands east to Samoa. When only seven birds survived, the species appeared doomed and there seemed to be good reason to abandon it and divert scarce resources to the more distinctive but more numerous kakapo or kiwi.

I have no doubt that the resources put into black robins were warranted. Can we abandon an endangered species and leave it to become extinct when people directly caused its plight? To date we have managed to avoid such ethical dilemmas, but in future we may not be so lucky. The black robin recovery, once under way, had an end point in sight. The factors that caused the species to become endangered – predators and habitat loss – had been neutralised, so that once the population recovered, management would no longer be needed. This compares favourably with black stilt management, which will be ongoing for at least the foreseeable future. The black robin is a charismatic species that lay people can relate to: television documentaries were made and the media followed the robins' recovery. This bird's plight is symptomatic of wide-ranging ecological problems and it was used to convey a conservation message to a huge audience, many of whom would have little other knowledge of endangered species and their predicament.

The science of conservation biology is in its infancy. The black robin recovery project was carefully documented and the success or failure of each measure was recorded.[46] Many new techniques were developed that have been adapted for use with other endangered species. The financial cost of saving the black robin has never been calculated, but because the entire operation was done on a shoestring budget and much of the labour was by volunteers, it was not huge. The species was saved and at least its short-term future seems secure.

The black robin is the most familiar of many endangered species endemic to the Chatham Islands. Eighteen of the non-marine birds that bred on the Chathams are now gone from those islands (see Chapter 5) and most of the survivors are endangered species now confined to a few small islands. Conservation programmes are under way to save the shore plover, parea, Forbes' parakeet, Chatham Island oystercatcher, taiko and Chatham petrel. These are not the only rare birds endemic to the Chatham Islands.

Are small populations viable?

Conventional wisdom in conservation biology suggested that inbreeding levels greater than one to three per cent are deleterious and, in order to prevent inbreeding, populations need to have at least 50 individuals in the short term and 500 for long-term viability. The figures were originally suggested just as guidelines, but in a discipline where rules are rare, these numbers had appeal and after a few repetitions the '50/500' guideline was soon adopted as one of conservation biology's few rules. Likewise, it has commonly been assumed that, under normal circumstances, animals avoid breeding with close kin.

These assumptions have been the basis of most population viability analyses and are widely used overseas. They have been adopted in reserve design and, in New Zealand, when assessing the suitability of islands for the translocation of endangered species. However, since 1990 these assumptions have been critically evaluated and found wanting.[47]

John Craig[48] has shown that many New Zealand birds violate these assumptions. Most native terrestrial birds remain in year-round territories, and young birds tend to settle close to the territories in which they were raised. Some of these species exhibit levels of inbreeding far higher than conservation geneticists have generally considered desirable. Craig noted that matings with siblings, parents, grandparents or cousins have been recorded in at least eight per cent of saddleback and 14 per cent of tui matings, and DNA analysis suggested that inbreeding could be even higher in blue ducks. The pukeko is one of our most abundant native birds, yet 70 per cent of their matings can be with close kin. Black robins, Forbes' parakeets, Campbell Island teal and Tiritiri Island bellbirds persisted for over a century with populations of fewer than 20 pairs, and island populations of various native birds are the result of translocations of as few as five individuals. No evidence of genetic defects is apparent.

Conservation biologists have long argued as to whether it is most desirable to manage rare species in one large population to avoid inbreeding, or in many small populations to spread the risk of local extinction. Common sense tells us that many large populations are best but failing this, Craig's analysis shows that, for some species at least, small populations have short- to medium-term viability. Recent research, not yet published, suggests that inbreeding may lower breeding success in some New Zealand birds.

Eradication of rats from Breaksea Island

In New Zealand, endangered-species conservation has relied heavily on those few islands that fortuitously remained free of rats and other introduced predators. The eradication of introduced mammals from islands has long been a conservation strategy in New Zealand, but most early eradications were of rabbits and feral animals such as cattle and goats.[49] Cats and rats pose a far greater threat to native species, but they are much more difficult to exterminate. Little more than a decade ago, the eradication of these predators from any but very small islands was considered impossible. The development of eradication techniques has been pivotal in wildlife conservation, and the expertise developed in New Zealand is now used overseas.

Cats have been eradicated from eight New Zealand islands, including large, rugged, bush-clad Little Barrier Island (3083 hectares). The cats were simultaneously hunted, poisoned and trapped into local extinction.

Eradicating cats is hard work, and Dick Veitch, who masterminded the Little Barrier campaign, has suggested that their success depended mostly on the enormous commitment the team had to achieving their goal.[50] The Little Barrier eradication took four years and involved 128 people for a total of 3880 person-days, using 950 traps (baited with fresh fish) plus 27,000 poisoned baits. The last few cats proved especially elusive: it took almost a year and 32,165 trap nights to catch the last five. The rewards have been immense. Hihi and Cook's petrels have increased in abundance, and it became possible to release saddlebacks, kokako and kakapo on Little Barrier.

Rats infest more islands that could otherwise provide refuge for rare species than do other introduced predators, but eradication was a daunting task. As recently as 1978, Dr Ian Atkinson and Dr Kazimierz Wodzicki, who at the time knew as much about rats as anyone, were convinced that eradicating them from even very small islands was impossible. In the early 1980s, safer, more effective poisons became available and two groups of scientists investigated how these could be used in rat eradication. A Wildlife Service (later DOC) team of Phil Moors, David Towns and Ian McFadden began trials on islands in the Hauraki Gulf and succeeded in removing rats from several islands, the largest being Double Island (32 hectares).[51]

Meanwhile, a second team, headed by Rowley Taylor and Bruce Thomas, was trying a very different approach. Rather than waging a long-drawn-out war until the last rat was killed, Taylor and Thomas proposed an intensive, short battle against a naive population and began their campaign on nine-hectare Hawea Island in Breaksea Sound, Fiordland.[52] They built bait stations that rats could enter but birds would not, and, weeks before eradication was attempted, these were set out at 40-metre intervals over the entire island. In April 1986, brodifacoum poison (Talon) was placed in every tunnel and replenished daily. Twelve days later the last rat was dead. Taylor and Thomas then turned their attention to nearby Breaksea Island. At 170 hectares, Breaksea was six times larger than any island from which rats had been eradicated; it was steep, rugged, covered with dense rainforest, and most people believed the task impossible. To everyone's amazement, the Breaksea campaign succeeded and about 21 days after the baits were first set out, in May 1988, all the island's rats were dead.

The secret to Taylor and Thomas's success was careful preparation and exposing a population of naive rats to a highly palatable poison, killing them all in a single 'hit' before they could develop a phobia of poisons, bait stations or people. Previous eradication programmes had involved repeated poisoning or trapping episodes until the last rat was killed. The first 90 per cent had always been easy to kill, but total eradication had foundered when the last few rats proved too expensive or too difficult.

This campaign was preceded by some basic but crucial research. Using captive rats, Taylor and Thomas discovered that the animals would eagerly devour a lethal dose of Talon in a single meal, even when other foods were available, and that even the shyest rats rapidly learned to associate the bait stations with food. They estimated the density of the rat population and identified the time of year (winter) when food was most scarce and rats had ceased breeding. They also determined how far rats would travel in search of food, so they knew how close together bait stations had to be set. With hindsight, they found the 40-metre spacing on Hawea Island was unnecessarily close, and on Breaksea the stations were set at 25- or 50-metre intervals on trap lines 100 metres apart.

The plan required massive preparation. Tracks were cut encircling Breaksea at 60-metre vertical intervals, then 743 bait stations were positioned, each of which had to be replenished daily throughout the 21-day campaign. Helicopters were used to set bulk-feeder stations on rock stacks and other inaccessible places. In all, it took 260 kilograms of Talon and 557 person-days (many by volunteers) to accomplish what up to then had been an impossible dream.

Rodents have now been eradicated from more than 60 islands,[53] the largest being Campbell Island (11,300 hectares). On some islands, Taylor and Thomas's meticulous though labour-intensive method of placing baits in fixed bait stations has been replaced by aerial broadcast of baits from helicopters. This has proven successful but uses more poison and there is a greater risk of accidentally poisoning other animals. On Breaksea Island, bird surveys were carried out before and after poisoning, and careful watch was kept for dead birds or baits that had been interfered with by birds. Two dead robins were found, kea interfered with some bait stations, and insects nibbled some baits, but the losses suffered by protected species were minor and numbers quickly rebounded once the rats were gone.[54] Five years later, Fiordland skinks, large weevils and saddlebacks had colonised the island, mostly with human assistance, and the vegetation was recovering from a century of rat impact.

Kokako: conservation of a mainland forest bird

The decline of the kokako was similar to that of many other forest birds. Forest clearance before European arrival reduced the range of both North and South Island kokako. Their decline was hastened after European settlement, so that by the mid-twentieth century the North Island kokako had a very restricted distribution and the South Island species was very rare, if not extinct. In the 1970s, probably fewer than 1400 North Island kokako remained. The situation was more dire than these figures suggest, as most of the 30 populations were in decline and about 20 of them contained fewer than 30 birds.[55] Great Barrier was the only island on which kokako occurred, but they were scarce and the last died in 1994.

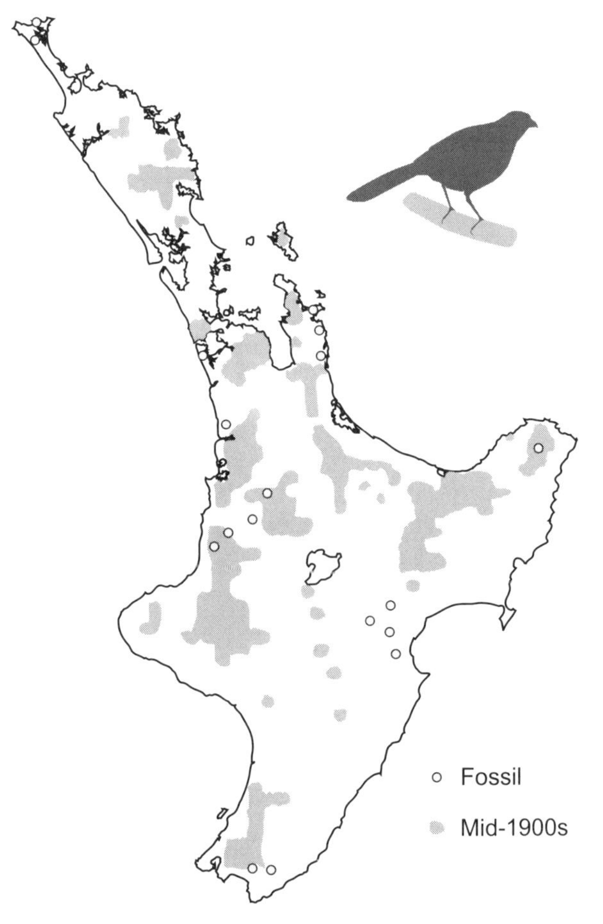

Figure 9.2 Distribution of North Island kokako

Adapted from Innes & Flux 1999

The buff weka is a subspecies that is extinct in its native eastern South Island range but has become an introduced pest on the Chatham Islands.

Photograph by Kerry-Jayne Wilson

Although Reischek's parakeet looks like a mainland red-crowned parakeet, molecular studies suggest that this Antipodes Island form is a distinct species.

Photograph by John Marris

Key to Plate 4 (overleaf): Present-day Oparara Valley, Karamea, with podocarp-beech rainforest and farmland.

Painting by Pauline Morse

Banished from North and South Islands by introduced predators, the little spotted kiwi survives only on islands to which it has been introduced and the Karori Wildlife Sanctuary (see below).

Photograph by Kerry-Jayne Wilson

This predator-proof fence surrounds the Karori Wildlife Sanctuary, a publicly accessible refuge in Wellington city for free-living native birds, including some endangered species.

Photograph by Kerry-Jayne Wilson

Kokako have subsequently been introduced to Kapiti, Little Barrier and Tiritiri Matangi Islands, but to date only the Little Barrier population (more than 120 adults) is self-sustaining.

The paucity of suitable islands and the sentimental belief that kokako should be maintained in their existing range inspired the development of pest-control strategies that enabled kokako to breed successfully in mainland forests. This has proven an important conservation breakthrough and the approaches developed for kokako can be adapted for the protection of other species.

Pureora and protesters

During the 1970s, Wildlife Service surveys had shown that Pureora State Forest held the largest remaining population of kokako as well as a good diversity of other native birds, and indicated that protection of Pureora was critical for the survival of the species.[56] Logging of native timber at Pureora had been vigorously opposed by conservation groups, but their views had gone unheeded. The dispute culminated in 1978 when protesters from the Native Forests Action Council climbed into the treetops and refused to come down until logging was halted. The standoff between protesters and loggers grabbed media attention and eventually a three-year moratorium was declared to allow scientists to determine the effects selective logging would have on the birds.

This research showed that logging and kokako were incompatible. Logging simplifies the forest structure and reduces the abundance and diversity of kokako foods. The large trees most desired for logging are of special importance to kokako. They support lianes and epiphytes that are important foods as well as providing vertical access through the forest, essential for a bird that flies so poorly. Forest gaps created by the removal of large trees were colonised by plants not used by kokako, resulting in short-term deterioration of kokako habitat.[57]

The findings resulted in the permanent protection of North Pureora Forest and provided important information for the conservation of kokako elsewhere. Without the treetop protest there would have been no moratorium, no research and no reserve. Perhaps the research and management detailed below would never have happened if protesters had not increased public awareness of the plight of this evocative songster. While the focus of both protesters and research was on kokako, Pureora supported most forest birds then left on the mainland, and recently short-tailed bats have been discovered there. Kokako were the focus of protest but not the only beneficiaries of the protective status that eventuated.

Kokako, possums and predators

Kokako now have to contend with a suite of introduced predatory mammals as well as herbivorous mammals that are potential competitors for food. Over time, views have changed on the relative importance of these various threats. Deer and goats take some of the same food plants as kokako, but their primary impact on the birds is long term because they browse seedlings and saplings, thus preventing regeneration. Possums, like kokako, forage at all levels in the forest, and rats take fruits and other nutritious plant foods. Studies on the foods taken by kokako and these introduced mammals showed extensive overlap in their diets, and it was thought that competition for favoured foods with possums and other browsing mammals posed a major threat to kokako.[58]

For effective conservation, it is necessary to establish which threats actually limit the population one is seeking to protect, and what level of control is needed to enable the protected species to increase in number. This was the aim of an eight-year research and management programme, led by John Innes, that set out to determine the cause of kokako decline, while at the same time developing management methods to neutralise the threats.[59] The specific aim of that study was to determine whether the control of predators and competitors would result in an improvement in kokako breeding success and, in the longer term, increase their numbers. This was management by scientific experimentation.

Three study sites were used – Mapara, Rotoehu and Kaharoa – all tawa-dominated forests in the central North Island. Scientific research would normally require more replicates and that the replicates be essentially identical, but the researchers had to make do with the limited number of kokako-inhabited forests available. These three forests differed in vegetation composition, size and topography. The programme was a major commitment of resources to control mammal pests in three forests, let alone the larger number scientific trials should require.

Another problematic factor that initially concerned the researchers was their inability to control predators without also affecting competitor populations. Rats ate kokako eggs as well as favouring kokako foods, and during the study it was discovered that the bird's primary competitor, the possum, was also a predator on eggs and chicks,[60] so that the distinction between predators and competitors was flawed anyway. The goal of this programme was kokako conservation, and research was a tool to achieve this – something other conservation scientists should remember more often. The paper describing this huge experiment[61] discusses these constraints and gives some useful tips on how to obtain scientific rigour from such necessarily simple experimental designs. It should be essential reading for anyone undertaking research on or management of threatened species.

The researchers applied simultaneous 'maximum practical' control of possums, rats and mustelids in two study sites, and cats were shot on sight at all three. Pest control was timed so that pest numbers were as low as possible during the kokako's breeding season. Pest control took place from 1989 to 1996 at Mapara, 1990–93 at Kaharoa, and 1994–96 at Rotoehu. The methods of controlling pests were refined during the course of the study, but all changes to control strategies were carefully documented. Kokako breeding success was recorded at all sites each year.

It was found that more pairs of kokako fledged young in managed blocks than in unmanaged blocks. At Kaharoa, 85 per cent of pairs fledged young after three years of management, but none was successful two years after management stopped. Conversely, only 14 per cent of nesting attempts succeeded at Rotoehu before management, but success increased once management began. At Mapara, only a quarter of all pairs attempted to breed in 1989–90 when the study began, and this increased during the course of the study. The early paucity of breeding attemps was due to an initial shortage of females in the population, presumably a result of prior predation on incubating females. As young females entered the population, the male/male pairs broke up and those males re-paired with females. After eight years of predator control the Mapara kokako population had tripled. The number of breeding pairs increased eightfold as the population changed from one with mostly old males to one dominated by young birds with a near-even sex ratio.

The study by Innes and his colleagues showed that predation was a more immediate threat to kokako than was competition. The most significant predators were rats, which took eggs and occasionally young chicks, and possums, which ate eggs, nestlings and possibly incubating females. In the light of these results, Innes suggested that the correlation between possum spread and kokako decline observed by earlier studies[62] was probably due to predation, not competition for food.

In a mainland situation, it is impossible to eradicate introduced mammals. The aim is to keep certain species at 'acceptably' low levels and to lower the rate of re-invasion by other species. This study showed that for more than half of the kokako-nesting attempts to succeed, by November each year possum abundance had to be no greater than one catch per hundred trap nights, and the tracking rate for rats, below one

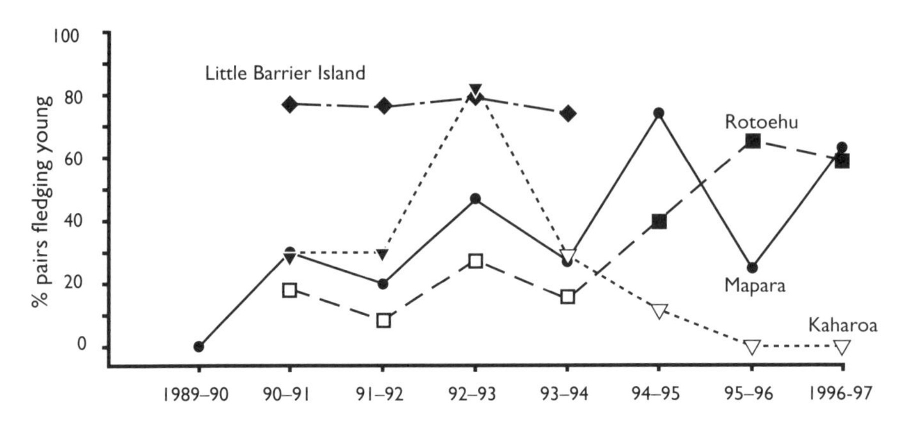

Figure 9.3 Breeding success of kokako on predator-free Little Barrier Island and at managed and unmanaged mainland sites

Source: Innes et al. 1999

per cent.[63] Rat numbers will recover to pre-poisoning levels in three to five months, even when poisoning reduces their numbers by over 90 per cent. Rats tend to avoid poisons and baits they have encountered previously, so kill rates are lower and the time to recovery shortens in second and subsequent years of poisoning. It was important to swap between the two favoured poisons (1080 and brodifacoum) every second year to prevent rats and possums becoming bait-shy.

Rats are known predators of kokako, they are easily killed and there are reliable methods of monitoring their numbers. Therefore, rat control is warranted. With mustelids the situation is different. They probably kill kokako, but not as many as do rats, harrier hawks or possums. Some mustelids are probably killed by eating poisoned rats, but there is no way to measure this. Because mustelid culls are expensive and their effectiveness is unknown, mustelid control was not recommended by Innes and his colleagues.

Kokako management has succeeded in part because the objectives were clearly stated and adhered to. Success was measured by the increase in kokako-breeding success, not by the numbers of rats and possums killed. In too many programmes the success of predator-control operations is measured by the numbers of pests killed, sometimes without even knowing which limits the breeding success of the protected species.

Kokako management – the way forward

> . . . this edition is a celebration of the success of kokako recovery planning so far. Since the 1980s when the first plan was written, our knowledge of kokako ecology has grown enormously. We have now identified key threats and developed successful pest-management regimes to counter them.
>
> – John Innes and Ian Flux (1999)

Thus opened the third edition of DOC's Kokako Recovery Plan. Few recovery plans begin so optimistically, and a decade ago optimism was unknown for species dependent on mainland reserves for their continued survival. The factors threatening kokako are now known, as are methods to neutralise those threats. In 1999, about 270 pairs of kokako remained and the few secure mainland populations were reliant on ongoing pest control. Most mainland populations were still declining and some comprised only one or two birds, so that certain genetic lineages were on the brink of extinction. The aims of the recovery plan are that, by 2020, there will be a thousand breeding pairs, at least 20 populations will have 50 pairs, management will be extended to very large forest blocks such as Te Urewera National Park, and kokako will be returned to the lower North Island.[64] These, Innes and Flux say, are achievable goals provided that pest control is financially, biologically and politically sustainable. They suggest that pest control could be pulsed (five years on, five years off), which would enable managers to rotate between multiple sites. The translocation of birds to unoccupied mainland sites presents an as-yet-untried challenge.

While the management described above was focused specifically on kokako, controlling possums and rats changed the vegetation composition and increased the abundance of other native birds and invertebrates. In the early 1990s, it became apparent that mainland reserves could be managed for the enhancement of a wide range of ecological values.[65] So was born the 'mainland island' concept that will be discussed later in this chapter.

Kiwi and kaka, mainland management – the next challenge

> The costs of managing a mainland population of kiwi are undoubtedly high and the pay-offs uncertain. For all that, we suggest that the costs of not doing anything are even greater.
>
> – John McLennan and Murray Potter (1992)

Until recently, kiwi and kaka had not been considered priorities for conservation. Although the numbers and range of kiwi and kaka taxa have declined since human settlement, all recognised taxa except the great spotted kiwi had populations on predator-free islands. Two decades ago no taxon was thought to be in immediate danger of extinction. However, recent studies have shown that kiwi and kaka are not as secure as once thought. In the 1980s, it was discovered that little spotted kiwi had become extinct on the South Island mainland and the population on D'Urville Island was on the very brink of extinction. The species was effectively restricted to Kapiti Island.[66] Another wake-up call came when a single dog may have killed over half the birds in one of the largest North Island brown kiwi populations during a six-week period.[67] Brown kiwi, previously considered to comprise a single species with three subspecies, was shown to be two separate species, each with a critically endangered, genetically distinct population.[68] Breeding success of mainland kiwi and kaka populations was shown to be too low to sustain the populations.[69]

For little spotted kiwi, translocation to predator-free islands was the logical solution, and new populations were established on four predator-free islands.[70] With four islands currently below carrying capacity, this is the only kiwi species whose numbers are currently increasing. In July 2000, little spotted kiwi were liberated in the predator-fenced Karori Wildlife Sanctuary in Wellington, the first step in returning the species to the mainland.

For other species of kiwi and kaka, conservation is problematic and the major conservation effort requires management on the mainland. North Island brown kiwi remain widespread in the northern half of the island, but most populations have declined by at least 90 per cent.[71] In Hawke's Bay, they have all but disappeared from lowland areas, including suitable forest habitats. The genetically distinctive Okarito and Haast kiwi populations are critically endangered. Great spotted kiwi are the least known of the kiwi and their status is uncertain. Lowland populations are declining, but those in more mountainous regions appear to be stable.

Predation is the main threat facing all mainland kiwi populations. While adult brown and great spotted kiwi are able to defend themselves against all introduced predators except dogs and ferrets, eggs and, in particular, juvenile birds are vulnerable to stoats and cats. Barring episodes such as the rogue dog in Waitangi Forest, mentioned above, only about three per cent of adult brown kiwi are lost to predators each year. However, with their natural longevity and low reproductive rates, even this low predation rate may contribute to a population decline. Traps set for possums are another major cause of adult kiwi mortality.

The most vulnerable stage in a kiwi's life are the weeks following hatching. Only 10 per cent of North Island brown kiwi chicks reach adulthood, and most die during the first hundred days after hatching. Introduced predators account for half those deaths. Mainland kiwi populations will survive only if the losses to predators during this critical period are cut by at least a third.[72] The level of predator control necessary to achieve this is not yet known, but studies to determine management requirements are under way. Captive rearing of young kiwi to be released into the wild once they are of adult size is an alternative strategy in use for some populations. Even with this level of protection, populations will remain stable only if adult survival remains high, and this means that dogs will have to be kept out of kiwi habitats. At present it is too expensive and too logistically demanding to conserve all remaining kiwi populations. Until conservation strategies are refined, the suggestion is that at least two populations in each region should be conserved. In the meantime, populations of the national symbol of which Kiwis of the human kind are so proud, will continue to become extinct.

The situation with kaka is slightly different. There are secure populations of both North and South Island kaka on offshore islands, but on the mainland both forms are now thinly distributed through a few large tracts of indigenous forest.[73] Mainland conservation is perhaps not essential to save the species, but it would be sad to see these magnificent large parrots lost from our national parks. The threats to kaka vary from place to place, but predators, food competition with introduced mammals and wasps, habitat loss, habitat fragmentation and logging have all contributed to their decline and pose threats to existing populations. Mainland kaka populations contain more males than females, a result of predation on nesting females.[74]

In the nearly monotypic beech forests of Nelson Lakes National Park, where the choice of foods is limited, kaka rely on honeydew as their main energy source and longhorn beetle larvae as their main protein source.[75] In beech forests, kaka may only breed in years when beech trees produce seed. Honeydew is a crucial food for many species of native birds (see Chapter 7), but very high densities of introduced wasps have reduced the availability of this vital food source. In the presence of wasps and introduced mammals, breeding success of the kaka was virtually nil; but was it competition for food with wasps or predation that limited their populations? The kaka were provided with high-energy foods, but still only two out of 20 nesting attempts were successful, and the main cause of nest failure was predation.[76] Although little direct evidence was obtained, Peter Wilson and his co-workers concluded that stoats were the primary culprits, as kaka were scarce on islands where stoats were present but healthy populations remained on islands with rats, possums and even cats. It was suggested that, on the mainland, kaka had more often persisted in beech forests than in other more diverse forest types because in beech forests kaka only bred in mast years while stoat numbers were low. Stoat numbers increase after the beech seeds have fallen, so they are high the following season. Thus, in beech forests nesting females are seldom killed by stoats, whereas in more diverse forest types, where kaka breed annually and fluctuations in stoat numbers are less marked, more nesting females are killed. Wilson concluded that kaka could be expected to survive on the mainland only if predators, particularly stoats could be controlled.

After Peter Wilson's study was completed, an adjacent part of Nelson Lakes National Park was declared a mainland island where predators and other pests were controlled, and this resulted in chicks fledging from eight of the 10 kaka breeding attempts monitored.[77] No females were killed once control was instigated. Predator control in this mainland island can raise the productivity of kaka, but if the young birds merely dispersed into the surrounding unprotected forests, this would not be effective conservation. Using a computer model, Ron Moorhouse suggested that the area of control needs to be twice the 825 hectares currently protected.

Kakapo, charismatic megaparrots

One of the most wonderful, perhaps, of all living birds.
– Dr Sclater, British Museum

In 2001, there were only about 60 kakapo, and their peculiar breeding strategy (see Chapter 3) makes them a most difficult species to save. Females produce a single clutch of two to four eggs once every two to four years, and males play no part in incubation or chick rearing.[78] The chicks are noisy and the nests acquire a pungent odour, so predatory mammals can easily find them. Eggs and chicks are left for hours at a time while the female forages, and even kiore can take these when the mother is absent. Other predatory mammals are an even greater threat.

Kakapo were once found through most of New Zealand and their decline began long before European settlement. By the late nineteenth century they were still common in South Westland and Fiordland, but their range was shrinking and by 1976 they were thought to be on the very brink of extinction. It was then believed that only about 15 male birds survived, in remote parts of Fiordland.[79] There was a reprieve when, in 1977, a new population was found in southern Stewart Island. Hope was short-lived because it was soon discovered the birds were falling prey to feral cats, and at the time cat control in this rugged, remote region was ineffective. Between 1982 and 1993, 61 kakapo (38 males and 23 females) – probably all the birds left alive at the time of capture – were transferred to Little Barrier, Codfish, Maud or Mana Islands.[80]

In order to increase productivity, some birds on Little Barrier Island and, later, birds on Codfish and Maud Islands were fed protein-rich plant foods to supplement their natural diet. Supplementary feeding did induce some males to display and a few females to lay, but productivity remained too low to compensate for adult mortality.[81]

Efforts to save the species by some of New Zealand's best conservation managers met with limited success. By 1996, only 50, mostly ageing birds survived and only eight of the remaining females had produced fertile eggs.[82] A review of kakapo management was commissioned. As a consequence, a National Kakapo Team was set up, headed by Paul Jansen, Don Merton and Graeme Elliott, each known for their innovative approaches to management or research. The ebbing kakapo tide had turned.

In 1996, 'Richard Henry' (the only surviving Fiordland kakapo, named in honour of that pioneer kakapo rescuer), and a female, 'Flossie', were transferred to Maud Island. Neither of them had bred during their sojourn on Little Barrier, but in 1998 they produced three chicks, including one female.[83]

The 1998–99 season proved unexpectedly fruitful. Most Codfish Island birds had been placed on Pearl Island in Port Pegasus, Stewart Island, while kiore were eradicated from Codfish, and five of these females unexpectedly bred, producing among them 14 eggs, which were artificially incubated to yield six chicks. (Pearl Island has rats, so the chicks could not be left with their mothers.) Some of the females who had their eggs taken for artificial rearing, then laid a second clutch – something kakapo had hitherto been thought unable to do. 'Lisa', a female assumed dead, was rediscovered on Little Barrier Island incubating three eggs, all of which produced female offspring.[84] By 1999, there were once again as many kakapo as when all known birds had been removed from Stewart Island, and the population included six young females soon to reach breeding age. In 2001, the National Kakapo Team predicted that during the following summer the rimu trees on Codfish Island would produce an especially heavy crop of berries, and this would stimulate kakapo to breed. During the 2001 winter, virtually all females were shifted to Codfish. The rimu fruited as predicted and 2002 was a bumper year for kakapo: by autumn more than 20 chicks had been fledged, yielding a 30 per cent increase in the kakapo population in a single season. For the first time in a century, recovery of this 'most wonderful' of birds appeared possible.

Black stilt

Today, black stilts only breed in the Mackenzie Basin, but a century ago they were widespread in many wetland habitats throughout the South Island and the southern North Island. As with so many species, predation and habitat loss contributed to their decline, but with black stilts there is a complicating factor: hybridisation with the closely related pied stilt. The conservation of black stilts has raised some pertinent issues about how we assign conservation priorities, what it is we seek to conserve and what sacrifices we are prepared to make to save endangered species.

The defensive behaviours used by other stilts to draw predators away from their nests were inappropriate in the mammal-free environment in which the black stilt evolved, and became lost over time. If left unmanaged in today's mammal-infested habitats, only one per cent of eggs laid by black stilts will survive to fledgling.[85] Black stilts will re-lay if early clutches are lost, so in 1981, spurred on by the success in cross-fostering black robins, Wildlife Service workers fostered early-laid black stilt eggs to pied or hybrid stilts. The foster parents successfully raised the black stilt chicks, but the birds produced did not recruit into the black stilt breeding population. Most hybrid and pied stilts leave the Mackenzie Basin to winter near the coast, whereas black stilts remain in the area all year round. The cross-fostered stilts migrated with their foster parents, and the few that did return tended to select pied or hybrid birds as mates. Consequently, there was a shift in management from the production of many fledglings, few of which recruited into the adult population, to the production of fewer, higher-quality fledglings.[86] Cross-fostering of stilts ceased in 1987.

In order to protect black stilt eggs from predators, the eggs were removed from the nest and the birds given ceramic eggs to incubate.

The real eggs were placed in incubators and returned to black stilts (not necessarily the real parents) just before hatching. As a result of artificial incubation and predator trapping around nests, about a third of eggs produced fledglings and the numbers of black stilts increased from 32 in 1982 to 70 in 1990. However, there was only a marginal increase in the number of breeding pairs.[87]

In 1986, a captive-breeding-and-release facility was built near Twizel, within the birds' current range. The goal was to hold six pairs in captivity, enough to produce 15–20 juveniles for release each year. In addition, each year about 30 wild-laid eggs were to be raised in captivity. Water was diverted through the aviaries to ensure that some natural food was available, although with the artificially high density of stilts supplementary feeding was necessary. Aviary-raised chicks had a chance to become familiar with natural foods while in the security of their cage, and growing up in a family group enabled young birds to learn appropriate behaviours and prevented them from imprinting on humans. Young birds were released into the predator-proof enclosure surrounding the aviaries so they had the opportunity to hone their foraging and social skills in a secure, familiar environment before dispersal into the wild. The braided Ohau River, good black stilt habitat, is within sight of the breeding facility. Between 1987 and 1991, a total of 23 aviary-reared birds were released and at least eight survived for more than a year.[88] In subsequent years more birds have been released and survival has been higher.

The immediate problem of nest predation has been controlled and captive rearing now boosts the number of juveniles entering the population, but this has not been enough to save the species. The black stilt is one of the most difficult of New Zealand birds to save. It is vulnerable to all introduced predators plus the native harrier hawk, strongly flighted and the small population is dispersed over a huge area of highly modified and possibly marginal habitat. In the foreseeable future, ongoing intensive management will be required just to maintain the present pitifully small population. The wisdom of devoting resources to a species that is reasonably similar to the widespread and common pied stilt, and for which the prognosis for recovery is poor, has often been questioned.

Black stilts, pied stilts and hybrids

Black stilts look quite different from pied stilts, yet the two forms hybridise and hybrid offspring can breed with either pied or black mates. Black stilts prefer to breed with other black birds or, failing that, with dark-coloured hybrids, but with so few birds spread over the million-hectare Mackenzie Basin, pied stilts are often the only mates available.[89]

Black and pied stilts differ in some behaviours and ecological preferences.[90] Black stilts prefer braided rivers but will use other wetland habitats, especially those close to braided rivers. Pied stilts are primarily birds of coastal estuaries or lakes but also utilise most other wetland habitats, including braided rivers. Black stilts are solitary nesters whereas pied stilts usually nest in loose colonies, and the black stilt chicks forage further from their parents than do pied chicks. Black stilts use anti-predator distraction displays only after their eggs have hatched, whereas pied stilts use them throughout the breeding cycle. All these factors make black stilts more vulnerable to introduced predators than are pied stilts.

Black stilts evolved in New Zealand from pied stilt ancestors, but they were not isolated from the ancestral species sufficiently long to evolve differences great enough to prevent the two forms from interbreeding when the ancestral species colonised New Zealand for the second time. Are their differences sufficient to justify their being recognised as a distinct species? As the conservation dollar is stretched ever further, can we justify spending so much money on a bird that is similar to, and closely related to, another common species?

In appearance, black stilts are more different from pied stilts than are the two species of tuatara, or a cluster of skink or gecko species. What is more, to most people black stilts have much greater aesthetic appeal than these reptiles. Conservation is, in part, about preserving beautiful life forms, and on these grounds there is a case for saving black stilts.

One of the scientific justifications for conservation is the preservation of genetic diversity. With a population of fewer than 80 pure black birds, there are now possibly more black stilt genes in the hybrid population than in the black population. A study of the DNA of black and pied stilts showed that, as a result of interbreeding, New Zealand pied stilts carry many black stilt genes and are more similar to black stilts than are Australian pied stilts.[91] A subsequent study by Brenda Greene showed that the genetic distance between pied and black stilts was small and similar to that between certain hybridising subspecies of overseas birds.[92] However, because black stilts selectively mate with black rather than pied birds, Greene argued that the two stilt forms should be considered separate species for conservation purposes.

When a population of animals becomes isolated from its parent stock, the animals change in response to prevailing environmental conditions, eventually becoming so different that they are unable to reproduce with their parent stock. At that point they are indisputably separate species. Black stilts were re-exposed to their parent pied stilt stock partway along this evolutionary process, at a time when human changes to the environment and newly introduced predators favoured the parental species.

There is no black-and-white answer to the stilt conundrum. The species concept is of square boxes into which the round stilt 'pegs' do not easily fit. Whether we should devote money to saving them is debatable. I think black stilts are pretty and I would be sad if I were unable to see them again. However, the rational view I might espouse with less familiar or less appealing species would be that we should devote our resources to more distinctive taxa.

To some extent the argument is irrelevant. Much of the money devoted to black stilt conservation has come from electricity-generating companies, who would be unlikely to devote money to species less directly impacted by hydro-electric development.

Conservation of braided-river habitats, but at what price?

Braided rivers are poorly represented in New Zealand's conservation portfolio. The headwaters of some rivers are in national parks, but few have any form of protection over their lower reaches. All braided rivers have been modified by hydro-electric development, stopbanks, gravel extraction, stock grazing, recreational activities or by the encroachment of vegetation (in particular legumes such as Russell lupin) on to riverbeds.[93] Because these are stabilised, introduced plants spread and leave progressively less habitat for breeding birds. Bird species that primarily nest on braided rivers prefer beds with low to moderate vegetation cover, not the dense vegetation that establishes on stabilised rivers. The tall vegetation also provides cover for predators so the remaining birds are subjected to increased predation.

Farmers whose land could be made more productive through irrigation see the rivers as a source of water, but this abstraction alters water quality, flow regimes and invertebrate numbers. Hydro-electric power development has greatly altered many braided rivers. Over much of its length, the once-braided Waitaki River was transformed into a series of lakes. In the Mackenzie Country, the Upper Waitaki Power Scheme has diverted water that used to flow down the braided sectors of the Tekapo and Ohau Rivers into canals. Raising Lakes Tekapo and Pukaki for hydro storage has flooded the deltas of rivers flowing into them, which were important foraging areas for black stilts[94] and presumably other species. Braided rivers are used for recreational activities such as off-road driving, not all of which are compatible with the needs of the birds. As riverbed habitat has been lost or degraded, the distribution of species most dependent on rivers has become more restricted; for instance, about a third of all black-fronted terns and almost all black stilts now breed on just a few rivers in the Upper Waitaki Basin.[95]

Some birds of braided rivers continue to decline in abundance. Counts were made on nine rivers in the Upper Waitaki Basin during the 1960s and again in the 1990s.[96] Trends varied between rivers and between species, but South Island pied oystercatcher, banded dotterel, black-fronted tern and black-billed gull numbers had declined on half or more of the rivers surveyed. Of the six specialist braided-river species, only wrybill and black stilt had not declined in abundance over that 30-year period. They were already threatened or endangered by the 1960s. The decline in black-fronted tern numbers appears to have been especially marked.[97]

Of the various threats faced by braided-river birds, the encroachment of vegetation is the most insidious, and Project River Recovery has restored the lower Ahuriri River and the Tekapo delta without compromising fishing and other recreational uses. Willows were removed by mechanical digger, left in piles to dry and then burned. Young willows, lupins, gorse and other weeds were sprayed with the herbicide Grazon, largely by helicopter. This herbicide was chosen because it is not toxic to people, stock, fish or bees, is broken down in soil or water, and does not accumulate in the food chain.

The programme was controversial – few people liked the idea of applying herbicides to rivers – but the alternative was worse. Without management the braided rivers would have degraded further and some birds might have become extinct. Weeding by hand was impractical over the large area concerned, and heavy machinery such as bulldozers would have caused unacceptable damage.

At the Tekapo delta site, willow clearance did enable birds to recolonise. Similar numbers of banded dotterels, black-fronted terns, pied stilts and South Island pied oystercatchers nested in areas cleared of willows as nested in areas that were previously free of willows.[98] Wrybills visited the delta after willows had been removed, but the cleared areas probably did not meet this species' specialised habitat requirements.

For black stilts and other braided-river birds, there is little option but to manage them in highly modified habitats, in the presence of the threats that made them rare and in places where human demands conflict with the birds' requirements.[99] Management of entire catchments is probably the only option for conserving these birds. It requires that we not only accept the ongoing monetary costs of conservation, but that New Zealanders reduce their demands on the environment and accept restrictions on some recreational activities.

Frogs, reptiles and bats: the forgotten fauna

Until recently, far more effort had been devoted to saving birds such as the black robin or black stilt (members of genera that occur in other countries) than to New Zealand's more highly distinctive tuatara, frogs, lizards and bats. Tuatara and frogs have long enjoyed protective legislation, but until 1990 there had been few attempts to actively manage non-bird wildlife. The last decade has seen a shift from passive protection, through legislation described by some as management by 'benign neglect' of these other species, to active management. There has been better documentation of continuing declines in reptile and amphibian populations (see Chapter 2) and the implementation of conservation programmes to increase populations of certain reptiles and frogs.[100] Although the last decade has seen extensive research on bats (see Chapter 3), hands-on management has not yet been attempted. DOC's Bat Recovery Plan seeks to protect existing populations, establish new populations on suitable islands and raise public awareness of bats.[101] In the following sections I discuss some of the management efforts devoted to nonbird species.

New homes for frogs

Now that Hamilton's frog and the Maud Island frog are considered separate species, the former, with a population of fewer than 300, is perhaps the world's rarest frog.[102] They presumably once inhabited the forest that covered most of Stephens Island, but when the forest was cleared they became confined to a 600-square-metre boulder bank near the island's summit (see Chapter 2). This area was virtually devoid of vegetation and the frogs survived in this marginal habitat by retreating deep below ground during hot weather, coming to the surface only on damp nights.

In 1951, the bank was fenced to exclude stock, and this enabled grasses and *Muehlenbeckia* vines to shade part of the frogs' refuge. Subsequently, trees were planted to provide better shade and a fence was built to exclude predatory tuatara from half the bank.[103]

About 50 metres from the bank there is a bush remnant that is climatically suitable for the frogs but lacked permanently humid retreats and an adequate prey resource. In 1991, a pit was excavated in this remnant and filled with 15 tonnes of rocks to create fissures and crevices similar to those on the frog bank. The newly constructed 'frog pit' was an irregular shape, with arms extending between trees and its surface contoured to provide a range of microclimates. Tuatara were excluded from the pit and a range of invertebrates introduced to provide food for the frogs.[104] Twelve frogs were introduced to the pit a year later. Survival was high and, four years after transfer, a juvenile frog was seen. A corridor of rock-filled pits and trenches beneath a planted forest canopy will be created to link the two habitats and allow frogs to move between the two populations.

Maud Island frogs, which until recently were restricted to a 10-hectare patch of forest, have been translocated: in 1984-85 to a second location on the island, then in 1997 to Motuara Island, also in the Marlborough Sounds.[105] Inter-island transfers of the Maud Island species are being undertaken not only for the conservation of that vulnerable species but to develop successful techniques for subsequent translocations of Hamilton's frog.

Translocations of lizards

The translocation of frogs and reptiles is in some ways simpler than the movement of birds, but it presents challenges not encountered with those transfers. One of the first lizard translocations took place in 1988, when 28 endangered Whitaker's skinks were taken from Middle Island to Korapuki Island after rats had been eradicated from the latter.[106] Over the next five years, more Whitaker's and three other skink species were liberated on Korapuki, the animals all being sourced from other islands in the Mercury group.

David Towns[107] described the particular challenges posed by lizard translocations. Whitaker's skink has a particularly low reproductive rate, first breeding when four years of age, then once every second year, and producing only two to four young per litter. During her lifetime, a female Whitaker's skink only produces about 16 offspring. This means that even under ideal circumstances newly established Whitaker's populations increase slowly. The reproductive rates for most other New Zealand lizards are higher, so they can establish viable populations more quickly. Towns suggested that on small islands species like Whitaker's skink should be introduced first so their populations can build up before the more fecund species arrive. Whitaker's skink, Duvaucel's gecko and tuatara, each of which has very low reproductive rates, could take decades to build up to modest densities, and it could be centuries or even millennia until the population attains the number the island could support.[108]

Lizards seldom move more than a few metres from their birth site, so they have limited ability to explore the island on which they are liberated. The success of a release depends on the lizards' finding suitable habitats and foods within a few metres of their release site. Thus, for a release to succeed, their habitat requirements must be known to enable an appropriate release point to be chosen. In contrast, birds can quickly explore over a far wider area to find the habitats and foods that best suit them, so their release point is less crucial. Lizards can survive on very small islands, so more islands are available for their conservation than for birds.

Where several lizard species were to be introduced, Towns[109] suggested that each be liberated on a different part of the island so populations could build up before they came into competition with one another. The addition of top predators such as tuatara should be delayed until their prey species were well established. Predators that extirpated tuatara and lizards from islands probably also removed certain large ground-dwelling invertebrates that were the reptiles' most common prey. Thus reptile re-introductions are likely to fail unless carried out in conjunction with the restoration of invertebrate communities. Restoration projects on Mana Island and the Mercury Islands set out to restore both reptile and invertebrate communities, along with the seabirds that import necessary nutrients to the restored ecosystems and whose burrows provide refuges for the smaller animals.[110]

Not all New Zealand lizards have such low reproductive and dispersal rates. Forty Fiordland skinks were translocated to Hawea Island in Breaksea Sound after Norway rats were eradicated.[111] The population built up quickly and in 1992, less than five years after their release, over 200 skinks were present, some more than 200 metres from the release site. A release of Fiordland skinks had been planned for Breaksea Island, but before this took place the skinks colonised the island naturally from a nearby rock stack. Since those initial translocations to Korapuki and Hawea Islands, a total of 12 lizard species have been re-introduced to seven different islands following rodent eradication.[112] On some of those islands, rodents were eradicated expressly to facilitate reptile conservation.

Reptile conservation on the mainland

Virtually all native reptiles are vulnerable to introduced predators and habitat change, and, as discussed in Chapter 2, mainland reptile communities have been greatly depleted. Perhaps in years to come restoration of certain mainland communities will be possible. In the meantime, efforts are being made to preserve the last mainland population of Whitaker's skink and to conserve the lizard community at Macraes Flat in Central Otago, for which there is no island replicate.[113]

Whitaker's skink was once found over much of the North Island but survives only on a few small islands and at one mainland site, Pukerua Bay, near Wellington, where about 300 remain in a half-hectare greywacke boulder bank. The lizards are effectively isolated in this tiny area where crevices between the rocks provide refuge from predators. Their survival is little short of miraculous because the habitat has been greatly modified by grazing animals and fire, the full suite of predatory mammals is present and the bank is adjacent to a popular walking trail. Management of this population is a difficult balancing act. Fencing to exclude stock has resulted in an increase in fire-prone long grass, while control of rabbits, rodents or mustelids could upset the existing predator/prey balance and cause the remaining predators to eat more lizards. Control of cats is especially problematic in this urban-fringe nature reserve, which is visited by local pets.

The threats faced by lizards at Macraes Flat were described in Chapter 2. At both these mainland sites the lizards face multiple threats, and management of any single threat can alter the intensity of others. However, there are some crucial differences between the two sites. At Macraes Flat, vestiges of little-modified habitat remain and all species are still sufficiently numerous that experiments to determine appropriate land-management procedures are possible. If good procedures can be devised, restoration of the Macraes lizard community should be possible, even in areas not currently occupied by lizards. At Pukerua Bay, Whitaker's skinks have no opportunity to expand in range and the population is so critically endangered that the options for effective conservation are severely limited.[114]

Tuatara – bad taxonomy can kill

During the nineteenth century, two species of tuatara and several subspecies were described, but for most of the following century only a single species, *Sphenodon punctatus*, was recognised. The vulnerability of tuatara to predators, including kiore, was widely acknowledged, but since it was believed that the same species occurred on 40 islands, one of which had an estimated 100,000 animals, the species was not regarded as threatened. However, in 1990 Charles Daugherty and his colleagues from Victoria University visited most tuatara islands, studied the morphological and allozyme variations between the populations, and discovered that the tuatara on North Brother Island in Cook Strait were sufficiently different from the rest to be considered a separate species. They reinstated the scientific name S. *guntheri* that had been given to these animals in 1877 but not used for a century.[115] There were only 300 Gunther's tuatara, all on this tiny, windswept island.

Daugherty also considered tuatara on other Cook Strait islands to be a different subspecies from those on islands off the eastern North Island. During the nineteenth century, the tuatara on Little Barrier Island had been considered to be a separate subspecies, but none could be found during Daugherty's survey. A few were located in 1991 and placed in a rat-free enclosure pending the eradication of kiore from that island. However, Little Barrier tuatara are no longer given subspecific status.[116]

Relegation of all tuatara to a single taxon gave a false sense of security for one of New Zealand's special zoological treasures. Because the genetic diversity among tuatara was not realised, one of only two species surviving from this Jurassic lineage was placed at unnecessary risk.[117]

There was an obvious need to increase the numbers and range of Gunther's tuatara, but this was no simple matter. One priority was to establish them on at least two further islands – but which? South Brother Island, a kilometre from their present refuge, was one option, but it is difficult of access and so seldom visited that it was not known whether tuatara were already present and, if so, which species. South Brother Island is pristine and the policy is to avoid liberations, even of endangered species, on the few remaining unmodified islands. Eggs were taken to establish a captive population as insurance against rats getting to North Brother Island, and to provide young animals to seed new populations.[118] Wild-caught adults plus captive-raised young have subsequently been released on two other islands in the Cook Strait region.[119] One of these is Somes Island in Wellington Harbour, which is open to the public, so as the tuatara population builds up, those who ultimately fund most conservation work will have an opportunity to see these fascinating creatures in the wild. Captive-hatched animals have also been used to speed up the recovery of northern populations when kiore were eliminated from tuatara islands. The numbers on some of these islands were so low that recovery was unlikely without such assistance.

In contrast to the intensive management of certain rare birds, little has been done to ensure the survival of tuatara, despite their extraordinary zoological interest. The conservation of birds has even proceeded to the possible detriment of tuatara. Kiwi, kakapo, saddleback, kokako and hihi have all been introduced to tuatara islands in efforts to save the birds, with little thought to possible impacts the birds could have on tuatara. Insects are even lower in the conservation pecking order. On both Stephens and North Brother Islands, endangered insects have declined, one species (the large carabid beetle *Mecodema puntellum*) to probable extinction. Other large insects have become locally extinct from both islands. The unusually high density of tuatara on these islands, possibly an artefact of human disturbance, has probably contributed to the insect's decline.[120]

Tiritiri Matangi, an open sanctuary

Most of the conservation case studies discussed so far in this chapter have been located on islands where public visits are prohibited, or in mainland locations where viewing of the animals is at best difficult. Tiritiri Matangi Island, or Tiri as it is affectionately known, is an open sanctuary where anyone may visit and view some rarely seen and endangered species.[121] Today, the island is mostly clad in regenerating coastal bush, yet in 1980 it was covered by pasture grass with four small, degraded forest remnants. Since 1984, over 280,000 trees have been planted in a regeneration programme that is the result of innovative science, the dedication of resident rangers Ray and Barbara Walter, and the hundreds of volunteers who have planted trees and erected the boardwalks that enable people to visit the forest and observe the birds without damaging the environment.

Tiri has a long history of human use and was extensively modified by burning and grazing. Maori had lived there for perhaps hundreds of years. A lighthouse was built in 1865 and the island inhabited by keepers, their families and their stock until the light was automated in the 1970s. In the 1960s, kiore, cats, goats, sheep and cattle were present. In 1971, stock were removed so that the island could regenerate into native forest. However, recovery was slow, hampered by the grass sward and tangles of bracken fern. Left to nature, regeneration would have taken a long time.

The concept of replanting the island using public labour and making it an open sanctuary was the brainchild of John Craig and Neil Mitchell of the University of Auckland. A nursery was set up on the island to provide seedlings grown from locally gathered seed, and the aim was to replant 60 per cent of the island over a 10-year period. The remaining area consisted of archaeological and historic sites, and ridges from which there were views of Auckland, the Coromandel Peninsula and the Hauraki Gulf islands.

Beginning in 1984, pohutukawa and a few other fast-growing pioneer species were planted to provide groundcover and suppress the grass and bracken to create suitable habitats for other tree species. Ten years later, all the areas to be planted were done. Some recently planted areas were still effectively grass sward with pohutukawa, *Coprosma*, cabbage tree and manuka seedlings, while in the first areas to be restored, cabbage trees and pohutukawa were three to four metres tall. Since then, trees have grown quickly and the forest is now losing its artificial appearance as natural seeding takes over.

From the beginning it was envisaged that animals would be introduced to Tiri. Red-crowned parakeets had been released in 1973, and next to arrive were saddlebacks, in 1984. It had been thought that saddlebacks needed mature forest, and they were released in the existing bush remnants on Tiri to determine whether they could live in regenerating forest and thus whether they could adapt to habitats on other predator-free islands. The saddlebacks thrived and have produced more chicks per pair than they have on islands with more mature forest. 'Surplus' Tiri saddlebacks have since been transferred to Mokoia Island in Lake Rotorua.

In 1991, the first takahe to reach Tiri were welcomed ashore by a crowd of over 500 people, with extensive media coverage. They were both males, but when they built a nest and showed other appropriate behaviours, a takahe egg was flown in from Maud Island. The male pair successfully raised the chick, which unfortunately died a year later. Subsequently, further takahe have been released on Tiri. Today, there are fewer than 180 takahe in the world, yet visitors to Tiritiri Matangi can view these endangered animals. Six other native bird species have been introduced (Table 9.1), and blue penguins, grey-faced petrels, bellbirds, tui and other native species are also present. Bellbirds from Tiri are now regular visitors to the adjoining mainland, to the great delight of residents at Whangaparaoa.

Kiore were eradicated from Tiri in 1993 by an aerial drop of brodifacoum. Before this, non-toxic pellets containing a fluorescent dye but

Table 9.1 Bird species translocated to Tiritiri Matangi Island and their approximate population size in the year 2000

	Date of translocation	Population in 2000
Red-crowned parakeet	1973	Hundreds
Saddleback	1984	400–500
Brown teal	1990	About 6
Whitehead	1989, 1990	Hundreds
Takahe	1991 and subsequently	19
North Island robin	1992, 1993	60–80
Little spotted kiwi	1993	21–24
Hihi	1995	>50
North Island kokako	1998	8

Source: Unpublished information supplied by J. Taylor

otherwise identical to the poison baits were spread over a small area to determine which birds were likely to take the baits. Brown teal, takahe and pukeko ate the pellets, so before the poison drop, takahe and teal were caught and penned in safety, to be released again once the poison had degraded. It was impossible to protect pukeko, and some did die, but since they are a common species throughout most of the country, their loss was seen as a small price to pay for the greater conservation good. To guard against the arrival of rats, 50 permanent poison-bait stations are set up on the wharf and on beaches around the island. With kiore eradicated, the introduction of lizards, tuatara and some rare invertebrates is planned.

From the outset Tiri has been a sanctuary with free public access, and a commercial launch provides transport for those who do not have a boat of their own. More than 13,000 people visit the island each year. Other island-restoration projects are under way, the two most notable being Somes and Mana Islands in Wellington Harbour.[122]

Kiwi in Wellington City

Three kilometres from Wellington's central business district is the Karori Water Supply Reserve, a 250-hectare watershed with significant areas of native forest and wetland. In 1992, an ambitious plan was hatched to fence the area, eradicate all introduced mammals, encourage forest recovery, and then introduce native birds, thus creating an urban sanctuary for rare species.[123]

It was no easy task. A predator-proof fence that could exclude even mice and cats had to be designed then built. This was completed in August 1999 and, after allowing a month for any domestic cats to leave (the fence being designed so cats could climb out but not in), all 14 species of introduced mammals present were eradicated. Never before had so many species been exterminated simultaneously, and this was achieved by a combination of poisoning and trapping. There was an initial operation to knock down the possum numbers before poison was laid to eradicate rats, mice and rabbits. Cats and stoats were eradicated by allowing them to eat the poisoned rat and rabbit carcasses. Last to be removed were hedgehogs, one rabbit and four possums that had survived the initial eradication operation.

Management does not stop once a fence is in place and the pests are eradicated. Ongoing monitoring is required to ensure re-invasion does not occur, but at Karori this costs less than on a mainland island. Tracking tunnels and bait stations have to be checked and the sanctuary is regularly patrolled by a trained dog to detect any re-invading mammals. The house mouse is the only introduced mammal that has managed to enter the enclosure. The fence is checked every week to ensure it has not been breached by tree-falls or vandals.

Little spotted kiwi, weka, brown teal and North Island robin have been liberated in the reserve, and plans are afoot to introduce a variety of other species, including whitehead, tomtit, rifleman, kaka and tuatara.

Flightless birds will be safely confined by the fence, but not so flighted birds. However, it is hoped that they will establish resident populations within the security of the sanctuary before they disperse over the fence. The weka are contained separately within the reserve to enable ground-nesting birds, reptiles and invertebrates to build up in numbers before they have to contend with weka.

Karori Wildlife Sanctuary was opened to the public in 2001 once walking tracks and other visitor facilities were in place. As it is operated by a private trust with no government funding, there is an entry charge for non-trust members.

Open sanctuaries like Tiritiri and Karori are more likely to be violated than remote islands with restricted access. A boat visiting Tiritiri could have rats on board; weed seed will be carried through the Karori fence; the fence itself may be damaged. Ideally, rare species will be liberated into open sanctuaries only after populations are established on more secure refuges, so no species depends solely on an open sanctuary. Although that guideline has mostly been followed, both these sanctuaries are home to little spotted kiwi, and Tiritiri has takahe – both species with insufficient other refuges for these sanctuary populations to be considered 'expendable'. The gains through advocacy are considered to outweigh any risk a few members of the species may be exposed to.

Mainland islands

So-called 'mainland islands' are areas of natural habitat that are intensively managed for two purposes: to conserve species that cannot be marooned on offshore islands, and to restore indigenous mainland ecosystems. They are islands of managed habitat, effectively ring-fenced in some way from the surrounding unmanaged areas. Some are islands of forest habitat in a 'sea' of pasture, where certain threats can be eliminated and re-invasion prevented. For other threats such as stoats and rats, control will be ongoing. At Trounson Kauri Park in Northland, for example, there is a grid of poison bait stations throughout the reserve to control possums and rodents, while a ring of traps around the forest margin catches cats and mustelids as they attempt to enter the reserve from the unmanaged matrix.[124]

Mainland islands are a new concept in conservation – the first were initiated in 1995 – and have required the development of a range of new management techniques. Efficient and effective methods to control undesirable species are still evolving. The Department of Conservation has six mainland islands ranging in size from 117 to 6000 hectares, including the Nelson Lakes National Park project (see page 292).[125] In addition, there are a number of other similar projects, some on private land and others undertaken by local communities.

There have been significant reductions in key pests and increases in indigenous species at all mainland islands, although the pests targeted and the species protected vary between the reserves.

The Northern Te Urewera team has developed some particularly innovative approaches to ecosystem management. The project began in 1995 by establishing a single core area of 1300 hectares, the idea being to get the techniques right before expanding management to eventually cover much of the northern section of Te Urewera National Park.[126] In subsequent years, new core areas have been added that sit within a much larger background area over which possums are controlled (although even for that species control is more intense in core areas). Nested within each core area are zones that receive different treatments; for instance, deer are excluded from the entire Otamatua core area, mustelids are controlled over half the area, and rats only in the central block.

So far, mainland island management projects are reliant on annual doses of poison to control rats, and there is public wariness of the poisons used. The Northern Te Urewera team have responded to the concerns of local iwi and have attempted to control rats by trapping. Early results look promising, with experimental trapping knocking rat populations down faster and for longer than in nearby blocks where less labour-intensive poisoning was used.[127] The trapped rats have been freeze-dried and used as bait in stoat traps. They retain the ratty smell that attracts stoats, and last longer than the traditionally used hen eggs.

After only four years, some significant conservation benefits had been realised. Within the managed areas, kokako and robin numbers had increased more than threefold; many more young kiwi now survive; and bellbird, kereru, tomtit and tui have doubled in number. With whitehead, the increase has been even more dramatic. There has been a corresponding recovery in vegetation. This is an exciting and ambitious project, and it is planned to eventually achieve ecosystem restoration over a 50,000-hectare area. Results to date suggest this might just be possible.

CHAPTER 10

Seeking solutions

> Pukekaroro and the bush walls behind the mill were brilliant with spring colour, clumps of rata red and kowhai gold and new fern greens. But nowhere now was there silence, and rarely could one hear the song of a solitary bird calling for its mate. Not a tui had been heard about the bay for years, and even the enterprising wekas had been driven further into the forests.
>
> – Jane Mander (1920)

The story that has unfolded in this book is not a happy one and there is no happy ending. During the last 2000 years, 64 endemic vertebrates, including 58 of New Zealand's 133 endemic birds, have become extinct. Fires and habitat loss during the last 800 years, and the tumultuous changes since 1800, have resulted in rapid and extensive habitat loss, including the near-total destruction of the biologically diverse eastern dryland forests. Over five per cent of the threatened or endangered birds listed by Birdlife International[1] are endemic to New Zealand, a high figure when you consider this country's comparatively small avifauna. Over 600 plant and animal taxa are at risk of extinction,* a similar number to that for the entire United States, which has a far larger tax base to support threatened species recovery.[2] Thanks only to intensive management, it is now nearly 40 years since a New Zealand native bird or mammal last became extinct. However, if management ceased, a frightening number of endangered birds would soon be extinct, threatened species endangered and common species threatened. Our conservation managers have their backs against the wall.

* Estimates of the number of threatened and endangered species vary. Molloy and Davis 1994 list fractionally over 400 taxa, but de Lange et al. 1999 list over 500 plant taxa, bringing the total to well in excess of 700.

Today, most New Zealanders live in landscapes that more closely resemble those of a distant continent than the landscapes present little more than a century ago. The animals and plants most people encounter day to day are introduced or have colonised since human-induced habitat change began. New Zealand once had one of the most distinctive biotas on the planet, but now has one of the world's most highly modified. Sadly, most people have little awareness of what has been lost, nor do they appreciate how rapid environment degradation has been. As long as conservation remains discretionary, protection of these taonga will continue to be compromised in favour of economic development.

The year 1948, when takahe were rediscovered, marked a major turning point in New Zealand's wildlife history. Until then the loss of native species had been considered regrettable but inevitable, and few attempts had been made to save native wildlife. With the rediscovery of takahe there was an expectation that the species should be saved, and virtually overnight this created a new role for the fledgling Wildlife Branch.

Since 1948, conservation management has done much to hold the fort. A few species have been saved, nine of the 13 national parks and all the marine reserves have been gazetted, and we now have a government department responsible for conservation. Yet, during that same half-century, vast areas of forest and wetland habitats have been lost, and many species, including kiwi, kaka, parakeets, mainland robins, blue duck, weka, tuatara, lizards and frogs, are now less common.

In this final chapter I review the changes that have occurred to New Zealand's vertebrate biota since first human contact. I also discuss how wildlife conservation has evolved since 1948 and how people's changing attitudes have influenced this, and I speculate on how future attitudinal change could affect the conservation of indigenous and introduced species.

The changing vertebrate fauna

In previous chapters I have discussed extinct and introduced species, but now I want to consider the net result of human-induced change to the New Zealand vertebrate biota. At first glance it appears that the fauna has been enriched, as more species have been introduced than have become extinct and habitat changes have enabled 14 more bird species to colonise New Zealand by trans-Tasman dispersal[3] (Table 10.1, page 312). Consequently, there are now 58 more species of vertebrates

Table 10.1 Extinct and introduced vertebrate species on the New Zealand mainland, including the land-bridge islands

Vertebrate taxa	Extinctions	Introductions	Recent colonists	Change
Freshwater fish	1	21	0	+20
Frogs	3	3	0	0
Reptiles	3	1	0	–2
Birds	45	39	14	+8
Mammals	1	33	0	+32
Total	–53	+96	+14	+58

Adapted from Wilson (1997) using the bird species list compiled by Holdaway et al. (2001)

in New Zealand (excluding the outlying islands) than there were 2000 years ago. Since there is no evidence that any species became extinct between the early Pleistocene and the arrival of kiore, there are probably more vertebrate species in New Zealand now than at any time during the last two and a half million years. Indeed, there may well be more vertebrate species now than ever before.

The 45 extinct bird species listed in Table 10.1 includes only those lost from the New Zealand mainland, including its land-bridge islands. Two of the species – black swan and shore plover – survived elsewhere and have been repatriated to the mainland. The shore plover survived on Rangatira Island in the Chatham group and has recently been reintroduced to Portland Island in Hawke Bay. If Trevor Worthy[4] is correct and the swan that became extinct in New Zealand was the Australian black swan, then the species was lost from New Zealand early in the Polynesian era, only to be re-introduced in 1864. In addition, 13 birds endemic to the Chatham Islands became extinct (Chapter 5), and the Phoenix petrel and a megapod were lost from the Kermadec Islands but survive in other countries.[5] Thus a total of 66 vertebrates, including 58 bird species, have been lost from the New Zealand region.

Of those 66 locally extinct vertebrates, all but the two birds lost from the Kermadec Islands and the re-introduced black swan were endemic to the New Zealand region, and 27 belonged to endemic orders, families or subfamilies (Table 10.2, opposite). In contrast, most introduced species and recent colonists have wide natural distributions and are common in their native range. Many were also introduced to other countries. Four of the introduced species are threatened in their natural range, and some introduced birds are now more common in New

Table 10.2 Status of species that belong to families and orders endemic to New Zealand

	Total species	Extinct	Threatened or endangered	%age extinct, threatened or endangered
Leiopelmatidae (New Zealand frogs)	7	3	4	100
Sphenodontida (tuatara)	2	–	2	100
Dinornithiformes (moa)	11	11	–	100
Apterygiformes (kiwi)	5	1	4	100
Aptornithidae (aptornis)	2	2	–	100
Strigopinae (kakapo)	1	–	1	100
Nestorinae (kea and kaka)	3	1	1	66
Acanthisittidae (New Zealand wrens)	6	4	–	66
Mohouinae (whitehead, etc.)	3	–	1	33
Turnagridae (piopio)	2	2	–	100
Callaeatidae (wattlebirds)	5	2	3	100
Mystacinidae (short-tailed bats)	2	1	1	100

Adapted from Wilson 1997 using the bird species list compiled by Holdaway et al. 2001

Table 10.3 Status of native birds that breed in the New Zealand region

	Non-marine birds	Seabirds
Breed in New Zealand	137	86
Endemic to New Zealand (full species only)	120 (87%)	38 (44%)
Extinct	53 (39%)	5 (6%)
Threatened or endangered	23 (17%)	20 (23%)

Note that threatened and endangered species are those taxa given category A or B status by Molloy & Davis 1994 that equate to full species as listed in Appendix 1

Zealand than in Britain, whence our populations were sourced. They will be discussed later in this chapter.

Of the birds lost, most have been terrestrial or wetland species (Table 10.3). Marine birds survived because most had populations on offshore islands, but huge ecological changes have occurred with their loss from mainland ecosystems (Chapter 8). As predators have spread to ever more islands and fisheries-related impacts have intensified, seabirds have declined and there are now almost as many threatened or endangered as there are threatened non-marine species.

Since first human contact, the New Zealand vertebrate biota has become increasingly cosmopolitan and dominated by species that are also common elsewhere. Two thousand years ago, over 80 per cent of the non-marine birds, 44 per cent of marine birds and all frogs, reptiles and bats were endemic, and a high proportion of them belonged to endemic families or orders. Today, most species belonging to those higher taxa are extinct or endangered. As we saw in Chapters 4 and 7, whole ecological guilds such as the flightless browsing birds were lost and whole ecosystems disrupted by the introduction of herbivorous and predatory mammals. Owing to habitat change, far more species have successfully colonised New Zealand by trans-Tasman dispersal during the last millenium than during previous millennia. Introduced species have now invaded virtually all indigenous ecosystems.

A home-grown approach to wildlife management

Just as New Zealand's isolation gave rise to the unique biota and ecological systems described earlier in this book, so the unusual challenges in this country encouraged the development of pioneering approaches to wildlife management (Chapter 9). New Zealanders had to save critically endangered endemic birds and control or eradicate introduced mammals at a time when few other nations had attempted such feats, and so had to develop appropriate techniques in isolation.

From the outset, conservation in New Zealand has had a focus on offshore islands. Translocation of vulnerable species to offshore islands was pioneered by Richard Henry, revived by early Wildlife Service field staff and perfected by subsequent workers long before the method was widely adopted elsewhere.[6] Predator-free islands were limited in number, so the need to eradicate introduced mammals from other islands stimulated wildlife managers to develop eradication techniques decades before this was even considered possible overseas. The crucial importance of islands in New Zealand conservation so strongly shaped the thinking of subsequent Department of Conservation (DOC) managers that the 'mainland island' concept was born. Although the techniques developed on offshore islands were of little practical application in mainland management, I doubt that the concept could have evolved in a country where islands had not had such an important focus for conservation.

New Zealand's home-grown techniques were developed by field-based workers, largely within the Wildlife Service, as practical solutions to local crises. This small, focused group of people, characterised by a 'can-do' attitude, was not scared to take risks, as with the black robin, or to attempt the impossible, as with cat eradication from Little Barrier Island. The history of takahe management[7] shows that although they were not always successful, they were extraordinarily persistent. Elsewhere in the world, wildlife management was perhaps more strongly influenced by academia: there were well-developed population theories that provided an understanding of the problems to be solved, and textbooks that offered a range of tried and appropriate techniques. Reading the case histories described in Graeme Caughley and Anne Gunn's *Conservation biology in theory and practice* will show just how unusual was the strongly pragmatic approach adopted in New Zealand.

Other countries have been slow to adopt the techniques developed here, and then mostly by contracting New Zealanders to do the job. For instance, Don Merton has worked with critically endangered birds in Mauritius, and New Zealanders have been contracted to eradicate introduced mammals from islands in other countries, including some with a long history of wildlife management.

Hawaii has experienced many of the same problems as New Zealand but, there, techniques and values have been applied that were appropriate for continental United States, with little success. A paper by Sherwin Carlquist[8] makes some interesting comparisons with the New Zealand situation. While the threats are similar in both places, in Hawaii most of the land is in private ownership and there are virtually no predator-free offshore islands. Thus many New Zealand management strategies are impractical.

Over the last 50 years, management of indigenous wildlife in this country has undergone great changes, and these are reviewed in the following sections.

Single-species management versus ecosystem management

Over time, management has changed from moving endangered species to threat-free islands, to treating those threats on site, and from saving single species to restoring the ecological communities of which they are a part.[9] The recent ecosystem focus in conservation is reflected in several recent government policy statements, including the New Zealand Biodiversity Strategy and DOC's vision document released in 2001.[10]

Figure 10.1 (pages 320–21) illustrates some of the trends and significant achievements of the half-century since takahe were discovered. For most of this time the focus has been on the management of single species such as saddleback or black robin, where the objectives can be clearly identified, results are measurable and management has the prospect of conclusion once the species is secure. This will remain a necessary first step in saving most species. However, focusing on a single species can lead to situations where a rare species is managed to the detriment of others. Translocating little spotted kiwi and kakapo to islands where they did not naturally occur averted their extinction but with unassessed trade-offs for the integrity of the recipient ecosystems. These species probably had little impact on the islands to which they were introduced, but saddlebacks, for example, are voracious predators and on some of the islands to which they have been introduced they may now threaten certain rare and endangered invertebrates. Wetapunga, the largest weta and probably the world's heaviest insect, now survives only on Little Barrier Island. Its numbers are declining and saddlebacks are highly efficient weta predators, which may play a part in their demise.[11]

In its native ranges, the buff weka is extinct and the North Island weka is critically endangered, but the survival of both was ensured (in both cases by accident rather than design) by introductions to islands outside their natural range. On the Chatham Islands, buff weka are now killed for food by islanders and by DOC staff to protect other species, including taiko and Chatham Island oystercatcher. Over 80 per cent of North Island weka now survive on a few small islands[12] and have been eradicated from others to protect species native to those islands. An attempt to re-establish them on the mainland failed owing to predation by ferrets and dogs, so islands are likely to remain important weka refuges.

Weka could be translocated to managed mainland reserves, but there they might pose a threat to equally threatened invertebrates or lizards. Since weka and these small animals last co-existed, a suite of new predators has been introduced. The addition of another predator, albeit a native species, could mean the difference between survival and extinction for some small animals. The re-introduction of buff weka to managed reserves on Banks Peninsula is feasible, but because of its geological and biological history (see Chapter 5), the peninsula has many endemic invertebrates, some of which are more critically endangered than the weka.[13] Before weka are re-introduced there or into other habitats where rare reptiles or invertebrates may be present, it would be prudent to ensure that they will not create new problems.

The debate concerning the relative value of single-species and ecosystem conservation is exemplified by the release of Campbell Island teal into the Codfish Island Nature Reserve, far outside its historical range. In their natural range, teal survive only on tiny, remote Dent Island. Teal were taken from there to the Mount Bruce Wildlife Centre, but rather than hold the progeny in captivity until rats were eradicated from Campbell Island, they were liberated onto a holding island.[14] There was no suitable island within the teal's native range, so they were released onto Codfish, the plan being to eventually relocate them all to Campbell Island. Some of the Codfish ducks will probably evade capture and it seems likely that this teal will remain on a nature reserve well outside its native range but within the natural range of the closely related brown teal, which is also endangered.

Ecosystems are difficult to define, boundaries are fuzzy, and they are complex, dynamic entities that are influenced by surrounding ecosystems. Our current poor understanding of ecosystem processes is a major constraint in managing them. Alan Saunders and David Norton[15] suggested that at best ecosystem management programmes 'may only ever be multi-species/multi-threat–focused rather than truly ecosystem-focused'.

One way of attaining community ends through a single-species focus is to select appropriate flagship or indicator species that will reflect changes occurring in the ecosystem.[16] Kokako are suitable for both these roles. They are charismatic birds, so their plight can help generate funding. They require a large area of healthy native forest, thus conservation actions that are good for kokako benefit other native species. At Mapara, rat and possum control to protect kokako also achieved an improvement in habitat quality and an increase in numbers of most other native species. For example, tui increased fivefold while kereru numbers more than doubled in response to control aimed at protecting kokako.[17] Black robins, while sharing the kokako's charisma, do not have habitat requirements that ensure the same ecological spin-offs for other species.

Since 1996, kakapo recovery has been carried out by the National Kakapo Team, a semi-autonomous group within DOC that sits outside the department's normal regional structure. This team has been successful in reversing the kakapo's decline, and one wonders why this model has been used only once. It could also ensure better management of sites with multiple conservation values. For instance, a Rangatira Island team could manage the island as something more than a refuge for endangered species. Rangatira supports more than 50 per cent of the global population of nine threatened or endangered birds, including the black robin, plus rare plants and invertebrates. Each endangered bird has been managed more or less independently of the other species. Management of the island as refuge for numerous species is obviously desirable, but that requires a much more detailed knowledge of the biology, not just of the rare species but equally importantly the common species; and not just of the animals but also the plant and vegetation changes. Multi-species management of this type requires very clearly defined goals and a sound understanding of the interactions between the various species and their environment. Site-based management would also give formal recognition to the importance of Rangatira Island as a breeding site for the locally abundant broad-billed prion and white-faced storm petrel.

An island-wide focus has been suggested for the management of rare species on Tiritiri Matangi Island, where it has been suggested that the single-species focus of the various recovery teams for each species may have impeded the progress of the restoration programme.[18]

Interventionist management

Over the last 50 years there has been a shift in the level of acceptable management. For example, during the first decade of takahe management it was acceptable to kill stoats and deer in the hope that this would benefit the birds, but not to manipulate takahe themselves. Later, takahe were taken into captivity and the wild birds left alone; subsequently, intensive management of wild birds begin (Chapter 9). The saddleback programme involved simply catching and releasing birds on to islands to establish new populations. Even this was controversial when, in 1964, the Royal Forest and Bird Protection Society did not support the translocation of saddlebacks at the time of the Big South Cape Islands rat invasion (Chapter 5).[19]

During the 1970s and 1980s, management of endangered species became much more hands-on as wildlife managers placed a greater value on the survival of the species than the costs imposed on the final few survivors.[20] This is exemplified by the black robin programme, where just about anything was acceptable in order to save this critically endangered species. The entire robin population was transferred to an island from which the tomtit, another native but a potential competitor, was eliminated. This was followed by a decade of intensive nest monitoring, cross-fostering and moving eggs and chicks from parent to foster parent and island to island (Chapter 9).

Intensive management of critically endangered species continues. Virtually all known pairs of Chatham petrels have been transferred into artificial nest boxes so the birds can be monitored and the burrows protected from broad-billed prions, the main cause of breeding failure.[21] All kakapo carry radio transmitters, which makes monitoring and research on these cryptic nocturnal birds possible. It is perhaps ironical that these flightless birds regularly fly (in aircraft) between nature reserves in DOC's efforts to save the species.

Captive management

Captive breeding can play a useful role in the conservation of rare species, and several, including Père David's deer and nene (Hawaiian goose), would now be extinct if individuals had not been taken into captivity. No New Zealand species has been saved by captive breeding, although captive rearing has played an important role in the recovery of some. One of the first actions taken to save takahe was to transfer birds to a captive-breeding facility at Mount Bruce. Success was limited and by the late 1970s the priority had shifted to management in the natural environment. Since 1985, captive rearing (rearing in captivity of chicks produced from wild-laid eggs, with release back into the wild once independent) has been important for takahe conservation (Chapter 9). Today, all the island birds and a quarter of all Fiordland takahe have been reared in captivity or are the progeny of birds reared in captivity.[22]

Captive management has played a role in black stilt conservation (Chapter 9). Eggs have been artificially incubated then returned to their wild parents just before hatching, and chicks have been raised in captivity then released into the surrounding predator-fenced area.[23] Captive breeding has been used in the management of brown teal and Campbell Island teal, the latter more successfully than the former. Over a thousand brown teal have been raised, most by private individuals, but very few have successfully established when released into the wild. The release of Campbell Island teal onto a predator-free island succeeded, whereas black stilt and brown teal have been released into habitats where the agents that made them rare (primarily predators) had not been neutralised.

Tuatara have also been bred in captivity. Gunther's tuatara were held in captivity until a second population of this critically endangered reptile could be established on Titi Island. The captive population was primarily insurance against the arrival of rats on their last island refuge, but both captive-born and wild animals were then used to seed the Titi Island population. On several of the islands inhabited by northern tuatara, the population was so low that without the injection of new stock through captive breeding the populations might not have rebounded following kiore eradication.[24] Research on captive animals helped develop methods to artificially incubate tuatara eggs and showed that egg production could be stimulated hormonally.

The Mt Bruce Wildlife Centre was originally set up for captive breeding of endangered birds, but today its primary role is advocacy – it displays to the public a variety of rare native animals in semi-natural surroundings. Kaka have been bred in captivity and released into the surrounding reserve. They are maintained through supplementary feeding and provide the public with a wonderful opportunity to experience free-flying kaka at close quarters. Even if they did not breed in the wild, their presence at Mt Bruce would still have an important advocacy function.

Island management

The survival into the post-takahe era of many species, including saddleback, Chatham petrel, tuatara, Duvaucel's gecko and Hamilton's frog, happened only because populations occurred on islands to which the predators they were vulnerable to did not gain access. Predator-free islands will continue to remain crucial for the survival of many endemic species. Islands offer a second important but sometimes overlooked advantage for conservation: most are Crown-owned, access is restricted and they are designated as reserves or sanctuaries. On the mainland, conservation must compete with other land uses.

The management of rare species on islands has largely relied moving endangered species to safe islands and eradicating predators. As of 1993, at least 50 species had been translocated in about 400 different transfers.[25] Since then, even more species have been shifted. Some of these projects failed, perhaps because all birds transferred were of the same sex or insufficient numbers were captured but, over time, translocations have become more sophisticated. The jury is still out on the best number of founder individuals to use and the appropriate mix of young and old, males and females to move to a new island.[26] Several releases of 20 or more individuals is the preferred method, although John Craig[27] has shown that some populations have successfully established from as few as five founding birds.

Even very small islands can support viable populations of lizards, so the options for island management of rare lizards are greater than for birds. However, unlike birds, which can quickly explore an entire island and locate suitable foods and habitats, New Zealand lizards tend to move over very short distances, so it is essential that they be released within a few metres of suitable habitat.[28] *Reintroduction biology of Australian and New Zealand fauna*, edited by Melody Serena, is an invaluable source of information on translocations and related procedures.

The scarcity of predator-free islands has been a major impediment to conservation. Probably the first successful mammal eradication was that of goats from Rangitira Island in 1916. As of 1994, 108 populations of 13 species of introduced mammals had been eradicated from 95 different islands.[29] Rodents had been eradicated from 62 of those islands.

The Wildlife Service saved saddlebacks and black robins on what were then the only places this was possible – remote islands beyond reach of both predators and people. Even today, most endangered species conservation is carried out in locations few people can visit, thus taxpayers, who foot the bill for conservation, are unable to view some of our most interesting species. The restoration of Tiritiri Matangi Island provided the first opportunity most people had to view certain rare species in the wild. Today, saddlebacks can be seen on Tiritiri, Mokoia, Motuara and Ulva Islands, and tuatara on Somes Island in Wellington Harbour. For other species, including black robins and kakapo, there are some major obstacles to be overcome before public access becomes a reality.

Management of endangered species on the mainland

Certain species such as black stilt, kiwi and kokako must be managed on the mainland alongside the threats that made them rare, and the development of strategies to achieve this has proven especially challenging. For these species, management mostly involves alleviating the threats, trusting that if this is achieved, the rare species will breed successfully.

Mainland management and island management differ in another crucial respect: the extent of conflict between conservation and other uses. On the mainland, management does not just have to contend with the threats that made the species rare: conservation often also requires economic or societal sacrifices. Protection of the Pureora kokako necessitated forgoing the income that could be derived from logging. As we saw in Chapter 9, conservation of braided-river birds is particularly fraught with difficulty. In addition to effective control of the full suite of predatory mammals on both public and private land, conservation will only be effective if people reduce their demands on the environment and restrict certain recreational activities.

The pulsed control of key pests that enables kokako to breed successfully and their populations to increase in the face of predators[30] can be readily adapted for the management of other species in other places. For most mainland situations, the immediate threat (usually predators) impacts on multiple species, so predator control can play a key role in the restoration of indigenous ecosystems. This is the basis for the mainland island programmes.

As well as DOC's six large mainland islands, there are a number of community based and private projects aimed at conserving habitat remnants or restoring forest or wetland habitats.[31] The mainland-island style of management is exciting and has provided good testing grounds for the development of ever more sophisticated *in situ* management techniques. Perhaps in future, management techniques will enable species that are at present island refugees to be repatriated to unfenced mainland reserves. The approach is most effective in relatively small, isolated habitat remnants, but as the techniques evolve, and provided that the resources are available, the methods can be applied to larger areas.

Mainland islands will succeed only if there is long-term and sustained control of threats, and this is not cheap: it cost DOC $1.8 million in 1998–99,[32] and there will always be some degree of uncertainty that the management will be continued. Unless it is sustained, the advances made in protecting these mainland ecosystems will be for nought. With community initiatives, there is an even greater risk that ongoing management will fail. Predator fences have much lower on-going costs but are currently feasible for only relatively small areas.

I am a keen advocate of mainland islands but acknowledge the risk that such high-profile ventures could siphon resources away from other projects. This could result in a form of triage where a few representative habitats containing an almost full range of species (or at least bird species) were carefully managed at the expense of habitats elsewhere. Others may argue that it is better to provide truly effective control at several key sites than partial protection over a larger area. There is a risk that future governments will withdraw support for these projects and all the good work will be undone. At present, management of mainland islands relies on the annual application of huge quantities of poisons, and there is growing opposition to toxin use.

A review of the ecological consequences of using poisons to control introduced mammals concluded that the conservation benefits generally outweighed the environmental costs.[33] However, there is some evidence to suggest that persistent use of brodifacoum may have long-term consequences and there is a need to better measure the side-effects of poisons and develop strategies that maximise benefits while reducing the quantity used. Such studies are under way.[34] The impact that 1080 possum control operations had on North Island robins has been quantified by Ralph Powlesland and co-workers.[35] In 1996, the baits used contained a high proportion of chaff, and that year about half of the robins died following the poison operation. The following year, improved baits were used and robin mortality was less than 10 per cent. After both operations, robins fledged nearly four times as many chicks in the poisoned blocks than in adjacent control blocks, so there was a net population increase even though the mortality in 1996 seemed unacceptably high.

Moreporks do not fare so well. About a fifth of those on Mokoia Island apparently died when mice were eradicated, presumably from eating poisoned rodents that were not yet dead.[36] The recovery time for this long-lived, slow-breeding bird will be longer than it was for the robins. Trials with rat control using traps instead of poisons are under way,[37] but mainland management will probably need to utilise large quantities of poisons for some years to come.

Effective management of rare species and ecosystems on the mainland has had another far-reaching spin-off. In the past, many conservation projects have made little use of biological knowledge. Too often managers consider that research is a luxury and, given the urgency of the situation, one that can be dispensed with. Alan Saunders and David Norton[38] suggested that the challenges posed by mainland management are stimulating a culture of management based on good science that is now permeating other projects. They further suggest that because field staff are involved with the research, it remains focused on the problems at hand and so is more likely to provide useful solutions. The mainland islands also promote a sense of stewardship where visitors and locals alike can see the results of good management.

Conservation and the molecular revolution

The last 20 years have seen the development of molecular technology that has dramatically changed our understanding of the evolution and systematics of birds in general and certain groups of New Zealand birds, reptiles and frogs in particular. It was formerly assumed that birds originated in the northern hemisphere, and that their main evolutionary radiation also took place there. There was even the suggestion that the Australasian region was a place where 'waifs and strays' took refuge from the 'advanced' fauna of the north. New Zealand was regarded as a sanctuary for the old, the slow and the simple.

However, the DNA-DNA hybridisation work[39] and subsequent molecular studies[40] (Chapter 3) showed that birds probably originated in Gondwana, and the Australasian region was where one of the major passerine radiations occurred. Thus the Australian and New Zealand faunas are important in understanding the evolution and ecology of birds.

Molecular studies of several genera of New Zealand vertebrates have revealed cryptic species (Chapter 2), and these studies have had dramatic conservation outcomes. Genetic studies have revealed previously unimagined diversity among New Zealand lizards.[41] They have turned up cryptic species in three of the families or orders endemic to New Zealand, in each case finding critically endangered species of tuatara, frogs or kiwi.[42] Molecular studies have resolved the long-debated status of orange-fronted and Forbes' parakeets, confirming the full species status of these critically endangered taxa, both of which had previously been designated subspecies of the more common yellow-crowned parakeet. The parakeet studies also provided insights into the biogeography of these birds (Chapter 3). They suggested that the genus originated in New Caledonia, spread to New Zealand via Norfolk Island, and from New Zealand colonised the Chatham and subantarctic islands.[43]

Two of these molecular studies have inspired ecological studies to test ideas generated by their sometimes unexpected findings. A behavioural study showed that both orange-fronted and yellow-crowned parakeets chose same-plumaged partners, in agreement with the genetic work that shows them to be separate species. Ecological studies of three cryptic species of skink that were once lumped together in a single subspecies showed that each had quite specific ecological preferences.[44]

Molecular techniques have proven to be powerful tools, enabling biologists to answer a variety of questions relating to the management of threatened species. For example, they have been used to sex animals, determine kinship and to determine whether there is gene flow between isolated populations.[45]

Conservation of introduced vertebrates: should we bother?

New Zealand now has several introduced species that are either globally rare or threatened in the country from which the local population was sourced. In this section I debate the responsibility New Zealand has for conserving these animals. The case of the Parma wallaby has been mentioned previously (Chapter 6). The brushtailed rock wallaby has disappeared from large parts of eastern Australia and remains common only at the northern end of its range. It is listed as vulnerable by both Environment Australia and by IUCN.[46] Brushtailed rock wallabies were introduced to Kawau, Motutapu and Rangitoto Islands. They were recently exterminated from the last two and eradication on Kawau is planned. The Himalayan tahr is threatened throughout its natural range and listed as vulnerable by IUCN. New Zealand has the only secure wild population. At present there is a global decline in amphibians, and each year species of frogs and other amphibians become extinct.[47] Both the green-and-golden bell frog and the golden bell frog have declined in their Australian homeland, and circumstantial evidence suggests that the decline may also be happening in New Zealand. Environment Australia lists both species as vulnerable, although the IUCN considers the golden bell frog to be endangered.

In Britain, at least 12 species of farmland birds have declined significantly since the 1970s,[48] and of these, song thrush, skylark, blackbird, starling, dunnock, house sparrow and yellowhammer are common in New Zealand. One other introduced species, the cirl bunting, is now so rare in Britain it is listed in the British Red Data Book. These European birds are still common on the European mainland and none is globally threatened.[49]

Does New Zealand have any global responsibility to conserve these species? My own view is that, as a small, not especially affluent nation with more than our fair share of conservation problems, we have no option but to devote our scarce conservation resources to saving endemic species. Sole responsibility for 5.3 per cent of the world's threatened and endangered birds is challenge enough for a country with just 0.06

per cent of the global human population. We also have sole responsibility for threatened reptiles, frogs, marine mammals, bats, invertebrates and plants, virtually all of which are endemic. While the situation is difficult enough in New Zealand, it is even worse in the small Pacific nations, which not only have more threatened endemic species per head of population but also lack the infrastructure and expertise to implement conservation programmes. Most threatened species and most endemic bird areas are in the less affluent nations.[50] The majority of the money and resources available for conservation is concentrated in a few affluent northern hemisphere nations that have very few globally threatened species. This imbalance between needs and resources is a major impediment to conservation.

The introduced farmland birds that are common in New Zealand but rare in their native Britain present no threat to our native species. I welcome European biologists who come here to study those birds in order to aid conservation in their native range, but I would resent New Zealand money being devoted to such research. Even the rarest of these species, the cirl bunting, is more numerous (in Europe but not Britain) than most of our endemic birds. Current research on the way farming practices influence skylark densities, and breeding success in Canterbury is of interest to European biologists who are trying to reverse the dramatic decline of that species in Britain and Western Europe.[51]

The Parma wallaby is no longer an issue, but tahr, brushtailed rock wallaby and bell frogs do present a quandary that has not been adequately addressed. New Zealand can easily ensure the survival of wild tahr. However, the species would thus be saved while indigenous alpine ecosystems and endemic plant species were threatened. In my opinion, conservation of those alpine communities and plants must have priority over the introduced tahr.

Both species of bell frog are essentially restricted to modified habitats and, except as possible vectors of chytrid fungi disease, pose no known threat to indigenous species. Study of the New Zealand populations may provide insights as to how to manage these species within their native range.[52] The status of these populations should be assessed and some perhaps be protected as insurance against further declines in Australia.

The case of the brushtailed rock wallaby is more of a quandary. All three islands on which it occurred have other conservation values that are impacted by the wallabies. Eradication from Rangitoto and Motutapu Islands in the late 1990s was a necessary precursor to ecological restoration of these islands.[53] Because Kawau Island has been highly modified already, perhaps there is some justification for retaining brush-tailed wallabies there.

Why should we conserve native species?

The world carries on quite happily without dodos and moa. Dinosaur fossils remind us that extinction is a natural phenomenon and even mass extinctions, though very rare, do happen from time to time. The forests did not collapse when kakapo became extinct throughout their natural range, and the ecosystems to which they were introduced had managed just fine without them. Few of the indigenous species discussed in this book play any crucial role in ecosystem function, and few are of any significant utilitarian value. There are exceptions: kereru, kokako, tui, bellbird, short-tailed bat and some lizards play important roles in pollination and seed dispersal (Chapter 7). Sooty shearwaters are among the few native birds some New Zealanders still eat, and paradise shelduck are one of the four species we can legally hunt. Dolphins, whales, penguins and some other species are important to our growing nature tourism industry.

One could be cynical and, given the overwhelming magnitude of the global environmental crisis, ask why we should bother to preserve native species and patch up tattered remnants of indigenous ecosystems. Obviously I think we should bother or I would not have written this book. I could simply argue that all species have an intrinsic right to exist, or that humans do not have an intrinsic right to condemn other species to extinction; but there are less esoteric justifications. At a national level our native biota is part of New Zealand's heritage and important to our national identity. The world would be a less interesting place without tui and kereru, and a more interesting place with moa and huia. For me, these are reasons enough but in addition there is an international obligation. Our biota is part of humanity's global heritage. Only New Zealand can conserve the endemic species and unusual ecosystems that make our country distinctive, but we do it on behalf of all humankind.

What is it we wish to conserve? This is no trivial question and there is no simple answer. Species can be saved in captivity, translocated to offshore islands or even to countries far outside their natural range, but while the species may then be secure, it is no longer part of the ecological community in which it evolved. In most parts of the world, conservation tends to be carried out in an ad hoc manner, lurching from crisis to crisis with little thought given to the over-arching goals. New Zealand is no exception, and David Towns and Murray Williams[54] have suggested that this piecemeal approach arises from the lack of a philosophical basis to underpin conservation. In their view, neither the legislation nor DOC's mission statement takes account of the context in which species should be conserved, nor does it recognise the dynamic processes that charaterise the natural world. They provocatively suggest that if endangered species were maintained in game farms, zoos and gardens, this would be a cheap and effective solution that was within the letter, if not the spirit, of the Conservation Act. Towns and Williams proposed that 'to preserve species as functioning members of a system of interacting organisms and their environment in which their essential nature is maintained' would provide an appropriate philosophical base to guide conservation in New Zealand. I would add that, where possible, species should be conserved within their natural range and within the ecological communities in which they evolved. According to Towns and Williams, greater emphasis should be placed on the community in which a species occurs rather than the prevalent emphasis on the species per se. Species are part of dynamic and changing ecosystems, but the Conservation Act and DOC's mission statement seek to conserve static entities.

Almost all species of indigenous vertebrates other than fishes are protected by law and can not be harvested in any way. The Conservation Act that confers this protective status on indigenous wildlife and protected lands effectively sets conservation apart from other human activities and does not see people as part of the indigenous ecosystems.[55] Total protection has not halted the ongoing loss of biodiversity, despite 30 per cent of the land area being 'locked up' in reserves, and some biologists have asked if sustainable utilisation of native species would result in more effective conservation. John Craig argued that because most native wildlife cannot be harvested, it has no recognised economic value; and because access to national parks and reserves is free, there is no mechanism for those using the parks to contribute to the costs of their ongoing protection. Government is essentially the sole funder of conservation, which is more strongly influenced by politics than by local communities. Craig[56] suggested that conservation of native species on private land is equally problematic. Because indigenous species cannot be used, landowners are better off replacing them with economically valuable exotic species.

New Zealand has few extant terrestrial birds that are large enough to bother harvesting, and these will not withstand harvest while their fragmented populations co-exist with introduced predators and competitors. Harvest of some marine birds and mammals is technically feasible, and I will return to this matter in the next section. Sustainable indigenous forestry is one way in which native forests could acquire economic value. In November 1999 the Labour government scrapped proposals to sustainably harvest large areas of beech forest on the West Coast. Ecologists were divided on the issue, and the range of views held by professional ecologists illustrates just how problematic sustainable exploitation really is. Some ecologists followed Craig's reasoning that predator control funded from the profits of timber extraction would enable conservation gains that were impossible while the forests produced no economic return. A few were philosophically opposed to exploitation. Many, including myself, were not convinced that selective logging would produce sufficient return to fund effective predator control over large enough areas to offset any disruption caused by logging. We were also concerned that removal of large trees would result in the loss of nesting and roosting holes used by threatened birds and bats. Philosophically I agree with Craig's reasoning; in practice I believe we need to know a lot more about the functioning of those forest ecosystems and the needs of the native animals before ecologically sustainable harvest can be guaranteed. The debate over the current preservationist approach to the conservation of indigenous species and sustainable harvest of those species will continue for years to come.

Cultural harvest

In the post-takahe era, the over-riding philosophy governing indigenous bird conservation has, with six exceptions, been one of total protection. Certain Maori iwi (tribes) retain the traditional right to harvest titi (sooty shearwaters) and oi (grey-faced petrels), and on payment of the appropriate game-licence fee anyone can hunt paradise shelduck, grey duck, shoveler and pukeko during the duck-hunting season. In pre-European times Maori hunted most native birds. Some species appear to have been harvested sustainably; others not. Currently, some Maori wish to resume harvest of certain native species, a right enshrined in the Treaty of Waitangi*, not because they need the food but in order to express

* The Maori version of the treaty gives Maori full access to natural resources, including indigenous wildlfe. The English version is more restrictive. Debate over this difference and which resources were included is at the core of Treaty politics.

their 'Maoriness' and to revive traditions with important cultural and spiritual significance. Rau Kirikiri and Graham Nugent[57] state that main motivation underpinning the desire to restore the traditional right to harvest is mana. The tribe's mana is enhanced by an ability to provide guests with sought-after traditional foods. The right to express rangatiratanga (sovereignty) over natural resources is, to the tribe, as important as the actual harvest. Maori desire greater involvement in the decision-making process, and Kirikiri and Nugent state that if Maori perceived the harvest to be unsustainable, it would not proceed. They further suggest that to some extent the opposition to increased Maori involvement in native species management may be driven by the unspoken desire of Pakeha to retain their current role as principal guardians of the nation's indigenous species.

At the heart of the debate is a fundamental difference in the way most Maori and most Pakeha view the conservation of indigenous species. Pakeha typically support total protection of the animals, while the traditional Maori view is conservation for human use.[58]

In pre-European times, we are told, the harvest of birds and other natural resources was regulated by tapu (religious restriction) and rahui (temporary bans) imposed by the tribal chiefs and tohunga (experts in natural resource management). Respectful use (manawhenua) of resources imposed on those users a responsibility to protect the resource for future generations (kaitiakitanga).[59] Pakeha legitimately question just how sustainable Maori harvest really was, and whether Maori could reestablish effective management of resources now that tribal structures, so important in the past, have broken down and those tohunga are long since dead. Pakeha can be quick to draw attention to examples where Maori management failed (for example, exploitation of moa) while conveniently forgetting our own poor record in environmental management. The right to harvest is not the right to expedite extinction; so is the sustainable harvest of native species feasible in the much-modified ecosystems of today? Some examples show how complex the debate is.

Kereru were traditionally harvested for food by Maori, and in some parts of New Zealand iwi wish to revive this tradition. Owing to habitat loss, introduced predators and competitors, there are now far fewer kereru than there were a century ago and breeding success is poor. However, kereru are adaptable and can thrive in fragmented landscapes if predators are controlled.[60] Although the current level of sustainable harvest is at best very low, it is technically possible to control predators and manage habitats to raise kereru numbers to a point where sustained harvest may again be possible. While some iwi wish to harvest kereru, other Maori and many Pakeha would find the killing of this handsome and much-loved bird abhorrent.

Rakiura Maori have retained the traditional right to harvest fledgling titi on islands around Stewart Island (see Chapter 8). In the past, the fledglings of many other species of petrel and albatross were taken as muttonbirds by tribes throughout New Zealand and the Chatham Islands. Properly managed, muttonbirding could be a biologically sensible way to harvest these seabirds, where adult mortality is naturally very low but a large proportion of young birds die in the weeks following fledging. Sustainable harvest would depend on a significant proportion of fat, healthy chicks escaping harvest. These, the birds best fitted to survive, are also the most sought-after for harvest. Other petrel species could also be harvested as long as the breeding islands were kept predator-free and adult birds strictly protected (as is the tradition of the Rakiura muttonbirders). In 1991, Chatham Island Moriori applied for a one-off take of a small number of albatross fledglings for ceremonial reasons.[61] There was vociferous public opposition, even from people who found shearwater harvest acceptable.

Harvest of New Zealand fur seals is technically feasible. Their populations crashed through overexploitation, first by Maori then by Europeans, but the population is now increasing and they are recolonising parts of their former range. Fur seal populations will decline rapidly if females are culled, but because the species is polygynous, a harvest of males could be sustained. Many of the most accessible fur seal colonies are male-only haul-outs.[62] However, there are other constraints to harvest. Hundreds of fur seals are killed each year as by-catch in trawl fisheries and, because a proportion of the animals taken are female, this amplifies the impact on the population. Each year thousands of New Zealanders and foreign tourists visit fur seal colonies or go swimming with the seals, which are now an important economic asset to Westport and Kaikoura. These tourists may be less-than-enthusiastic supporters of a scheme to harvest the species.

There are biological constraints to the exploitation of any species, and no harvest of indigenous wildlife should be contemplated until biological studies determine the take will truly be sustainable. For most native birds, these biological constraints are such that sustainable harvest is impossible in today's predator-filled, highly modified habitats. For kereru and some marine birds and mammals, it is probably technically feasible to meet the biological constraints of harvest, but the question still remains whether harvest is socially acceptable today. The cultural harvest issue raises some thorny and as-yet-unresolved questions about the rights Maori as tangata whenua (indigenous people, literally 'people of the land') have over the Pakeha majority. Many Pakeha families, my own included, have now been in this country for well over a century and we too regard ourselves as people of this land. Kirikiri and Nugent,[63] while promoting Maori rights to customary harvest, acknowledge the biological and social barriers that would need to be overcome. They conclude by suggesting that restoring certain traditional rights would increase Maori links to and commitment to the natural environment and could have considerable conservation benefits.

The cultural harvest debate has raised two important issues that hinder effective management. First, there is the typical Pakeha view of conservation by full protection versus the typical Maori view of conservation for sustainable harvest. Second, there is a lack of Maori involvement, brought about in part by the paucity of Maori with tertiary qualifications in resource management and ecology. One way ahead is co-management, where Maori and Pakeha are jointly responsible for the management of resources, species or protected areas.[64] Todd Taiepa discusses the issues in depth, giving examples to show how it can result in effective and inclusive management, and introduces a joint University of Otago and Rakiura Maori study of the muttonbird harvest. This has not only yielded top-quality research but, perhaps even more importantly, has broken down the barriers between Maori and Pakeha and encouraged iwi involvement in the research.

Setting priorities

> Conservation efforts are directed almost exclusively at the highly endangered species – kakapo and takahe, black robin and black stilt, kokako and little spotted kiwi. But the monocular concern with these species is allowing others to enter the self-same category.
>
> – Murray Williams (1986)

Conservation is too often crisis-driven and results in a kneejerk response when a species is found to be on the brink of extinction. Conservation managers cannot hope to save each threatened vertebrate species, let alone the hundreds of other threatened taxa one by one. In 1994, Janice Molloy and Alison Davis[65] listed all New Zealand's endangered, threatened or vulnerable animals and plants, taking many criteria into consideration to determine which were in most dire need of management. This document has been used by DOC to set priorities for species conservation The list determined which species were most likely to become extinct soonest, rather than which would most benefit from management. Thus, if used alone, it would result in most money being devoted to a minority of species, usually the most critically endangered with the lowest chances of recovery, while other species moved ever closer to this category.[66] At the time of writing (late 2001), DOC was revising its method of ranking conservation priorities. Several documents were under preparation, only one of which was available and then only in draft form.[67] This new system will, where possible, integrate threatened species management into site-based management programmes, reducing the need for a species-focused priority-setting system.

An alternative strategy is a form of triage[68] where species are assigned to one of three categories:

1. Those that are in no dire need for conservation action
2. Those for which management will achieve positive results
3. Those that are past the stage where management is likely to be effective.

If such a triage were to be applied, it would result in most conservation resources being directed towards vulnerable or threatened species where the prospects for recovery were still good. With these species there would be time to carry out research then develop sound long-term strategies. It is a rational approach, but if it had been adopted in New Zealand, the black robin, kakapo and black stilt would probably now be extinct. On the other hand, less-threatened species such as yellow-eyed penguin, kaka and North Island kiwi would have received management while they were still relatively common, and this would probably have been cheaper and more effective than the management measures needed for these now-endangered species.

Triage may be a pragmatic approach but raises some thorny ethical issues. In effect, we would abandon Category 3 species to extinction, and do we have the moral right to do that? On what grounds would we decide that a species was past the point where it could be saved, and, if this became accepted practice, would we see an incremental lowering of the triage bar?

Perceptions, values and conservation

People only conserve what they value and, as we saw in Chapters 6 and 9, the way New Zealanders value wildlife has changed and will continue to change. Not all people value animals, plants and wild places in the same ways, and these different value judgements influence conservation. Kiore is the only introduced mammal that remains on Little Barrier Island. This island is one of New Zealand's most important nature reserves: the only large land-bridge island that has never supported introduced browsing mammals. The iwi who are the traditional owners of the island wish to retain kiore and maintain Little Barrier Island in its pre-European condition. Presumably, this also means the removal of kiwi that have been translocated to the island by Pakeha. The retention of kiore will inevitably mean the final extinction of wetapunga (a giant weta which survives nowhere else) and the local extinction of tuatara, chevron skink and probably other reptiles and insects unless these animals can be translocated to other islands. The skink currently survives on only one other island. The continued presence of kiore also precludes using the island to conserve other kiore-susceptible species. Eradication of kiore has been proposed but is currently on hold. Pakeha are typically disdainful of the Maori desire to conserve rats, yet many of those same Pakeha wish to conserve certain introduced species including trout, deer and tahr, all of which are currently managed to the ecological detriment of the indigenous habitats they occupy.

Periodically, conflict arises between people who value the Kaimanawa wild horses (Chapter 6) and those who value the distinctive subalpine communities they inhabit and the rare and endemic plants present. Management compromises that would restrict the horses' range or relocate them to less sensitive areas entail regular horse culls (which incur the wrath of horse lovers) or would result in loss of the herd's 'wild character'.[69] With both the Little Barrier kiore and the Kaimanawa wild horses, opposing value judgements are in direct conflict.

Wildlife conservation in New Zealand has had a strong focus on birds. Even tuatara, arguably of greatest zoological interest, has received little practical conservation. By the time its status was finally assessed in the late 1980s, 10 populations had become extinct during the preceding century and one species had been placed at unnecessary risk of extinction (Chapter 9). Conservation is political and agitation by pressure groups can result in scarce resources being devoted to just a few high-profile species, usually those that are most expensive to manage. Public involvement in and support for conservation is vital but uninformed support can be counterproductive. As Harry Recher[70] has pointed out, most environmental pressure groups have narrow agendas and their focus is not always on the most critical issues. An example of this can be seen in the growth of opposition, both in New Zealand and overseas, to the use of the poison 1080 (sodium monofluoroacetate). This is driven by the ideology that applying poison to the environment, in particular to 'pristine' forests and World Heritage Areas, is bad. These critics fail to weigh up the perceived environmental costs with the conservation gains made by the use of 1080.

The importance of public concern in influencing conservation outcomes was evident with the Pureora kokako. The Wildlife Service had documented the importance of Pureora for kokako survival and presented a strong case to stop logging, but facts alone did not convince the government to save the forest: it was public opinion that provided the mandate for conservation .

The takahe and black robin have played important roles in raising the public's awareness of the plight of New Zealand's native birds, and flagship species such as these can successfully convey the conservation message to almost all sectors of society. The takahe rediscovery and the saving of the black robin are entrenched in New Zealand folklore and the stories have been retold many times. The trials, tribulations and successes of the black robin recovery were kept in the public eye – a brave move given the uncertainties of the outcome. The Natural History Unit (of Television New Zealand) produced documentaries on various New Zealand species, including two programmes that bought the black robin story into the living rooms of most New Zealanders. The Wildlife Service enjoyed a level of public support unmatched by DOC, largely a result of its high-profile publicity of the birds and its actions to save them.[71]

Today, having attained international stature as one of the best makers of nature films in the world, Natural History New Zealand is making documentaries for the more lucrative North American and European markets. In these days of globalisation, the New Zealand market is too small to be served by our own documentary makers, and advocacy for conservation has suffered accordingly. Despite the Internet and the unprecedented access to information it provides, and open sanctuaries such as Tiritiri Matangi Island and Karori, few New Zealanders are familiar with the ground-breaking conservation work carried out in this country. Consequently, the people in the field receive far less support from the public than they deserve. Most of those who visit sanctuaries are already supporters of conservation and already have some knowledge about the country's fauna and flora. Television takes the message to a far larger segment of the population. If we wish the next generation to value the environment and manage it in an ecologically sustainable way, natural history must be given as much prominence in schools as mathematics, and the state of the environment considered more important than the state of the economy. Conservation must be seen as a public good and for this to happen the general public must value the species and ecosystems being conserved. One of the challenges facing conservation advocates is how to use those species that people have an affinity for, such as birds and trees, to convey to the public a conservation message that explains the importance of conserving less charismatic groups.

What of the future?

New Zealanders in the late nineteenth century could have saved the huia but effectively chose not to, thus denying subsequent generations that part of their heritage. The Wildlife Service chose to save saddlebacks, little spotted kiwi and black robins at the only locations where that was then possible, on islands away from predators and people. Because Wildlife made that choice, it was possible later to re-introduce saddlebacks to Tiritiri Matangi Island and Karori Sanctuary. Will future generations thank us for the choices we make during our era of guardianship?

Conservation will become even more challenging. Unless the nation tackles the underlying problems of ever-increasing resource use and a growing, increasingly affluent population, conservation will never progress beyond the 'band aid' stage. So far we have largely been insulated from these problems by our large land area relative to our small population, but these halcyon days have now gone.[72] To date, conservation has been a discretionary activity. To be truly effective it must become a mainstream activity, as important to society as health or education.

During the post-takahe era, tremendous advances have been made in the management of rare species. Given the skills, commitment and initiative of the conservation workers in this country, I firmly believe we can prevent the extinction of any more indigenous species and that the environment 10 years hence could be greatly improved over that of today. However, I am not convinced that New Zealanders have the commitment to ensure this will happen. Conservation of indigenous species will come at a cost – not just monetary.

At present, conservation relies on annual applications of huge quantities of poisons to national parks and other reserves. If public perceptions change and this becomes socially unacceptable, the good work will quickly be undone. No one likes using poisons, and while most conservation professionals consider that the gains outweigh the ecological costs, some factions in society disagree.[73] If they had their way, we would soon see birds such as kokako and kiwi rejoin the slippery slide towards extinction. With our present technology, there is no end in sight to this sort of management on the New Zealand mainland. Will we as a nation accept this and be prepared to meet the financial costs in 10, 20, let alone a hundred years? If not, we will revert back to the pre-takahe days when many species survived only on islands. Countering the growing public resistance to the use of toxins is one of the great challenges facing DOC, and reducing dependence on toxins is a priority for research.

Perhaps, 10 years hence, poisons will no longer be needed. Fifteen years ago, the eradication of rats on any but very small islands was impossible; now, most years, DOC eradicates rodents from islands, including large ones. Conserving kokako looked hopeless in 1990, but now control of its enemies is almost routine and the techniques developed are being adapted for other species. The Te Urewera and other mainland island projects are developing more effective and cost-efficient controls, including the use of traps rather than poisons.

The black robin project showed that it was possible to rescue a species from the very brink of extinction. Conservation authorities still approach critically endangered species with the mind-set that recovery is possible whatever the prognosis. Underfunding is always an issue and, as the public sector becomes more cautious, perhaps we will see more resistance to spending on high-risk species rescues for which success at best comes slowly.

Miracles do happen: some god or other is sometimes on our side. It is worth holding the fort even when that is all we are doing. I expect that in 20 years' time the management methods described in this book will have been superseded by much more sophisticated and cost-effective controls. Perhaps by then black robins will have been returned to a predator-free Chatham mainland, and some summers the booming calls of kakapo will resonate from Karori across Wellington suburbs. Perhaps cats, rats and possums will have been eradicated from Stewart Island and we will have our first national park that is free of introduced predatory mammals.

APPENDIX I

A checklist to the amphibians, reptiles, mammals and birds of New Zealand

This appendix lists all native species of amphibians, reptiles, birds and mammals known to have been present during the Holocene (since the end of the ice ages). Migratory species that breed elsewhere but regularly visit New Zealand are listed, but vagrants (species not recorded in New Zealand every year) are not. Subspecies are listed only if endemic to New Zealand. All introduced species that had truly wild breeding populations somewhere in New Zealand in the 1990s are listed. The scientific and common names given here and the status indicated follows: amphibians and reptiles (Bell et al. 1998b, Daugherty et al. 1994, 1999, Patterson and Daugherty 1996); mammals (King 1990, Hill and Daniel 1985); birds (OSNZ 1990, Holdaway et al. 2001 Appendix 2). The higher level taxonomy for birds mostly follows Christidis and Boles (1994), who adopted many of the higher level taxonomic changes suggested by Sibley & Ahlquist (1990). The higher level taxonomy for the passerines adopted by Christidis and Boles (1994) and used in this book differs from that used by Holdaway et al. (2001). The only subfamilies listed are those that are endemic to New Zealand.

Key

C Endemic to the Chatham Islands.
E Endemic to the New Zealand region but after breeding may disperse to other parts of the world.
H Colonised New Zealand by its own powers of dispersal since human contact.
I Introduced to New Zealand.
Is Previously occurred on North or South Islands but now restricted (or nearly so) to offshore islands.
K Breeds on the Kermadec Islands but not on the New Zealand mainland.
M Migratory species that do not breed in the New Zealand region.
S Breeds on the Snares, Auckland, Campbell, Bounty or Antipodes Islands but not on the New Zealand mainland.
T Threatened or endangered species.
t Introduced species that is threatened or endangered in its native range.
V Breeds elsewhere but regularly occurs in New Zealand or New Zealand seas.
X Endemic to the New Zealand region and now extinct.
x Extinct in New Zealand but survives elsewhere.

ORDER ANURA

Native frogs – Family Leiopelmatidae

Leiopelma archeyi	Archey's frog	E T
L. hamiltoni	Hamilton's frog	E Is T
L. pakeka	Maud Island frog	E Is T
L. hochstetteri	Hochstetter's frog	E T
L. auroraensis	Aurora frog	E X
L. markhami	Markham's frog	E X
L. waitomoensis	Waitomo frog	E X

Introduced frogs – Family Hylidae

Litoria aurea	Green-and-golden bell frog	I t
L. ewingii	Whistling frog	I
L. raniformis	Golden bell frog	I t

ORDER SPHENODONTIDA (= RHYNCHOCEPHALIA)

Tuatara – Family Sphenodontidae

Sphenodon guntheri	Gunther's tuatara	E Is T
S. punctatus	Tuatara	E Is T
S. punctatus punctatus	Northern tuatara	
S. punctatus subsp.	Cook Strait tuatara	

ORDER SQUAMATA

Skinks – Family Scincidae

Cyclodina aenea	Copper skink	E
C. alani	Robust skink	E Is T
C. macgregori	McGregor's skink	E Is T
C. oliveri	Marbled skink	E Is
C. ornata	Ornate skink	E
C. whitakeri	Whitaker's skink	E Is T
C. northlandi	Northland skink	E X
Oligosoma acrinasum	Fiordland skink	E Is
O. chloronoton	Green skink	E
O. fallai	Falla's skink	E Is T
O. gracilicorpus	Narrow-bodied skink	E X
O. grande	Grand skink	E T
O. homalonotum	Chevron skink	E Is T
O. inconspicuum	Cryptic skink	E
O. infrapuctatum	Speckled skink	E
O. lineoocellatum	Spotted skink	E
O. maccanni	McCann's skink	E
O. microlepis	Small-scaled skink	E T

O. moco	Moko skink	E Is
O. nigriplantare nigriplantare	Chatham Island skink	C
O. nigriplantare polychroma	Common skink	E
O. notosaurus	Southern skink	E Is T
O. otagense form 'otagense'	Otago skink	E T
O. otagense form 'waimatense'	Scree skink	E T
O. smithi	Shore skink	E
O. stenotis	Small-eared skink	E Is T
O. striatum	Striped skink	E T
O. suteri	Egg-laying skink	E Is
O. zelandicum	Brown skink	E
Plus two as-yet-undescribed species		
Lampropholis delicata	Rainbow skink	I

Geckos – Family Gekkonidae

Hoplodactylus chrysosireticus	Goldstripe gecko	E T
H. delcourti	Kawekaweau	E X
H. duvaucelii	Duvaucel's gecko	E Is
H. granulatus	Forest gecko	E
H. kahutarae	Black-eyed gecko	E T
H. maculatus	Common gecko	E
H. nebulosus	Cloudy gecko	E Is T
H. pacificus	Pacific gecko	E
H. rakiurae	Harlequin gecko	E T
H. stephensi	Striped gecko	E Is T
Plus 15 as-yet-undescribed species		
Naultinus elegans	Common green gecko	E
N. gemmeus	Jewelled gecko	E
N. grayii	Northland green gecko	E
N. manukanus	Marlborough green gecko	E
N. rudis	Rough gecko	E
N. stellatus	Nelson green gecko	E
N. tuberculatus	West Coast green gecko	E

ORDER MARSUPIALIA

Wallabies and kangaroos – Family Macropodidae

Macropus eugenii	Dama wallaby	I
M. rufogriseus	Bennett's wallaby	I
M. parma	Parma wallaby	I t
M. dorsalis	Black-striped wallaby	I
Petrogale penicillata	Brushtailed rock wallaby	I t
Wallabia bicolor	Swamp wallaby	I

Possums – Family Phalangeridae

Trichosurus vulpecula	Brushtailed possum	I

ORDER INSECTIVORA

Hedgehogs – Family Erinaceidae

Erinaceus europaeus	Europeon hedgehog	I

ORDER CHIROPTERA

Bats – Family Vespertilionidae

Chalinolobus tuberculatus	Long-tailed bat	E T

Short-tailed bats – Family Mystacinidae

Mystacina tuberculata	Lesser short-tailed bat	E T
M. tuberculata aupourica	Kauri forest short-tailed bat	
M. tuberculata rhyacobia	Volcanic Plateau short-tailed bat	
M. tuberculata tuberculata	Southern short-tailed bat	
M. robusta	Greater short-tailed bat	E X

ORDER LAGOMORPHA

Rabbits and hares – Family Leporidae

Oryctolagus cuniculus	European rabbit	I
Lepus europaeus	Brown hare	I

ORDER RODENTIA

Rats and mice – Family Muridae

Rattus exulans	Kiore	I
R. norvegicus	Norway rat	I
R. rattus	Ship rat	I
Mus musculus	House mouse	I

ORDER CARNIVORA

Fur seals and sealions – Family Otariidae

Arctocephalus forsteri	New Zealand fur seal	
Phocarctos hookeri	Hooker's sea lion	E S T

True seals – Family Phocidae

Mirounga leonina

Southern elephant seal		S Is

Dogs – Family Canidae

Canis familiaris	Kuri/dog	I

Stoats and ferrets – Family Mustelidae

Mustela erminea	Stoat	I
M. nivalis	Weasel	I
M. furo	Ferret	I

Cats – Family Felidae

Felis catus	House cat	I

ORDER MYSTICETI

Baleen whales – Family Balaenopteridae

Balaenoptera musculus	Blue whale	M T
B. physalus	Fin whale	M
B. borealis	Sei whale	M
B. edeni	Bryde's whale	V
B. acutorostrata	Minke whale	M
Megaptera novaeangliae	Humpback whale	M T

Right whales – Family Balaenidae

Balaena glacialis	Southern right whale	M T
Caperea marginata	Pygmy right whale	V?

ORDER ODONTOCETI

Sperm whales – Family Physeteridae

Physeter macrocephalus	Sperm whale	M
Kogia breviceps	Pygmy sperm whale	V
K.simus	Dwarf sperm whale	V?

Beaked whales - Family Ziphiidae

Berardius arnouxi	Arnoux's beaked whale	T?
Hyperoodon planifrons	Southern bottlenose whale	T?
Ziphius cavirostris	Goose-beaked whale	
Tasmacetus shepherdi	Shepherd's beaked whale	T?
Mesoplodon layardi	Strap-toothed whale	
M. bowdoini	Andrews' beaked whale	
M. grayi	Scamperdown whale	
M. ginkgodens	Ginkgo-toothed whale	
M. hectori	Hector's beaked whale	T

Dolphins and killer whale – Family Delphinidae

Globicephala melaena	Long-finned pilot hale	
G. macrorhynchus	Short-finned pilot whale	
Orcinus orca	Killer whale	M
Pseudorca crassidens	False killer whale	
Peponocephala electra	Melon-headed whale	V?

Grampus griseus	Risso's dolphin	
Tursiops truncatus	Bottlenose dolphin	
Stenella caeruleoalba	Striped dolphin	V
S. attenuata	Spotted dolphin	V
Delphinus delphis	Common dolphin	
Lagenorhynchus obscurus	Dusky dolphin	
L. cruciger	Hourglass dolphin	
Cephalorhynchus hectori	Hector's dolphin	E T
Lissodelphis peronii	Southern right whale dolphin	
Porpoises – Family Phocoenidae		
Australophocoena dioptrica	Spectacled porpoise	V?

ORDER PERISSODACTYLA

Horses – Family Equidae		
Equus caballus	Feral horse	I

ORDER ARTIODACTYLA

Pigs – Family Suidae		
Sus scrofa	Feral pig	I
Cattle, sheep and goats – Family Bovidae		
Bos taurus	Feral cattle	I
Rupicapra rupicapra	Chamois	I
Hemitragus jemlahicus	Himalayan tahr	I t
Capra hircus	Feral goat	I
Ovis aries	Feral sheep	I
Deer – Family Cervidae		
Cervus elaphus scoticus	Red deer	I
C. elaphus nelsoni	Wapiti	I
C. nippon	Sika deer	I
C. unicolor	Sambar	I
C. timorensis	Rusa deer	I
Dama dama	Fallow deer	I
Odocoileus virginianus	White-tailed deer	I
Alces alces	Moose	I

ORDER DINORNITHIFORMES

Moa – Family Emeidae		
Subfamily Anomalopteryginae		
Anomalopteryx didiformis	Little bush moa	E X
Megalapteryx didinus	Upland moa	E X

Pachyornis elephantopus	Heavy-footed moa	E X
P. australis	Crested moa	E X
P. mappini	Mappin's moa	E X
Subfamily Emeinae		
Euryapteryx geranoides	Stout-legged moa	E X
E. curtus	Coastal moa	E X
Emeus crassus	Eastern moa	E X

Moa – Family Dinornithidae

Dinornis struthoides	Slender bush moa	E X
D. novaezealandiae	Large bush moa	E X
D. giganteus	Giant moa	E X

ORDER APTERYGIFORMES

Kiwi – Family Apterygidae

Apteryx mantelli	Northern brown kiwi	E T
A. australis	Tokoeka	E T
A. new species	Eastern kiwi	E X
A. owenii	Little spotted kiwi	E Is T
A. haastii	Great spotted kiwi	E T

ORDER PODICIPEDIFORMES

Grebes – Family Podicipedidae

Podiceps cristatus	Southern crested grebe	T
Poliocephalus rufopectus	New Zealand dabchick	E T
Tachybaptus novaehollandiae	Australasian little grebe	H

ORDER PROCELLARIIFORMES

Albatrosses – Family Diomedeidae

Diomedea epomophora	Southern royal albatross	E S
D. sanfordi	Northern royal albatross	E T
D. *exulans*	Wandering albartross	S
D. chionoptera	Snowy albatross	V
Thalassarche melanophris	Black-browed mollymawk	H S
T. impavida	Campbell mollymawk	E S T
T cauta	Shy mollymawk	S
T. salvini	Salvin's mollymawk	E S
T. eremita	Chatham mollymawk	C T
T. chrysostoma	Grey-headed mollymawk	S T?
T. chlororhynchos	Atlantic yellow-nosed mollymawk	V
T. bulleri	Buller's mollymawk	E S
T. new species. (*platei*)	Pacific mollymawk	C
Phoebetria palpebrata	Light-mantled sooty albatross	S

Shearwaters, petrels and prions – Family Procellariidae

Puffinus carneipes	Flesh-footed shearwater	
P. pacificus	Wedge-tailed shearwater	K
P. bulleri	Buller's shearwater	E T
P. griseus	Sooty shearwater	
P. tenuirostris	Short-tailed shearwater	V
P. gavia	Fluttering shearwater	E Is
P. huttoni	Hutton's shearwater	E T
P. spelaeus	Scarlett's shearwater	E X
P. kermadecensis	Kermadec little shearwater	E K
P. haurakiensis	North Island little shearwater	E Is
P. elegans	Subantarctic little shearwater	
Pelecanoides urinatrix	Northern diving petrel	Is
P. exsul	Richdale's diving petrel	Is
P.georgicus	South Georgian diving petrel	
Procellaria cinerea	Grey petrel	S
P. parkinsoni	Black petrel	E Is T
P. westlandica	Westland petrel	E T
P. aequinoctialis	White-chinned petrel	S
Lugensa brevirostris	Kerguelen petrel	V
Daption capense		
D. capense capense	Cape pigeon	V
D. capense australe	Snares Cape pigeon	E S
Fulmarus glacialoides	Antarctic fulmar	V
Macronectes giganteus	Southern giant petrel	V
M. halli	Northern giant petrel	S
Pachyptila turtur	Fairy prion	Is
P. crassirostris		
P. crassirostris crassirostris	Fulmar prion	S
P. crassirostris pyramidalis	Chatham fulmar prion	C T
P. belcheri	Thin-billed prion	V
P. desolata	Antarctic prion	S
P. salvini	Salvin's prion	V
P. vittata	Broad-billed prion	Is
Halobaena caerulea	Blue petrel	V
Pterodroma pycrofti	Pycroft's petrel	E T
P. leucoptera	Gould's petrel	V
P. cookii	Cook's petrel	E Is
P. nigripennis	Black-winged petrel	
P. axillaris	Chatham petrel	C T
P. inexpectata	Mottled petrel	E Is
P. cervicalis	White-napped petrel	E K
P. alba	Phoenix petrel	K x

P. neglecta	Kermadec petrel	K
P. macroptera		
P. macroptera gouldi	Grey-faced petrel	Is
P. magentae	Taiko	C T
P. new species	Unnamed	C X
P. lessonii	White-headed petrel	S
P. mollis	Soft-plumaged petrel	H S

Storm petrels – Family Hydrobatidae

Oceanites oceanicus	Wilson's storm petrel	V
O. maorianus	New Zealand storm petrel	X E
Garrodia nereis	Grey-backed storm petrel	Is
Pelagodroma marina		
P. marina maoriana	New Zealand white-faced storm petrel	E Is
P. marina albiclunis	Kermadec storm petrel	K T
Fregetta tropica	Black-bellied storm petrel	S
F. grallaria	White-bellied storm petrel	K

ORDER SPHENISCIFORMES

Penguins – Family Spheniscidae

Aptenodytes patagonicus	King penguin	V
Megadyptes antipodes	Yellow-eyed penguin	E T
Eudyptula minor	Blue penguin (white-flippered penguin)	
Eudyptes filholi	Eastern rockhopper penguin	
E. pachyrhynchus	Fiordland crested penguin	E T
E. robustus	Snares crested penguin	E S
E. sclateri	Erect crested penguin	E S T
E. new species	Chatham crested penguin	C X

ORDER PELECANIFORMES

Tropicbirds – Family Phaethontidae

Phaethon rubricauda	Red-tailed tropic bird	K

Gannets and boobies – Family Sulidae

Morus serrator	Australasian gannet	
Sula leucogaster	Brown booby	V
Sula dactylatra	Masked booby	K

Shags – Family Phalacrocoracidae

Phalacrocorax carbo	Black shag	
P. varius		
P. varius varius	Pied shag	

P. sulcirostris	Little black shag	H
P. melanoleucos		
P. melanoleucos brevirostris	Little shag	
Leucocarbo carunculatus	King shag (Stewart Island shag)	E T
L. onslowi	Chatham Island shag	C T
L. ranfurlyi	Bounty Island shag	E S
L. colensoi	Auckland Island shag	E S
L. campbelli	Campbell Island shag	E S
Stictocarbo punctatus		E
S. punctatus punctatus	Spotted shag	
S. punctatus steadi	Blue shag	
S. featherstoni	Pitt Island shag	C

ORDER CICONIIFORMES

Herons and bitterns – Family Ardeidae

Ardea novaehollandiae	White-faced heron	H
Egretta alba	White heron	
Egretta garzetta	Little egret	V
Egretta sacra	Reef heron	
Bubulcus ibis	Cattle egret	V
Nycticorax caledonicus	Nankeen night heron	V
Botaurus poiciloptilus	Australasian bittern	H
Ixobrychus novaezelandiae	New Zealand little bittern	E X

Ibises and spoonbills - Family Threskiornithidae

Plegadis falcinellus	Glossy ibis	V
Platalea regia	Royal spoonbill	H

ORDER ANSERIFORMES

Swans, geese and ducks – Family Anatidae

Cygnus olor	Mute swan	I
C. atratus	Black swan	I x*
Branta canadensis	Canada goose	I
Anser anser	Feral goose	I
Cnemiornis calcitrans	South Island goose	E X
C. gracilis	North Island goose	E X
Tadorna variegata	Paradise shelduck	E
T. new species	Chatham Island shelduck	C X
T. tadornoides	Chestnut-breasted shelduck	V
Hymenolaimus malacorhynchos	Blue duck	E T

* The swan that became extinct during the Polynesian era is now considered to be the Australian black swan that was introduced to New Zealand in the nineteenth century.

Anas platyrhynchos	Mallard	I
A. superciliosa		
A. superciliosa superciliosa	Grey duck	
A. gracilis	Grey teal	
A. chlorotis	Brown teal	E T
A. aucklandica	Auckland Island teal	E S
A. nesiotis	Campbell Island teal	E S T
A. rhynchotis	New Zealand shoveler	H
Pachyanas chathamica	Chatham duck	C X
Euryanas finschi	Finsch's duck	E X
Malacorhynchus scarletti	Scarlett's duck	E X
Aythya novaeseelandiae	New Zealand scaup	E
Mergus australis	Auckland Island merganser	E X
Biziura delautouri	de Lautour's duck	E X

ORDER FALCONIFORMES

Eagles and hawks – Family Accipitridae

Circus approximans	Australasian harrier	H
C. eylesi	Eyles's harrier	E X
Harpagornis moorei	Haast's eagle	E X

Falcons – Family Falconidae

Falco novaeseelandiae	New Zealand falcon	E T
Falco cenchroides	Nankeen kestrel	V

ORDER GALLIFORMES

Partridges, quails, pheasants and turkeys – Family Phasianidae

Callipepla californica	California quail	I
Colinus virginianus	Bobwhite quail	I
Alectoris rufa	Red-legged partridge	I
A. chukar	Chukor	I
Coturnix novaezelandiae		
C. novaezelandiae novaezelandiae	New Zealand quail	E X
Synoicus ypsilophorus	Brown quail	I
Phasianus colchicus	Ring-necked pheasant	I
Pavo cristatus	Peafowl	I
Meleagris gallopavo	Wild turkey	I
Numida meleagris	Tufted guineafowl	I

Megapodes – Family Megapodiidae

Megapodius sp.	Unknown species	K x

ORDER GRUIFORMES

Aptornis – Family Aptornithidae

Aptornis otidiformis	North Island aptornis (adzebill)	E X
A. defossor	South Island aptornis (adzebill)	E X

Rails, gallinules and coots – Family Rallidae

Gallirallus phillippensis		
G. philippensis assimilis	Banded rail	E
G. dieffenbachii	Dieffenbach's rail	C X
G. australis	Weka	E T
G. australis greyi	North Island weka	
G. australis australis	Western weka	
G. australis hectori	Buff weka	
G. australis scotti	Stewart Island weka	
Capellirallus karamu	Snipe-rail	E X
Cabalus modestus	Chatham Island rail	C X
Diaphorapteryx hawkinsi	Giant Chatham Island rail	C X
Dryolimnas *muelleri*	Auckland Island rail	E S T
Porzana tabuensis	Spotless crake	
P. pusilla		
P. pusilla affinis	Marsh crake	E
Gallinula hodgenorum	Hodgen's rail	E X
Porphyrio melanotos	Pukeko	H
P. mantelli	North Island takahe	E X
P. hochstetteri	South Island takahe	E T
Fulica atra		
Fulica atra australis	Australian coot	H
F. prisca	New Zealand coot	E X
F. chathamensis	Chatham Island coot	C X

ORDER CHARADRIIFORMES

Oystercatchers – Family Haematopodidae

Haematopus finschi	South Island pied oystercatcher	E
H. unicolor	Variable oystercatcher	E
H. chathamensis	Chatham Island oystercatcher	C T

Stilts and avocets – Family Recurvirostridae

Himantopus leucocephalus	Australasian pied stilt	H
H. novaezelandiae	Black stilt	E T

Dotterels and plovers – Family Charadriidae

Charadrius obscurus	New Zealand dotterel	E T
C. bicinctus	Banded dotterel	E
C. exilis	Auckland Island banded dotterel	E S

C. ruficapillus	Red-capped dotterel	V
C. melanops	Black-fronted dotterel	H
C. leschenaultii	Large sand dotterel	M
C. mongolus	Mongolian dotterel	M
C. novaeseelandiae	Shore plover	E Is T
Anarhynchus frontalis	Wrybill	E T
Pluvialis fulva	Pacific golden plover	M
P. squatarola	Grey plover	M
Vanellus miles	Spur-winged plover	H
Arenaria interpres	Turnstone	M
Snipe, sandpipers and godwits – Family Scolopacidae		
Coenocorypha barrierensis	North Island snipe	E X
C. iredalei	Stewart Island (and South Island) snipe	E X
C. chathamica	Extinct Chatham Island snipe	E X
C. pusilla	Chatham Island snipe	C E
C. huegeli	Snares Island snipe	E S T
C. aucklandica	Auckland Island snipe	E S T
C. meinertzhagenae	Antipodes Island snipe	E S
C. new species	Campbell Island snipe	E S T
Calidris canutus	Lesser knot	M
C. alba	Sanderling	M
C. ferruginea	Curlew sandpiper	M
C. acuminata	Sharp-tailed sandpiper	M
C. melanotos	Pectoral sandpiper	M
C. ruficollis	Red-necked stint	M
Numenius madagascariensis	Eastern curlew	M
N. phaeopus variegatus	Asiatic whimbrel	M
N. phaeopus hudsonicus	American whimbrel	M
N. minutus	Little whimbrel	M
Limosa lapponica	Bar-tailed godwit	M
L. limosa	Black-tailed godwit	M
L. haemastica	Hudsonian godwit	M
Tringa incana	Wandering tattler	M
T. brevipes	Siberian tattler	M
T. hypoleucos	Common sandpiper	M
T. nebularia	Greenshank	M
T. stagnatilis	Marsh sandpiper	M
T. flavipes	Lesser yellowlegs	M
T. terek	Terek sandpiper	M
Gulls and terns – Family Laridae		
Catharacta lonnbergi	Brown skua	Is

C. maccormicki	South Polar skua	V
Stercorarius parasiticus	Arctic skua	M
S. pomarinus	Pomarine skua	M
Larus dominicanus	Southern black-backed gull	
L. novaehollandiae		
L. novaehollandiae scopulinus	Red-billed gull	E
L. bulleri	Black-billed gull	E
Chlidonias leucopterus	White-winged black tern	M
Gelochelidon nilotica	Gull-billed tern	V
Sterna albostriata	Black-fronted tern	E T
S. caspia	Caspian tern	
S. striata	White-fronted tern	
S. fuscata	Sooty tern	K
S. vittata		
S. vittata bethunei	New Zealand Antarctic tern	E
S. nereis		
S. nereis davisae	New Zealand fairy tern	E T
S. albifrons	Little tern	M
S. paradisaea	Arctic tern	M
Anous stolidus	Common noddy	K
A. minutus	White-capped noddy	K
Procelsterna cerulea	Grey noddy	K
Gygis alba	White tern	K

ORDER COLUMBIFORMES

Pigeons and doves – Family Columbidae

Hemiphaga novaeseelandiae	Kereru	E T
H. chathamensis	Parea	C T
Columba livia	Rock pigeon	I
Streptopelia roseogrisea	Barbary dove	I
S. chinensis	Spotted dove	I

ORDER PSITTACIFORMES

Cockatoos – Family Cacatuidae

Cacatua galerita	Sulphur-crested cockatoo	I
C. roseicapilla	Galah	I

Parrots – Family Psittacidae

Kakapo – Subamily Strigopinae

Strigops habroptilus	Kakapo	E Is T

Kaka and kea – Subfamily Nestorinae

Nestor meridionalis	Kaka	E T

Scientific name	Common name	Status
N. meridionalis septentrionalis	North Island kaka	
N. meridionalis meridionalis	South Island kaka	
N. new species	Chatham Island kaka	E X
N. notabilis	Kea	E T

Other parrots – Subfamily Platycercinae

Scientific name	Common name	Status
Platycercus elegans	Crimson rosella	I
P. eximius	Eastern rosella	I
Cyanoramphus unicolor	Antipodes Island parakeet	E S
C. novaezelandiae		
C. novaezelandiae cyanurus	Kermadec parakeet	E K
C. novaezelandiae novaezelandiae	Red-crowned parakeet	E
C. novaezelandiae chathamensis	Chatham Island red-crowned parakeet	C T
C. erythrotis hochstetteri	Reischek's parakeet	E S
C. auriceps	Yellow-crowned parakeet	E
C. forbesi	Forbes' parakeet	C T
C. malherbi	Orange-fronted parakeet	E T

ORDER CUCULIFORMES

Cuckoos – Family Cuculidae

Scientific name	Common name	Status
Chrysococcyx lucidus		
C. lucidus lucidus	Shining cuckoo	E
Eudynamys taitensis	Long-tailed cuckoo	E

ORDER STRIGIFORMES

Owls – Family Strigidae

Scientific name	Common name	Status
Ninox novaeseelandiae		
N. novaeseelandiae novaeseelandiae	Morepork	E
Sceloglaux albifacies		E X
S. albifacies rufifacies	North Island laughing owl	
S. albifacies albifacies	South Island laughing owl	
Athene noctua	Little owl	I

ORDER CAPRIMULGIFORMES

Owlet-nightjars – Family Aegothelidae

Scientific name	Common name	Status
Aegotheles novaezealandiae	New Zealand owlet-nightjar	E X

ORDER CORACIIFORMES

Kingfishers - Family Alcedinidae

Dacelo novaeguinae	Kookaburra	I
Todiramphus sanctus	Sacred kingfisher	
T. sanctus vagans	New Zealand kingfisher	

ORDER PASSERIFORMES

New Zealand Wrens – Infraorder Acanthisittides

Family Acanthisittidae

Acanthisitta chloris		E
A. chloris granti	North Island rifleman	
A. chloris chloris	South Island riflleman	
Xenicus longipes		E X
X. longipes stokesii	North Island bush wren	
X. longipes longipes	South Island bush wren	
X. longipes variabilis	Stead's bush wren	
X. gilviventris	Rock wren	E
Traversia lyalli	Stephens Island wren	E X
Pachyplichas yaldwyni	Stout-legged wren	E X
Dendroscansor decurvirostris	Long-billed wren	E X

PARVORDER CORVIDA

Australasian warblers –- Family Pardalotidae

Gerygone igata	Grey warbler	E
G. albofrontata	Chatham Island warbler	C

Honeyeaters – Family Meliphagidae

Notiomystis cincta	Hihi (stitchbird)	E Is T
Anthornis melanura		E
A. melanura obscura	Three Kings bellbird	
A. melanura oneho	Poor Knights bellbird	
A. melanura melanura	Bellbird	
A. melanocephala	Chatham Island bellbird	C X
Prosthemadera novaeseelandiae		E
P. novaeseelandiae novaeseelandiae	Tui	
P. novaeseelandiae chathamensis	Chatham Island tui	T

Australasian robins – Family Petroicidae

Petroica toitoi	North Island tomtit	E
P. macrocephala	South Island tomtit	E
P. macrocephala chathamensis	Chatham Island tomtit	C

P. dannefaerdi	Snares Island tomtit	E
P. marrineri	Auckland Island tomtit	E
P. longipes	North Island robin	E
P. australis	South Island robin	E
P. australis rakiura	Stewart Island robin	
P. traversi	Black robin	C T

Whistlers and Allies – Family Pachycephalidae

Whitehead and allies – Subfamily Mohouinae

Mohoua albicilla	Whitehead	E
M. ochrocephala	Mohua (yellowhead)	E T
M. novaeseelandiae	Brown creeper	E

Australasian flycatchers – Family Dicruridae

Rhipidura fuliginosa		
R. fuliginosa placabilis	North Island fantail	E
R. fuliginosa fuliginosa	South Island fantail	E
R. fuliginosa penita	Chatham Island fantail	C

Woodswallows and Australian magpies – Family Artamidae

Gymnorhina tibicen	Australian magpie	I

Crows and ravens – Family Corvidae

Corvus frugilegus	Rook	I
C. new species	New Zealand raven	E X
C. moriorum	Chatham Island raven	C X

Piopio – Family Turnagridae

Turnagra tanagra	North Island piopio	E X
T. capensis	South Island piopio	E X

New Zealand wattlebirds – Family Callaeatidae

Callaeas wilsoni	North Island kokako	E T
C. cinerea	South Island kokako	E X
Philesturnus rufusater	North Island saddleback	E Is T
P. carunculatus	South Island saddleback	E Is T
Heteralocha acutirostris	Huia	E X

PARVORDER PASSERIDA

Larks – Family Alaudidae

Alauda arvensis	Skylark	I

Wagtails and pipits – Family Motacillidae

Anthus novaeseelandiae	New Zealand pipit	E
A. chathamensis	Chatham Island pipit	C

A. *aucklandicus*	Subantarctic pipit	E S
A. *aucklandicus aucklandicus*	Auckland Island pipit	
A. *aucklandicus steindachneri*	Antipodes Island pipit	
Sparrows and grass finches – Family Passeridae		
Passer domesticus	House sparrow	I
Finches – Family Fringillidae		
Fringilla coelebs	Chaffinch	I
Carduelis chloris	Greenfinch	I
C. *carduelis*	Goldfinch	I
C. *flammea*	Redpoll	I
Buntings – Family Emberizidae		
Emberiza citrinella	Yellowhammer	I
E. *cirlus*	Cirl bunting	I
Swallows and martins – Family Hirundinidae		
Hirundo tahitica	Welcome swallow	H
H. *nigricans*	Australian tree martin	V
H. *ariel*	Fairy martin	V
Bulbuls – Family Pycnonotidae		
Pycnonotus cafer	Red-vented bulbul	I
Old World warblers – Family Sylviidae		
Bowdleria punctata		E
B. *punctata vealeae*	North Island fernbird	
B. *punctata punctata*	South Island fernbird	
B. *punctata stewartiana*	Stewart Island fernbird	T
B. *punctata wilsoni*	Codfish Island fernbird	T
B. *caudata*	Snares Island fernbird	E S
B. *rufescens*	Chatham Island fernbird	E X
White-eyes – Family Zosteropidae		
Zosterops lateralis	Silvereye	H
Accentors – Family Prunellidae		
Prunella modularis	Dunnock	I
Thrushes and blackbird – Family Muscicapidae		
Turdus merula	Blackbird	I
T. *philomelos*	Song thrush	I
Starlings – Family Sturnidae		
Sturnus vulgaris	Starling	I
Acridotheres tristis	Myna	I

APPENDIX 2

Plant, invertebrate, marine, foreign bird and foreign mammal species mentioned in the text

Plants

Beech, Southern	*Nothofagus* spp.
Black maire	*Nestegis cunninghamii*
Black tree fern	*Cyanthea medullaris*
Bottle gourd	*Lagenaria siceraria*
Boxthorn	*Lycium ferocissimum*
Bracken	*Pteridium esculentum*
Broadleaf	*Griselinia littoralis*
Bush lawyer	*Rubus cissoides*
Cabbage tree	*Cordyline australis*
Chatham Island korokio	*Corokia macrocarpa*
Cherry laurel	*Prunus laurocerasus*
Creeping pohuehue	*Muehlenbeckia axillaris*
Feathery tutu	*Coriaria angustissima*
Five-finger	*Pseudopanax arboreus*
Flax, New Zealand	*Phormium tenax*
Fuchsia, tree	*Fuchsia excorticata*
Hall's totara	*Podocarpus hallii*
Heather	*Calluna vulgaris*
Hinau	*Elaeocarpus dentatus*
Horopito	*Pseudowintera axillaris*
Inaka	*Dracophyllum longifolium*
Kahikatea	*Dacrycarpus dacrydioides*
Kaikomako	*Pennantia corymbosa*
Kamahi	*Weinmannia racemosa*
Kanono	*Coprosma grandifolia*
Kanuka	*Kunzia ericoides*
Karaka	*Corynocarpus laevigatus*
Karamu	*Coprosma robusta*
Kauri	*Agathis australis*
Kawakawa	*Macropiper excelsum*
Kiekie	*Freycinetia banksii*
Kohekohe	*Dysoxylum spectabile*
Kowhai	*Sophora* spp

Kumara	*Ipomoea batatas*
Lancewood	*Pseudopanax crassifolius*
Lemonwood	*Pittosporum eugenioides*
Mahoe	*Melicytus ramiflorus*
Mangeao	*Litsea calicaris*
Manuka	*Leptospermum scoparium*
Marram grass	*Ammophila arenaria*
Matai	*Prumnopitys taxifolia*
Mingimingi	*Coprosma propinqua*
Miro	*Prumnopitys ferruginea*
Mistletoes	*Peraxilla colensoi, P. tetrapetala, Alepis flavida*
Mount Cook lily	*Ranunculus lyallii*
Muttonbird scrub	*Senecio reinoldii*
New Zealand cedar	*Libocedrus* spp.
Ngaio	*Myoporum laetum*
Nikau	*Rhopalostylis sapida*
Oak	*Quercus robur*
Paper mulberry	*Broussonetia papyrifera*
Pate	*Schefflera digitata*
Pepperwood, *see* Horopito	
Pigeonwood	*Hedycarya arborea*
Pingao	*Desmoschoenus spiralis*
Pohutukawa	*Metrosideros excelsa*
Pokaka	*Elaeocarpus hookerianus*
Pukatea	*Laurelia novae-zelandiae*
Putaputaweta (marbleleaf)	*Carpodetus serratus*
Puriri	*Vitex lucens*
Quintinia	*Quintinia acutifolia* and *Q. serrata*
Rata, northern	*Metrosideros robusta*
Rata, southern	*Metrosideros umbellata*
Rata, scarlet	*Metrosideros fulgens*
Raukawa	*Pseudopanaz edgerleyi*
Raurekau	*Coprosma australis*
Red mapou	*Myrsine australis*
Rewarewa	*Knigntia excelsa*
Rimu	*Dacrydium cupressinum*
Snow totara	*Podocarpus nivalis*
Stinkwood	*Coprosma foetidissima*
Supplejack	*Ripogonum scandens*
Taraire	*Beilschmieda tarairi*
Taro	*Colocasia esculenta*
Tawa	*Beilschmieda tawa*
Tawapou	*Planchonella novo-zelandica*

Ti	*Cordyline fruticosa*
Titoki	*Alectryon excelsus*
Toro	*Myrsine salicina*
Totara	*Podocarpus totara*
Tree daisy	Shrubs and trees in the genera *Olearia*, *Senecio* and *Brachyglottis*
Tree ferns	*Dicksonia* and *Cyathea* spp.
Weeping mapou	*Myrsine divaricata*
Wineberry	*Aristotelia serrata*
Wood rose	*Dactylanthus taylori*
Yam	*Dioscorea* spp.

Terrestrial and freshwater invertebrates

Carove's giant dragonfly	*Uropetala carovei*
Cave weta	Species in the family Raphidophoridae
Chirping cicada	*Amphipsalta strepitans*
Common wasp	*Vespula vulgaris*
Cook Strait click beetle	*Amychus granulatus*
Cook Strait (Stephens Island) giant weta	*Deinacrida rugosa*
Freshwater crayfish	*Paranephrops planifrons*
German wasp	*Vespula germanica*
Huhu grub	*Prionoplus reticularis*
Longhorn	*Ochrocydus huttoni*
Sixpenny scale insect	*Ctenochiton viridis*
Sooty beech scale	*Ultracoelostoma assimile*
Stephens Island weevil	*Anagotus stephenensis*
Stick insects	Phasmatodea
Tree weta	*Hemideina* spp
Wetapunga	*Deinacrida heteracantha*

Fish and marine invertebrates

Anchovy	*Engraulis australis*
Antarctic krill	*Euphausia superba*
Ahuru	*Auchenoceros punctatus*
Arrow squid	*Nototodarus sloanii*
Barracouta	*Thyrsites atun*
Bluefin tuna	*Thunnus maccoyii*
Hoki	*Macruronus novaezelandiae*
Jack mackerel	*Trachurus declivis*
Octopus	*Octopus maorum*
Pilchard	*Sardinops neopilchardus*

Foreign birds

Arctic tern	*Sterna paradisaea*
Bald eagle	*Haliaeetus leucocephalus*
Black-tailed native hen	*Gallinula ventralis*
Brown creeper (North American)	*Certhia familiaris*
Emperor penguin	*Aptenodytes forsteri*
Emu	*Dromaius novaehollandiae*
Great tit	*Parus major*
Grey partridge	*Perdix perdix*
Java sparrow	*Padda oryzivora*
Kagu	*Rhynochetos jubatus*
Kakerori	*Pomarea dimidiata*
King penguin	*Aptenodytes patagonicus*
Linnet	*Carduelis cannabina*
Long-legged bunting	*Emberiza alcoveri*
Nene (Hawaiian goose)	*Nesochen sandvicensis*
Red grouse	*Lagopus lagopus*
Ostrich	*Struthio camelus*

Foreign mammals

Alpaca	*Lama pacos*
Chinchilla,	*Chinchilla laniger*
Llama	*Lama glama*
Père David's deer	*Elaphurus davidiensis*

Chapter 1

1. Diamond 1990
2. Birdlife International 2000
3. Collar et al. 1994, Stattersfield et al. 1998
4. Holdaway 1999a, b
5. Stevens et al. 1988
6. Stevens 1991, Stevens et al. 1998
7. Ibid.
8. Cooper & Milliner 1993, Stevens 1991
9. Copper et al. 2001
10. Long 1998.
11. Chambers et al. 2001, Cooper & Millener 1993, Stevens 1991
12. Wardle 1984
13. www.tepapa.govt.nz/communications/press_releases/pr_snake.html.20 July 20002
14. Stevens et al. 1998, Stevens 1991
15. Kennedy et al. 1999
16. Stevens et al. 1998
17. Cooper & Cooper 1995, Cooper & Millener 1993
18. Chambers et al. 2001, Cooper & Cooper. 1995
19. Stevens et al. 1988
20. Ibid., Stevens 1991
21. Cooper et al. 1993
22. Wilson 1997
23. Stevens et al. 1988
24. Ibid.

Chapter 2

1. Daugherty et al. 1994
2. Bell et al. 1985, McIntyre 1997
3. Cree & Butler 1993, Dawbin 1982
4. Green & Cannatella 1993
5. Chambers et al. 2001, Daugherty et al. 1993
6. Long 1998, McKee & Wiffen 1998
7. Bell 1978
8. Bell 1994, Bell et al. 1985
9. Bell et al. 1998a
10. Holdaway & Worthy 1996, Worthy 1997a
11. Towns and Daugherty 1994
12. Thurley & Bell 1994
13. Bell et al. 1998b, Worthy 1987a
14. Bell 1982, Bell et al. 1985
15. Bell 1994
16. Bell et al. 1998b
17. Holyoak et al. 1999
18. Worthy 1987b, Bell 1994
19. Bell et al. 1985, Bell et al. 1998b
20. Green 1994
21. Bell 1978, 1982, Bell et al. 1985
22. Thurley & Bell 1994
23. Bell 1982, 1985a
24. Bell 1982, 1985a, Thurley & Bell 1994
25. Bell 1994, Bell et al. 1998b
26. Worthy & Holdaway 1994a
27. Cree & Butler 1993, McIntyre 1997
28. Daugherty et al. 1990a
29. Cree et al. 1995
30. Newman & McFadden 1990
31. Daugherty et al. 1990a
32. McIntyre 1997
33. Towns et al. 2001
34. Cree & Butler 1993, Gaze 2001
35. Newman and McFadden 1990, Newman 1987b
36. Crook 1975, Worthy & Holdaway 1994a
37. Newman 1987b
38. Newman et al. 1994, Cree 1994
39. Cree et al. 1992
40. Newman 1987a
41. Thompson et al. 1992
42. Newman 1977, Ussher 1999, Walls 1981
43. Carmichael et al. 1989
44. Ussher 1999
45. Newman 1987b
46. Crook 1975, Cree et al. 1992
47. Daugherty et al. 1990b
48. Daugherty et al. 1994
49. Gill & Whitaker 1996

50. Worthy 1991a, Daugherty et al. 1994
51. Chambers et al. 2001, Daugherty et al. 1994
52. Bauer 1990
53. Chambers et al. 2001
54. Hardy 1977
55. Patterson & Daugherty 1995
56. Daugherty et al. 1993
57. Bauer 1990, Cooper & Millener 1993
58. Cooper and Millener 1993
59. Chambers et al. 2001
60. Bell et al. 1998b, Daugherty et al. 1990a
61. Daugherty et al. 1994, Hitchmough 1997
62. Gleeson et al. 1999, Boon et al. 2000, Baker et al. 1995
63. Hitchmough 1997
64. Chambers et al. 2001
65. Boon et al 2000, Greene 1999
66. Patterson & Daugherty 1995
67. Towns et al. 1985
68. Patterson 1992, Patterson & Daugherty 1990
69. Freeman 1997a
70. Bauer and Russell 1986
71. Worthy 1997a
72. Whitaker 1994
73. Thomas 1985
74. Whitaker 1984
75. Whitaker 1994, Freeman 1997a
76. Towns & Elliott 1996
77. Whitaker 1982
78. Thompson et al. 1992
79. Whitaker 1968b
80. Porter 1987
81. Gill 1976
82. Patterson 1992
83. Whitaker 1984, 1997
84. Barwick 1959
85. Cree 1994, Cree & Guillette 1995
86. Cree 1994
87. Ibid.
88. Whitaker 1968a
89. Thompson et al. 1992
90. Bannock et al. 1999
91. Tocher 1998
92. Towns et al. 1985, Whitaker 1968b, 1973, Towns 1994
93. Towns 1994
94. Whitaker 1968b
95. Towns and Daugherty 1994
96. Crook 1973, Whitaker 1973)
97. Towns 1991, 1994
98. Towns 1994, Towns et al. 2001
99. Patterson 1997
100. Tocher 1998
101. Patterson 1992, Towns et al. 1985
102. Towns and Daugherty 1994
103. Worthy 1998a
104. Towns and Daugherty 1994
105. Thomas 1985, Taylor and Thomas 1993
106. Towns and Daugherty 1994
107. Freeman 1993, 1997b
108. Bell 1985b, Tocher & Newman 1997
109. Tocher 1998
110. Whitaker 1996, Towns et al. 2001
111. Towns et al. 2001

Chapter 3

1. Schrodde & Faith 1991
2. Cooper & Millener 1993
3. Cooper and Millener 1993, Cooper et al. 1992
4. Sibley et al. 1982, Cooper & Penny 1997
5. Schrodde & Faith 1991
6. Cooper & Millener 1993, Stevens 1991
7. Christidis et al. 1996
8. Chambers et al. 2001
9. Wilson 1997
10. Fleming 1979, Baker 1991
11. Ibid.
12. Fleming 1976, *HANZAB*
13. Taylor 1985
14. Boon et al. 2001a, b
15. Taylor 1985, 1998
16. Boon et al. 2000, 2001a
17. Baker 1991
18. OSNZ 1990, see also Appendix 1, Foggo et al. 1997
19. Stattersfield et al. 1998
20. Cooper & Penny 1997
21. Edwards & Boles 2002
22. Sibley & Ahlquist 1990, Sibley 1991
23. Johnson 2001
24. Holdaway 1988, 1991
25. Christidis & Boles 1994
26. Sibley & Alquist 1990
27. Cooper & Millener 1993, Cooper et al. 1992, 2001

28. Cooper et al. 2001
29. Ibid.
30. Sibley et al. 1982
31. Raikow & Bledsoe 2000
32. Ibid.
33. Johnson 2001
34. Edwards & Boles 2002
35. Ibid.
36. OSNZ 1970
37. Kennedy et al. 1999, Kirsch et al. 1998
38. Hand et al. 1998
39. Daniel 1979
40. Kirsch et al. 1998
41. O'Donnell 2000a
42. Williams 1991a
43. Williams et al. 1991
44. Ibid
45. Sagar & O'Donnell 1982, *HANZAB*
46. Watt 1975
47. Diamond 1981
48. Ibid., Livezey 1990
49. *HANZAB*
50. Livezey 1992
51. Carlquist 1965, Trewick 1996
52. Trewick 1996
53. Trewick & Worthy 2001
54. Diamond 1991, Trewick 1997a, b
55. Diamond 1981, Trewick 1997b
56. Livezey 1990
57. Williams et al. 1991
58. Millener & Worthy 1991, Millener 1989
59. Rando et al.1999
60. Begon et al. 1986, Ford 1989
61. *HANZAB*, Merton et al. 1984
62. Warham 1990
63. Ibid.
64. Fenwick & Browne 1975
65. Richdale 1957
66. Gosler 1993
67. Knetmans & Powlesland 1999
68. Butler and Merton 1992
69. Saul et al. 1998
70. Trewick 1996, *HANZAB*
71. Niethammer 1970
72. Stevens 1990
73. Williams et al. 1991
74. Flack 1976
75. Gill et al. 1983
76. Wilson 1990, *HANZAB*, Powlesland et al. 1992, Elliott et al. 1996
77. *HANZAB*
78. O'Donnell 1996a, b
79. Williams 1991b
80. Triggs et al. 1991
81. Merton et al. 1984, Powlesland et al. 1992, Merton & Empson 1989
82. Castro et al. 1996
83. Christidis et al. 1996
84. McLennan 1988
85. Reid & Williams 1975
86. McLennan 1988, Jolly 1989
87. McLennan & McCann 1991
88. Reid 1977
89. McLennan 1988, Jolly 1989
90. Jamieson & Spencer 1996, Moorhouse 1996
91. Darwin 1871, Maynard-Smith 1978
92. Jamieson & Spencer 1996, Moorhouse 1996
93. Phillipps 1963, Gill & Martinson 1991
94. Frith 1997
95. See review by Craig 1991a
96. e.g. by Guthrie-Smith 1910, 1914; Stead 1932; Richdale 1965
97. Ford 1989
98. McLean & Gill 1988, Gill & McLean 1992
99. Sherley 1990
100. Young 1978
101. Craig & Jamieson 1990
102. Craig 1991
103. *HANZAB*
104. Veitch 1977
105. *HANZAB*
106. Maloney 1999
107. Pierce 1983
108. Pierce 1979
109. *HANZAB*
110. Beer 1966
111. O'Donnell 2001
112. Hill & Daniel 1985
113. Worthy et al. 1996
114. O'Donnell et al. 1999
115. O'Donnell 2001
116. O'Donnell 2000a
117. O'Donnell 2000b
118. Daniel 1979
119. Lloyd 2001
120. O'Donnell et al. 1999, Webb et al. 1998
121. Arkins et al. 1999, Daniel 1976, 1979

122. Lloyd 2001
123. Sedgeley 2001
124. Daniel 1979, Lloyd 2001
125. Lloyd 2001, O'Donnell et al. 1999, Webb et al. 1998
126. O'Donnell et al. 1999, O'Donnell & Sedgley 1999
127. O'Donnell & Sedgley 1999
128. O'Donnell 2001
129. Sedgeley & O'Donnell 1999, O'Donnell et al. 1999
130. Sedgeley 2001, Lloyd 2001
131. Molloy 1995, O'Donnell 2000a

Chapter 4

1. Worthy 1990
2. Atkinson & Millener 1991
3. Worthy & Holdaway 1993, 1994a, 1995, 1996, 2000; Worthy 1997a, 1998a, b, c, d
4. Worthy 1998b presents a concise summary of the bird species associated with particular vegetation types in the South Island. The other papers on Quaternary South Island fossil faunas (Worthy & Holdaway 1993, 1994a, 1995, 1996 and Worthy 1997a, 1998a) present detailed accounts of regional and habitat variations in vertebrate faunas.
5. Holdaway 1989
6. Worthy & Holdaway 1993, 1994, Holdaway & Worthy 1996
7. Worthy & Holdaway 1996, Worthy 1998a,b
8. Holdaway & Worthy 1997
9. Holdaway 1989, Holdaway & Worthy 1997
10. Ibid.
11. Worthy & Holdaway 1995, Worthy 1997a
12. Worthy & Holdaway 1994a, b, Holdaway & Worthy 1996
13. Holdaway & Worthy 1996
14. Worthy 2001
15. Worthy & Holdaway 1996, Worthy 1997a
16. Worthy & Holdaway 1994b
17. Worthy & Holdaway 1996, Worthy 1997a
18. Worthy & Holdaway 1994b, Holdaway & Worthy 1996
19. Worthy & Holdaway 1995
20. Atkinson & Millener 1991
21. Worthy & Holdaway 1996, 1997, Worthy 1998a
22. Worthy & Holdaway 2000
23. Powlesland 1981
24. Worthy & Mildenhall 1989, Millener & Worthy 1991
25. Heath 1986
26. Millener 1989
27. Millener & Worthy 1991, Worthy 1998b, Worthy & Holdaway 1994a
28. Atkinson & Millener 1991, Worthy & Holdaway 1993, 1994a
29. Buller, in Turbott 1967
30. Pierre 2000
31. Buller, in Turbott 1967, Potts 1882
32. Worthy & Holdaway 1993, 1994a
33. Worthy & Holdaway 1994a
34. Holdaway & Worthy 1997, Worthy 1997a
35. Worthy et al. 1996
36. Holdaway & Worthy 1997
37. Mills et al. 1984
38. Worthy 1990
39. Worthy 1998b
40. Worthy 1991b (These estimates of body weight differ from those given by Atkinson and Millener 1991)
41. Burrows 1980, Burrows et al. 1981
42. Worthy 1991b, Burrows 1980
43. Worthy 1990, 1991b
44. Worthy & Holdaway 2000
45. Worthy 1990, 1998b
46. Mills et al. 1984
47. Beauchamp & Worthy 1988
48. Holdaway & Worthy 1997
49. Worthy & Holdaway 1996, Worthy 1998c
50. Worthy 1997b, Worthy & Holdaway 1994a, 1996
51. Worthy 1998c
52. Greenwood & Atkinson 1977, Atkinson & Greenwood 1989
53. Burrows 1980
54. McGlone & Webb 1981
55. Atkinson & Greenwood 1989
56. Best 1984, Best & Powlesland 1985
57. Worthy & Holdaway 1994a, 2000
58. O'Donnell & Dilks 1994
59. Beggs & Wilson 1987
60. O'Donnell & Dilks 1994

61. Wilson & Brejaart 1992, *HANZAB*
62. Diamond and Bond 1999
63. Taylor 1985
64. Powlesland 1987, Leathwick et al. 1983
65. Clout and Hay 1989
66. Worthy & Holdaway 1993, Worthy 1998c
67. Gravatt 1971
68. O'Donnell & Dilks 1994
69. Worthy & Mildenhall 1989, Worthy & Holdaway 1993
70. Worthy & Holdaway 1993
71. Worthy & Holdaway 1995, 1996, Worthy 1997a
72. Worthy & Holdaway 1995
73. Worthy 1997a
74. Worthy & Holdaway 1996, Worthy 1997a
75. Worthy 1998a
76. Worthy & Holdaway 1996, Worthy 1998a
77. Worthy & Holdaway 1995
78. Worthy & Holdaway 1994, Worthy 1997a
79. Worthy & Holdaway 1995
80. Worthy 1998c
81. Worthy 1997b, 1998b
82. Worthy & Holdaway 1996, Worthy 1995, 1998c
83. Worthy 1998c
84. Worthy & Holdaway 1996
85. Worthy 1998c
86. Worthy 1995
87. Worthy 1997b, Worthy & Holdaway 1994a, 1996
88. Holdaway & Worthy 1997
89. Worthy 1998c
90. *HANZAB*
91. Kear and Scarlett 1970
92. Livezey 1989

Chapter 5

1. Worthy & Holdaway 1994a
2. Holdaway 1999b
3. Parrish & Williams 2001
4. Diamond 1992, Flannery 1994
5. Flannery 1994
6. Diamond 1992
7. Ibid.
8. Ibid., James 1995
9. Steadman 1991b, 1997
10. James 1991
11. Wragg & Weisler 1994
12. Steadman 1997
13. Martin & Klein 1984, Flannery 1994
14. Wilson 1988
15. Stork 1997
16. Lawton & May 1995
17. Wilson 1988, Ehrlich 1995
18. Worthy et al. 1991
19. Atkinson 1985
20. Holdaway 1996, 1999a
21. Atkinson & Towns 2001
22. Holdaway & Beavan 1999
23. Atkinson & Towns 2001
24. King 1990
25. Atkinson 1985, Holdaway 1999b
26. Worthy & Holdaway 1994a, b, 1996, Worthy 1997
27. Holdaway 1999b, Booth et al. 1996
28. Atkinson & Towns 2001
29. Wilson 1999
30. Holdaway 1999a
31. Holdaway 1999b, Worthy 1997
32. Smith 1989, Gill 1998
33. McGlone 1989, Anderson 1989b
34. Holdaway 1999b
35. Anderson 1983
36. Trewick & Worthy 2001
37. Holdaway 1999b
38. McGlone et al. 1994
39. McGlone 1989
40. Worthy 1998c
41. McGlone & Wilmshurst 1999, Ministry for the Environment 1997
42. Anderson 1983, 1989a, b, Anderson et al. 1996
43. Holdaway & Jacomb 2000
44. Smith 1989
45. Higham et al. 1999
46. McGlone et al. 1994
47. McGlone et al. 1994
48. Anderson 1989b
49. Holdaway & Jacomb 2000
50. Burrows 1982
51. Holdaway 1999b
52. Worthy & Holdaway 1994a
53. Beauchamp 1999
54. James 1995
55. McGlone 1989
56. East & Williams 1984
57. Holdaway 1999b
58. Lever 1994

59. Beauchamp 1997a, b
60. Atkinson 1973, King 1984, 1990
61. King 1984
62. Worthy 1997a
63. Lever 1994
64. King 1990
65. Fitzgerald & Veitch 1985
66. Millener 1989
67. Fitzgerald & Veitch 1985
68. King 1984
69. Veitch 1985
70. King 1984
71. Holdaway 1999b
72. King 1984
73. Wilson et al. 1998
74. Challies 1999
75. Wilson 1999
76. M. King 1981
77. Cox 1994
78. Worthy 1993, Wilson 1998
79. Trotter 1975, McCulloch 1987, Challis 1995, Worthy unpub.
80. Wilson 1998
81. Dennis 1986
82. Stead 1927
83. Potts, 1873, 1874
84. Stead 1927
85. Buller, in Turbott 1967
86. King 1984, Holdaway 1999b
87. Stead 1927
88. Wilson et al. 1998
89. Stead 1927
90. Ibid., Turbott 1969
91. Dawson & Creswell 1949, Bull et al. 1985, Crossland 1996
92. Turbott 1969, Bull et al. 1985
93. Stead 1927, Oliver Hunter unpub. notes, K-J Wilson pers. obs.
94. Stead 1927
95. Trotter 1975, Worthy unpub.
96. Stead 1927, Crossland 1996
97. Challies 1999
98. *Chatham Islands: Heritage and conservation*
99. Millener 1996, Tennyson & Millener 1994
100. Millener 1996
101. Tennyson & Millener 1994
102. Veitch 1985
103. Guthrie-Smith 1925, Bell 1978
104. Bell 1978
105. Watt 1975
106. Bell 1978
107. Wilson 1997
108. Dewar 1984, Diamond 1992, James 1995
109. James 1991, 1995, Steadman 1991b
110. James1995, Steadman 1991a, Wragg & Weisler 1994

Chapter 6

1. Holdaway 1996
2. Davidson 1984
3. McDowall 1994, Lever 1992, 1994
4. Lamb 1964
5. Galbreath 1989
6. Atkinson 1978
7. King 1984
8. *HANZAB*
9. Crosby 1986
10. McDowall 1994. Histories of provincial acclimatisation societies include Lamb 1964 (Canterbury), Ashby 1967 (Auckland) and Sowman 1981 (Nelson).
11. Lamb 1964
12. Lamb 1964, Drummond 1907
13. Ashby 1967
14. Thomson 1922
15. Ibid., McDowall 1994
16. Thomson 1922
17. McDowall 1994
18. Thomson 1922
19. Hutton & Drummond 1905, McDowall 1994
20 Druett 1983
21. *The Press*, 9 January 1883
22. McDowall 1994
23. Dawson & Bull 1970
24. Porter et al. 1994.
25. Lever 1994, Flux 1994
26. King 1990, Gibb & Williams 1994
27. Thomson 1922
28. Flux 1994, Gibb & Williams 1994
29. Lever 1994
30. King 1984
31. King 1990, Gibb & Williams 1994
32. King 1984
33. The voluminous literature on rabbits and rabbit control in New Zealand is reviewed by Gibb & Williams 1994.
34. Parliamentary Commissioner for the Environment 1998
35. Norbury 2001

36. Guthrie-Smith 1969
37. King 1990
38. Taylor and Tilley 1984
39. King 1990
40. King 1990, Strachan 1995
41. Cockayne 1926, Kirk 1920
42. King 1990
43. Nugent et al. 2001
44. Sowman 1981
45. McDowall 1994
46. This population cycle, common to all the introduced herbivores, is described by Clarke 1976, Challies 1985 and, in a lighter vein, by Caughley 1983.
47. Caughley 1983, Riney 1956
48. McSaveney & Whitehouse 1989
49. Gage 1980
50. McSaveney & Whitehouse 1989, Adams 1980. The latter gives 15,000 tonnes per square kilometre per year but notes that erosion rates vary enormously place to place.
51. Mark 1989, Stewart & Burrows 1989
52. Challies 1991, Caughley 1983
53. Challies 1977
54. Nugent et al. 2001, Challies 1991
55. King 1984
56. Challies 1991
57. Forsyth & Tustin 2001
58. McDowall 1994
59. King 1990
60. *HANZAB*
61. Nugent 1992
62. Ibid.
63. *HANZAB*
64. Lever 1994
65. Caithness 1982, Williams 1981
66. Ogle & Cheyne 1981
67. Stephenson 1986
68. King 1990
69. Lever 1994
70. Bump & Robbins 1966
71. Williams 1974. Distribution maps in Simpson and Day (1993) show 10 introduced bird species present in Tasmania
72. Simpson & Day 1993, Strahan 1995, Lever 1994
73. Veltman et al. 1996, Duncan 1997
74. Duncan et al. 1999
75. King 1990
76. Siriwardena et al. 1998
77. Rudge 1982
78. Ibid.
79. Regnault, in Whitaker & Rudge 1976
80. Rudge 1983
81. Parkes 2001
82. Rogers 1991
83. Department of Conservation 1995
84. Rogers 1991

Chapter 7

1. Allen et al. 1984, Mark 1989, Stewart & Burrows 1989
2. Rose et al. 1992
3. Stewart & Burrows 1989
4. Ladley & Kelly 1995, 1996, Wilson 1984
5. Nugent et al. 1997
6. King et al. 1996b
7. Innes 1995
8. King 1983
9. King et al. 1996b
10. King 1990
11. Taylor 1984
12. Taylor 1975
13. Innes 2001
14. Daniel 1973
15. Innes 2001
16. Clout et al. 1995
17. Innes et al. 1999, Bradfield & Flux 1996
18. Badan 1986, Campbell et al. 1984
19. Fitzgerald et al. 1996
20. King & Moody 1982a
21. Murphy 1996
22. King et al. 1996a, b, Murphy et al. 1998
23. Brockie 1992
24. Gillies 2001
25. Fitzgerald & Karl 1979
26. King et al. 1996a
27. King &Moody 1982b
28. Murphy & Dowding 1995, Murphy et al. 1998
29. King et al. 1996a
30. Murphy et al. 1998
31. Murphy & Dowding 1994
32. King et al. 2001, Cuthbert et al. 2000, King & Moody 1982b
33. King et al. 1996,a Murphy et al. 1998
34. Smith et al. 1995
35. Clapperton 2001
36. Ramsay 1978, Watt 1975, John

Marris, pers. comm.
37. Meads et al. 1984
38. Bremner et al. 1984
39. King 1984
40. Innes 2001
41. Brockie 1992
42. Murphy & Dowding 1995
43. O'Donnell et al. 1996
44. McLennan et al. 1996
45. Clapperton 2001, King et al. 2001
46. Wilson et al. 1998
47. Flack & Lloyd 1978, Flack 1976
48. Brown 1997
49. Murphy & Bradfield 1992
50. Murphy et al. 1998, Brown & Alterio 1996
51. Clapperton 2001, King et al. 2001
52. Gillies 2001
53. O'Donnell 1996a
54. See *NZ Journal of Ecology*, vol. 23, no. 3
55. Elliot 1996a, b
56. O'Donnell 1996b
57. King 1983
58. Murphy & Dowding 1995, Fitzgerald et al. 1996
59. Clout & Hay 1981
60. Brockie 1992
61. Hunt & Gill 1979, Gill 1980
62. O'Donnell & Dilks 1994
63. Brockie 1992
64. O'Donnell & Dilks 1994
65. Brockie 1992
66. O'Donnell & Dilks 1994
67. Gill 1980
68. Brockie 1992, Gill 1980, O'Donnell & Dilks 1994
69. Brockie 1992
70. *HANZAB*
71. Haw & Clout 1999
72. *HANZAB*
73. Lloyd 1985
74. Castro & Robertson 1997
75. Lloyd 1985, Ecroyd 1993, Lord 1991
76. Daniel 1976, Lloyd 2001, Lord 1991
77. Lee et al. 1991
78. O'Donnell & Dilks 1994, McCann 1963
79. Whitaker 1987
80. Burrows 1994b
81. Lee et al. 1988, 1991
82. Whitaker 1987
83. Burrows 1994b
84. Williams & Karl 1996
85. Burrows 1980
86. Williams et al. 2000
87. Daniel 1973
88. Atkinson 1973
89. Allen et al. 1994
90. Williams et al. 2000
91. Clout & Hay 1989 , Burrows 1994a & b , Lee et al. 1991 & Whitaker 1987
92. Ladley & Kelly 1995, 1996
93. Ladley et al. 1997
94. Ladley & Kelly 1995
95. Gaze & Clout 1983
96. Wilson et al. 1998
97. Brockie 1992
98. Stewart & Burrows 1989
99. Nugent et al. 1997
100. Norton 1995
101. Stewart & Burrows 1989
102. Forsyth et al. 2000, Forsyth & Hickling 1998
103. Parkes & Thomson 1995
104. Forsyth & Hickling 1998
105. Forsyth & Tustin 2001
106. Ibid.
107. Nugent 1994
108. O'Donnell 1994, Cowan 1991
109. Stewart & Rose 1988
110. Campbell 1990
111. Cowan 2001
112. Brown et al. 1993
113. Brockie 1992
114. Leathwick et al. 1983
115. Bradfield & Flux 1996, Innes et al. 1999
116. A. N. D. Freeman 1999
117. Hunt & Gill 1979
118. Saunders & Hobbs 1991
119. G. Howling & K-J. Wilson unpub. data
120. Potter 1990
121. Molloy 1995
122. Gibb 2000
123. Gill 1989
124. Day 1995
125. A. B. Freeman 1997c

Chapter 8

1. Sagar & Warham 1993, Robertson 1993

2. Barnett 1997
3. Harcourt 2001
4. The biology of seals is described by J. E. King 1983, and information on the New Zealand species is reviewed in C. M. King 1990 and Harcourt 2001. Crawley & Wilson 1976 describe the natural history of the New Zealand fur seal.
5. Numata et al. 2000
6. *HANZAB*
7. Walker 1995
8. Weimerskirch & Sagar 1996
9. Warham & Wilson 1982
10. West & Nilsson 1994
11. Robertson & van Tets 1982, Taylor 1982
12. Knox 1975
13. Warham 1996
14. A. N. D.Freeman 1997, Freeman et al. 1997, 2001
15. Stahl & Sagar 2000a, b
16. Wood 1990
17. Knox 1975
18. More detailed accounts of New Zealand hydrology and bathymetry are given by Knox 1975 and Bradford & Roberts 1978.
19. Knox 1975
20. Bradford & Roberts 1978
21. Ibid., Bradford et al. 1986
22. Bradford et al. 1986
23. Bartle 1990, Bradford & Roberts 1978
24. Gaskin 1968
25. Warham 1996
26. Knox 1975, R. H. Taylor et al. 1995, Wilson 1981
27. Harcourt 2001
28. Warham 1996
29. Ibid., *HANZAB*
30. Warham 1977
31. Warham 1996
32. *HANZAB*
33. Imber 1999, Nicholls 1994, Robertson & Nicholls 2000
34. Imber 1999, Walker & Elliott 1999
35. Walker et al. 1995.
36. Walker & Elliott 1999
37. Stahl & Sagar 2000a, b
38. Robertson & Nicholls 2000
39. Robertson et al. 2000
40. Stahl & Sagar 2000b
41. Bartle 1974 lists the birds present; *HANZAB*, Harper et al. 1985 and Warham 1996 provide information on their foods and feeding methods.
42. Bradford et al. 1986, Bartle 1974
43. Imber 1973, Warham 1996
44. Harper 1983
45. Weimerskirch & Sagar 1996
46. Kitson et al. 2000
47. A. N. D. Freeman et al. 1997, 2001
48. *HANZAB*, Wragg 1985
49. Wingham 1985
50. The vegetation of some seabird islands and the impacts petrels have on vegetation and soils have been described by Warham 1996, Fineran 1973 and Gilham 1957, 1960.
51. Hawke et al. 1999
52. J. P. C. Watt 1975
53. Thomas 1985
54. Cox et al. 1967
55. Beach et al. 1997
56. Gaze 2000
57. Best 1977
58. Richdale 1946, Howard 1940, Anderson 1980.
59. Davidson 1984
60. Lyver & Moller 1999
61. Warham 1982
62. Lyver 2000
63. Lyver et al. 1999
64. Moore 2001, G. A. Taylor 2000
65. Slooten & Dawson 1995
66. Wilson 1992
67. Robertson 1991
68. G. A. Taylor 2000
69. Gardner & Wilson 1999
70. G. A. Taylor 2000 & unpub.
71. Warham 1996
72. Davidson 1984, Smith 1989
73. Wilson 1981
74. Smith 1989, Gill 1998
75. R. H. Taylor 1982, 1992
76. Harcourt 2001
77. Sorensen 1969
78. Ibid.
79. Crawley & Wilson 1976
80. J. E. King 1983
81. Carey 1991, Harcourt 2001
82. Challies 1999
83. Jones 2000, Lyver et al. 2000

84. Wilson 1999, 2000
85. Lyver et al. 2000
86. Imber 1975
87. Booth et al. 1996
88. Fitzgerald & Veitch 1985
89. Cuthbert et al. 2000
90. *HANZAB*, Cooper et al. 1986, St Clair & St Clair 1992
91. A. N. D. Freeman 1998
92. Murray et al. 1993, Slooten & Dawson 1995
93. Bartle 1991, Gales 1997
94. Lyver et al. 1999
95. Bartle 1990
96. Croxall & Gales 1997
97. Slooten & Dawson 1995
98. Tennyson 1992
99. Alexander et al. 1997
100. Brothers 1991
101. Murray et al. 1993
102. Gales 1997, Robertson & Nunn 1997
103. G. A. Taylor 2000
104. Sagar et al. 1994, 1999
105. Sagar et al. 2000
106. Gales 1997
107. G. A. Taylor 2000
108. Croxall et al. 1990
109. Walker & Elliott 1999
110. Murray et al. 1993
111. Gales 1997, Croxall & Gales 1997
112. Croxall & Gales 1997, Robertson 1997, G. A. Taylor 2000
113. Robertson 1997
114. Croxall 1997
115. Harcourt 2001
116. Ibid., Slooten & Dawson 1995
117. Harcourt 2001
118. G. A. Taylor 2000, Warham 1996
119. Slooten & Dawson 1995
120. Ibid., G. A. Taylor 2000
121. Wodzicki et al. 1984
122. Wilson 1981
123. Warham & Wilson 1982, West & Nilsson 1994

Chapter 9

1. Andrews 1986
2. Galbreath 1989
3. M. King 1981
4. Galbreath 1989
5. Potts 1882, Galbreath 1989
6. Ibid.
7. Hill & Hill 1987
8. Guthrie-Smith 1914, 1936
9. Galbreath 1993
10. Caughley 1983
11. Galbreath 1993
12. Balance 2001
13. Lee & Jamieson 2001
14. Lee 2001
15. Galbreath 1993
16. Lee & Jamieson 2001, Maxwell 2001
17. Eason & Willans 2001
18. Ibid., Maxwell 2001
19. Clout & Craig 1994
20. Eason & Willans 2001
21. Mills & Mark 1977
22. Mills et al. 1991, Mills & Mark 1977
23. Lee 2001
24. Mills & Mark 1977
25. Mills et al. 1982b
26. Bamford 1985–86, Mills et al. 1982a
27. Bunin & Jamieson 1995
28. Lee 2001
29. Maxwell 2001
30. Mills et al. 1984
31. Beauchamp & Worthy 1988
32. Bunin & Jamieson 1995
33. Trewick & Worthy 2001
34. Gray & Craig 1991, Caughley 1994
35. Mills et al. 1982b, Jamieson & Ryan 2001
36. Jamieson & Ryan 2001
37. Bunin et al. 1997
38. Merton 1975
39. Brian Bell 1978
40. Galbreath 1993
41. Lovegrove 1996
42. Butler & Merton 1992, Merton 1992
43. Butler & Merton 1992
44. Ardern & Lambert 1997
45. Craig 1991b
46. Butler & Merton 1992, Merton 1992
47. See Caughley 1994 for a review of the relevant literature and a critique of the application of these ideas in conservation biology.
48. Craig 1991b
49. Veitch & Bell 1990
50. Veitch 1985, 1995, Veitch & Bell 1990
51. Moors 1985, Veitch & Bell 1990
52. Taylor & Thomas 1989, 1993

53. Veitch 1995
54. Taylor & Thomas 1993
55. Innes et al. 1999, Innes & Flux 1999
56. Hay 1984
57. Ibid.
58. Leathwick et al. 1983
59. Innes et al. 1999
60. Brown et al. 1993
61. Innes et al. 1999
62. Leathwick et al. 1983
63. Innes et al. 1999, Innes & Flux 1999
64. Innes & Flux 1999
65. Saunders & Norton 2001
66. Jolly & Colbourne 1991
67. Taborsky 1988
68. Baker et al. 1995
69. McLennan et al. 1996, Wilson et al. 1998
70. Jolly & Colbourne 1991, Colbourne & Robertson 1997
71. McLennan et al. 1996, McLennan & Potter 1992
72. McLennan & Potter 1992, McLennan et al. 1996
73. O'Donnell & Rasch 1991
74. Wilson et al. 1998, Greene & Fraser 1998
75. Beggs & Wilson 1987, 1991, Wilson et al. 1998
76. Wilson et al. 1998
77. R. Moorhouse et al. unpub.
78. Powlesland et al. 1992
79. Powlesland et al. 1995, Clout & Craig 1994.
80. Lloyd & Powlesland 1994, Powlesland et al. 1995
81. Powlesland & Lloyd 1994, Clout & Craig 1994
82. Cresswell 1996
83. Merton & Clout 1999
84. Ibid.
85. Reed et al. 1993a, Reed & Merton 1991
86. Reed et al. 1993b
87. Reed & Merton 1991, Reed et al. 1993a
88. Reed et al. 1993a
89. Pierce 1984b, Greene 1999
90. Pierce 1985, 1986
91. Reed et al. 1993a
92. Greene 1999
93. Maloney 1999, Maloney et al. 1997
94. Sanders 1999
95. Maloney et al. 1997
96. Maloney 1999
97. Maloney et al. 1997
98. Maloney et al. 1999
99. Reed et al. 1993a
100. Towns et al. 2001
101. O'Donnell 2001, J. Molloy 1995
102. Bell et al. 1998b
103. Brown 1994
104. Ibid.
105. Tocher & Newman 1997
106. Towns 1994
107. Ibid.
108. Towns et al. 2001
109. Towns 1994
110. Towns et al. 2001
111. Thomas & Whitaker 1994
112. Towns et al. 2001
113. Ibid.
114. Ibid.
115. Daugherty et al. 1990
116. Cree et al. 1994
117. May 1990
118. Cree et al. 1994, Cree & Butler 1993
119. Nelson & Daugherty 1997
120. Gibbs 1999
121. Galbraith & Hayson 1995
122. Timmins et al. 1987
123. http://www.sanctuary.org.nz
124. Saunders & Norton 2001
125. Ibid.
126. Burns et al. 2000
127 Ibid.

Chapter 10

1. Birdlife International 2000
2. Towns & Williams 1993
3. Wilson 1997
4. Worthy 1998c
5. Worthy & Brassey 2000, Holdaway et al. 2001
6. Hill & Hill 1987, Merton 1965, Williams 1977, Armstrong & McLean 1995
7. Lee & Jamieson 2001
8. Carlquist 1990
9. Mansfield 1996
10. For details, see Saunders & Norton 2001
11. Gibbs 1999
12. Beauchamp & Chambers 2000, Beauchamp et al. 2000

13. Wells et al. 1983
14. Gummer & Williams 1999
15. Saunders & Norton 2001
16. Towns & Williams 1993
17. Bradfield & Flux 1996
18. Craig & Stewart 1994
19. Brian Bell 1978
20. Reed & Merton 1991
21. Gardner & Wilson 1999, Sullivan & Wilson 2001
22. Eason & Willans 2001
23. Reed et al. 1993a
24. Towns et al. 2001
25. Saunders 1995
26. Armstrong & McLean 1995
27. Craig 1991b
28. Towns 1994
29. Veitch 1995
30. Innes et al. 1999
31. Saunders & Norton 2001
32. Ibid.
33. Innes & Barker 1999
34. A special issue of the *New Zealand Journal of Ecology* (vol. 23, no. 2) presented a range of papers that discuss the risks, benefits and current research on the use of toxins in New Zealand.
35. Powlesland et al 1999
36. Stephenson et al 1999
37. Burns et al 2000
38. Saunders & Norton 2001
39. Sibley & Ahlquist 1990
40. e.g. Cooper & Penny 1997
41. Daugherty et al. 1994
42. Baker et al. 1995, Bell et al. 1998b, Daugherty et al. 1990a
43. Boon et al 2000, 2001a
44. Patterson 1992, A. B. Freeman 1997a
45. Lambert & Millar 1995
46. www.environment.gov.au, www.redlist.org, Strahan 1995
47. Waldman & Tocher 1998
48. Siriwardena et al. 1998
49. Birdlife International 2000, www.redlist.org
50. Birdlife International 2000, Stattesfield et al. 1998
51. Thomsen et al. 2001
52. Pyke & White 1997
53. Sadleir & Warburton 2001, Miller et al. 1994
54. Towns & Williams 1993
55. Craig et al. 2000
56. Ibid.
57. Kirikiri & Nugent 1995
58. Roberts et al. 1995 provide a good introduction to Maori perspectives on conservation.
59. Kirikiri & Nugent 1995
60. Clout et al 1995
61. Robertson 1991
62. Crawley & Wilson 1976, Wilson 1981
63. Kirikiri & Nugent 1995
64. A detailed discussion of co-management is beyond the scope of this book, but interested readers should see Taiepa et al. 1997
65. Molloy & Davis 1994
66. Towns & Williams 1993
67. Molloy et al. 2001
68. Towns & Williams 1993
69. Linklater et al. 2000
70. Recher 1996
71. Galbreath 1993
72. Ministry for the Environment 1997
73. Craig et al. 2000

REFERENCES

Adams, J. 1980. High sediment yields from major rivers of the western Southern Alps, New Zealand. *Nature* 287: 88–89.

Alexander, K., G. Robertson and R. Gales. 1997. *The incidental mortality of albatrosses in longline fisheries*. Australian Antarctic Division, Hobart.

Allen, R. B., I. J. Payton and J. E. Knowlton. 1984. Effects of ungulates on structure and species composition in the Urewera forests as shown by exclosure. *New Zealand Journal of Ecology* 7: 119–30.

Allen, R. B., W. G. Lee and B. D. Rance. 1994. Regeneration in indigenous forest after eradication of Norway rats, Breaksea Island, New Zealand. *New Zealand Journal of Botany* 32: 429–39.

Anderson, A. J. 1980. Towards an explanation of protohistoric social organisation and settlement patterns amongst the southern Ngai Tahu. *New Zealand Journal of Archaeology* 2: 3–23.

Anderson, A. 1983 Faunal depletion and subsistence change in the early prehistory of Southern New Zealand. *Archaeology in Oceania* 18: 1–10

Anderson, A. 1989a. *Prodigious birds: Moas and moa hunting in prehistoric New Zealand*. Cambridge University Press, Cambridge.

Anderson, A. 1989b. Mechanics of overkill in the extinction of New Zealand moas. *Journal of Archaeological Science* 16: 137–51.

Anderson, A., T. Worthy and R. McGovern–Wilson. 1996. Moa remains and taphonomy, pp. 200–13 in A. Anderson, B. Allingham and I. Smith (eds), *Shag River mouth: The archaeology of an early southern Maori village*. Research Papers in Archaeology and Natural History 27, Australian National University, Canberra.

Andrews, J. R. H. 1986. *The southern ark: Zoological discovery in New Zealand 1769–1900*. Century Hutchinson, Auckland.

Archer M. 1984. Systematics: an enormous science rooted in instinct, pp. 125–50 in A. Archer and G. Clayton (eds), *Vertebrate zoogeography and evolution in Australia*. Hesperian Press, Marrickville.

Archey, G. 1941. *The moa: A study of the Dinornithiformes*. Bulletin of the Auckland Institute and Museum 1: 1–119.

Ardern, S. L. and D. M. Lambert. 1997. Is the black robin in genetic peril? *Molecular Ecology* 6: 21–28.

Arkins, A. M., A. P. Winnington, S. Anderson and M. N. Clout. 1999. Diet and netarivorous foraging behaviour of the short-tailed bat (*Mystacina tuberculata*). *Journal of Zoology (London)* 247: 183–87.

Armstrong, D. P. and I. G. McLean. 1995. New Zealand translocations: Theory and practice. *Pacific Conservation Biology* 2: 39–54.

Ashby, C. R. 1967. *The centenary history of the Auckland Acclimatisation Society 1867–1967*. Auckland Acclimatisation Society, Auckland.

Ashmole, N. P. 1971. Seabird ecology and the marine environment, pp. 223–85 in D. S. Farmer and J. R. King (eds), *Avian Biology*, vol. 1. Academic Press, New York.

Atkinson I. A. E. 1973. Spread of the ship rat (*Rattus r. rattus* L.) in New Zealand. *Journal of the Royal Society of New Zealand* 3: 457–72.

Atkinson, I. A. E. 1978. Evidence for effects of rodents on the vertebrate wildlife of New Zealand islands, pp 7–30 in P. R. Dingwel, I. A. E. Atkinson and C. Hay (eds), *The ecology and control of rodents in New Zealand nature reserves*. Department of Lands and Survey Information Series 4.

Atkinson I. A. E. 1985. The spread of commensal species of *Rattus* to Oceanic Islands and their effects on island avifaunas, pp. 35–81 in *ICBP Technical Publication No. 3*.

Atkinson, I. A. E., and R. M. Greenwood. 1989. Relationships between moas and plants. *New Zealand Journal of Ecology* 12 (supplement): 67 96.

Atkinson, I. A. E. and P. R. Millener 1991. An ornithological glimpse into New Zealand's pre-human past. *Acta XX Congressus Internationalis Ornithologici*: 129–92.

Atkinson, I. A .E. and D. R. Towns 2001. Advances in New Zealand mammalogy 1900–2000: Pacific rat. *Journal of the Royal Society of New Zealand* 31: 99–109.

Badan, D. 1986. Diet of the house mouse (*Mus musculus* L.) in two pine and a kauri forest. *New Zealand Journal of Ecology* 9: 137–141.

Baker, A. J. 1991. A review of New Zealand ornithology. *Current Ornithology* 8: 1–76.

Baker, A. J., C. H. Daugherty, R. Colbourne and J. L. [*sic*] McLennan. 1995. Flightless brown kiwis of New Zealand possess extremely subdivided population structure and cryptic species like small mammals. *Proceedings of the National Academy of Science of the USA* 92: 8254–58.

Baker, A. N. 1983. *Whales and dolphins of New Zealand and Australia*. Victoria University Press, Wellington.

Balance, A. 2001. Takahe: The bird that twice came back from the grave, pp. 18–22 in W. G. Lee and I. G. Jamieson (eds), *Fifty years of conservation management and research*. University of Otago Press, Dunedin.

Bamford, J. C. 1985/86. Takahe management – wandering in circles. *New Zealand Wildlife* 10: 4–5.

Bannock, C. A., A. H. Whitaker and G. J. Hickling. 1999. Extreme longevity of the common gecko, (*Hoplodactylus maculatus*) on Motunau Island, Canterbury, New Zealand. *New Zealand Journal of Ecology* 23: 101–03.

Barnett, A. 1997. Pouting whales are suckers for squid. *New Scientist*, 1 February 1997: 17.

Bartle, J. A. 1974. Seabirds of eastern Cook Strait, New Zealand, in autumn. *Notornis* 21: 135–66.

Bartle, J. A. 1990. Sexual segregation of foraging zones in procellariiform birds: Implications of accidental capture on commercial fishery longlines of grey petrels (*Procellaria cinerea*). *Notornis* 37: 146–50.

Bartle, J. A. 1991. Incidental capture of seabirds in the New Zealand subantarctic squid trawl fishery, 1990. *Bird Conservation International* 1: 351–59.

Barwick, R. E. 1959. The life history of the common New Zealand skink *Leiolopisma zelandica* (Gray 1843). *Transactions of the Royal Society of New Zealand* 86: 331–80.

Bauer, A. M. 1990. *Phylogenetic systematics and biogeography of the Carphodactylini (Reptilia: Gekkonidae)*. Bonner Zoologische Monographien 30.

Bauer, A. M. and A. P. Russell. 1986. *Hoplodactylus delcourti* n.sp. (Reptilia: Gekkonidae), the largest known gecko. *New Zealand Journal of Zoology* 13: 141–48.

Beach, G. S., K-J. Wilson and C. A. Bannock. 1997. *A survey of the birds, lizards and mammals of Motunau Island, Canterbury, New Zealand*. Lincoln University Wildlife Management Report No. 14.

Beauchamp, A. J. 1997a. The decline of the North Island weka (*Gallirallus australis greyi*) in the East Cape and Opotiki regions, North Island, New Zealand. *Notornis* 44: 27–35.

Beauchamp, A. J. 1997b. Sudden death of weka (*Gallirallus australis*) on Kawau Island, New Zealand. *Notornis* 44: 165–70.

Beauchamp, A. J. 1999. Weka declines in the north and north-west of the South Island, New Zealand. *Notornis* 46: 461–69.

Beauchamp A. J. and R. Chambers. 2000. Population density changes of adult North Island weka (*Gallirallus australis greyi*) in the Mansion House Historic Reserve,

Kawau Island, in 1992–1999. *Notornis* 47: 82–89.

Beauchamp, A. J. and T. H. Worthy. 1988. Decline in distribution of the takahe *Porphyrio* (= *Notornis*) *mantelli*: A re-examination. *Journal of the Royal Society of New Zealand* 18: 103–18.

Beauchamp A. J., G. C. Staples, E. O. Staples, A. Graeme, B. Graeme and E. Fox. 2000. Failed establishment of North Island weka (*Gallirallus australis greyi*) at Karangahake Gorge, North Island, New Zealand. *Notornis* 47: 90–96.

Beer, C. G. 1966. Adaptations to nesting habitat in the reproductive behaviour of the black-billed gull *Larus bulleri*. *Ibis* 108: 394–410.

Beggs, J. R. and P. R. Wilson. 1987. Energetics of South Island kaka (*Nestor meridionalis meridionalis*) feeding on the larvae of kanuka longhorn beetles (*Ochrocydus huttoni*). *New Zealand Journal of Ecology* 10: 143–47.

Beggs, J. R. and P. R. Wilson. 1991. The kaka *Nestor meridionalis*, a New Zealand parrot, endangered by introduced wasps and mammals. *Biological Conservation* 56: 23–38.

Begon, M, J. C. Harper and C. R. Townsend. 1986. *Ecology: Individuals, populations and communities*. Blackwell, Oxford.

Bell, Brian D. 1978. The Big South Cape Islands rat irruption, pp. 33–40 in P. R. Dingwell, I. A. E. Atkinson and C. Hay (eds), *The ecology and control of rodents in New Zealand nature reserves*. Department of Lands and Survey Information Series 4.

Bell, Ben D. 1978. Observations on the ecology and reproduction of the New Zealand leiopelmid frogs. *Herpetologica* 34: 340–54.

Bell, B. D. 1982. New Zealand frogs. *Herpetofauna* 14: 1–21.

Bell, B. D. 1985a. Development and parental-care in the endemic New Zealand frogs, pp. 268–78 in G. Grigg, R. Shine and H. Ehmann (eds), *The biology of Australasian frogs and reptiles*. Surrey Beatty, Chipping Norton.

Bell, B. D. 1985b. Conservation status of the endemic New Zealand frogs, pp. 449–58 in G. Grigg, R. Shine and H. Ehmann (eds), *The biology of Australasian frogs and reptiles*. Surrey Beatty, Chipping Norton.

Bell, B. D. 1994. A review of the status of New Zealand *Leiopelma* species (Anura: Leiopelmatidae), including a summary of demographic studies in Coromandel and on Maud Island. *New Zealand Journal of Zoology* 21: 341–49.

Bell, B. D., D. G. Newman and C. H. Daugherty. 1985. The ecological biogeography of the archaic New Zealand herpetofauna, pp. 99–106 in G. Grigg, R. Shine and H. Ehmann (eds), *The biology of Australasian frogs and reptiles*. Surrey Beatty, Chipping Norton.

Bell, B. D., C. H. Daugherty and R. A. Hitchmough. 1998a. The taxonomic identity of a population of terrestrial *Leiopelma* (Anura: Leiopelmatidae) recently discovered in the northern King Country, New Zealand. *New Zealand Journal of Zoology* 25: 139–46.

Bell, B. D., C. H. Daugherty and J. M. Hay. 1998b. *Leiopelma pakeka*, n.sp. (Anura: Leiopelmatidae), a cryptic species of frog from Maud Island, New Zealand, and a reassessment of the conservation status of *L. hamiltoni* from Stephens Island. *Journal of the Royal Society of New Zealand* 28: 39–54.

Best, E. 1977. *Forest Lore of the Maori*. Government Printer, Wellington.

Best, H. A. 1984. The foods of kakapo on Stewart Island as determined from their feeding sign. *New Zealand Journal of Ecology* 7: 71–83.

Best, H. and R. Powlesland. 1985. *Kakapo*. John McIndoe, Dunedin.

Birdlife International. 2000. *Threatened birds of the world*. Lynx Editions and Birdlife International, Barcelona and Cambridge.

Boon, W. M., J. C. Kearvell, C. H. Daugherty and G. K. Chambers. 2000. Molecular systematics of New Zealand *Cyanoramphus* parakeets: Conservation of orange-fronted and Forbes' parakeets. *Bird Conservation International* 10: 211–39.

Boon, W. M., J. C. Kearvell, C. H. Daugherty and G. K. Chambers. 2001a. *Molecular systematics and conservation of kakariki* (Cyanoramphus) *spp.* Science for Conservation

176. Department of Conservation, Wellington.
Boon, W. M., C. H. Daugherty and G. K. Chambers. 2001b. The Norfolk green parrot and New Caledonian red-crowned parakeet are distinct species. *Emu* 101: 113–21.
Booth, A. M., E. O. Minot, R. A. Fordham and J. G. Innes. 1996. Kiore (*Rattus exulans*) predation on the eggs of the little shearwater (*Puffinus assimilis haurakiensis*). *Notornis* 43: 147–53.
Bradfield, P. and I. Flux. 1996. *The Marpara kokako project 1989–1996: A summary report*. Department of Conservation, Hamilton.
Bradford, J. M. and P. E. Roberts. 1978. Distribution of reactive phosphorus and plankton in relation to upwelling and surface circulation around New Zealand. *New Zealand Journal of Marine and Freshwater Research* 12: 1–15.
Bradford, J. M., P. P. Lapennas, R. A. Murtagh, F. H. Chang and V. Wilkinson. 1986. Factors controlling summer phytoplankton production in greater Cook Strait, New Zealand. *New Zealand Journal of Marine and Freshwater Research* 20: 253–79.
Bremner, A. G., C. F. Butcher and G. B. Patterson. 1984. The density of indigenous invertebrates on three islands in Breaksea Sound, Fiordland, in relation to the distribution of introduced mammals. *Journal of the Royal Society of New Zealand*, 14: 379–86.
Brockie, R. 1992. *A living New Zealand forest*. David Bateman, Auckland.
Brothers, N. 1991. Albatross mortality and associated bait loss in the Japanese longline fishery in the Southern Ocean. *Biological Conservation* 55: 255–68
Brown, D. 1994. Transfer of Hamilton's frog *Leiopelma hamiltoni*, to a newly created habitat on Stephens Island, New Zealand. *New Zealand Journal of Zoology* 21: 425–30.
Brown, K. P. 1997. Predation at nests of two New Zealand endemic passerines: Implications for bird community restoration. *Pacific Conservation Biology* 3: 91–98.
Brown, K. and N. Alterio. 1996. Secondary poisoning of ferrets (*Mustela furo*) and poisoning of other mammalian pests by the anticoagulant poison brodifacoum, pp. 34–37 in *Ferrets as vectors of tuberculosis and threats to conservation*. Royal Society of New Zealand Miscellaneous Series No. 36.
Brown, K. P., J. G. Innes and R. M. Shorten. 1993. Evidence that possums prey on and scavenge birds' eggs, birds and mammals. *Notornis* 40: 1–9.
Brown, R. G. B., W. R. P. Bourne and T. R. Wahl. 1978. Diving by shearwaters. *Condor* 80: 123–25.
Bull, P. C., P. D. Gaze and C. J. R. Robertson. 1985. *The atlas of bird distribution in New Zealand*. Ornithological Society of New Zealand, Wellington.
Buller, W. L. 1872–73. *A history of the birds of New Zealand*. John van Voorst, London.
Buller, W. L. 1888. *A history of the birds of New Zealand*. The author, London.
Bump, G. and C. S. Robbins 1966. The newcomers, pp. 343–53 in A. Stefferud and A. L. Nelson (eds), *Birds in our lives*. US Department of the Interior, Fish and Wildlife Service, Washington, DC.
Bunin J. S. and I. G. Jamieson. 1995. New approaches toward a better understanding of the decline of takahe (*Porphyrio mantelli*) in New Zealand. *Conservation Biology* 9: 100–06.
Bunin J. S. and I. G. Jamieson and D. Eason. 1997. Low reproductive success of the endangered takahe *Porphyrio mantelli* on offshore island refuges in New Zealand. *Ibis* 139: 144–51.
Burns, R., A. Harrison, J. Hudson, G. Jones, P. Rudolf, P. Shaw, C. Ward, D. Wilson and L. Wilson. 2000. *Northern Te Urewera ecosystem restoration project: Summary annual report July 1999 to June 2000*. Department of Conservation, Gisborne.
Burrows, C. J. 1980. Some empirical information concerning the diet of moa. *New Zealand Journal of Ecology* 3: 125–30.
Burrows, C. J. 1982. On New Zealand climate within the last 1000 years. *New Zealand Journal of Archaeology* 4: 157–67.
Burrows, C. J. 1994a. Fruit types and seed dispersal modes of woody plants in the Ahuriri

Summit Bush, Port Hills, western Banks Peninsula, Canterbury, New Zealand. *New Zealand Journal of Botany* 32: 169–81.

Burrows, C. J. 1994b. The seeds always know best. *New Zealand Journal of Botany* 32: 349–63.

Burrows, C. J., B. McCulloch and M. M. Trotter. 1981. The diet of moas based on gizzard contents samples from Pyramid Valley, North Canterbury, and Scaifes Lagoon, Lake Wanaka, Otago. *Records of the Canterbury Museum* 9: 309–36.

Butler, D. and D. Merton. 1992. *The black robin: Saving the world's most endangered bird.* Oxford University Press, Auckland.

Caithness, T. 1982. *Game bird hunting: Problems, questions and answers.* Wetland Press, Wellington.

Campbell, D. J. 1990. Changes in structure and composition of a New Zealand lowland forest inhabited by brushtail possums. *Pacific Science*, 44: 277–96.

Campbell, D. J., H. Moller, G. W. Ramsay and J. C. Watt. 1984. Observations on foods of kiore (*Rattus exulans*) found in husking stations on northern offshore islands of New Zealand. *New Zealand Journal of Ecology* 7: 131–38.

Carey, P. W. 1991. *Fish prey species of the New Zealand fur seal* (Arctocephalus forsteri *Lesson*). Science and Research Internal Report 115. Department of Conservation, Wellington.

Carlquist, S. 1965. *Island life: A natural history of the islands of the world.* The Natural History Press, Garden City, New York.

Carlquist, S. 1990. Worst-case scenarios for island conservation: The endemic biota of Hawaii, pp. 207–12 in D. R. Towns, C. H. Daugherty, and I. A. E. Atkinson (eds), *Ecological restoration of New Zealand islands.* Conservation Sciences Publication 2, Department of Conservation, Wellington.

Carmichael, C. K., J. C. Gillingham and S. N. Keall. 1989. Feeding ecology of the tuatara (*Sphenodon punctatus*) on Stephens Island based on niche diversification. *New Zealand Journal of Zoology* 16: 269.

Castro, I. and A. W. Robertson. 1997. Honeyeaters and the New Zealand forest flora: The utilisation and profitability of small flowers. *New Zealand Journal of Ecology* 21: 169–79.

Castro, I., E. O. Minot, R. A. Fordham and T. R. Birkhead. 1996. Polygynandry, face to face copulation and sperm competition in the hihi *Notiomystis cincta* (Aves: Meliphagidae). *Ibis* 138: 765–71.

Caughley, G. 1983. *The deer wars: The story of deer in New Zealand.* Heinemann, Auckland.

Caughley, G. 1994. Directions in conservation biology. *Journal of Animal Ecology* 63: 215–44.

Caughley, G. and A. Gunn. 1996. *Conservation biology in theory and practice.* Blackwell, Oxford.

Challis, A. J. 1995. *Ka Pakihi Whakatekateka o Waitaha: The archaeology of Canterbury in Maori times.* Science and Research Series 89. Department of Conservation, Wellington.

Challies, C. N. 1977. Effects of commercial hunting on red deer densities in the Arawata Valley, South Westland, 1972–76. *New Zealand Journal of Forestry Science* 7: 263–73.

Challies, C. N. 1985. Establishment, control and commercial exploitation of wild deer in New Zealand, pp. 23–26 in P. F. Fenessy and K. D. Drew (eds), *Biology of deer production.* Royal Society of New Zealand Bulletin 22.

Challies, C. N. 1991. Status and future management of the wild animal recovery industry. *New Zealand Forestry*, May 1991: 10–17.

Challies, C. N. 1999. Changing fortunes of the white–flippered penguin. *Notornis* 46: 410.

Chambers, G. K., W. M. Boon, T. R. Buckley and R. A. Hitchmough. 2001. Using molecular methods to understand the Gondwanan affinities of the New Zealand biota: Three case studies. *Australian Journal of Botany* 49: 377–87.

Chatham Islands, The: Heritage and conservation. 1996. Canterbury University Press, Christchurch.

Chilton, C. (ed.) 1909. *The subantarctic islands of New Zealand*. Philosophical Institute of Canterbury, Christchurch.

Christidis, L. and W. Boles. 1994. *The taxonomy and species of birds of Australia and its territories.* Royal Australasian Ornithologists Union Monograph 2.

Christidis, L., P. R. Leeton and M. Westerman. 1996. Were bowerbirds part of the New Zealand fauna? *Proceedings of the National Academy of Sciences of the USA* 93: 3898–901.

Clapperton, B. K. 2001. Advances in New Zealand mammalogy 1900–2000: Feral ferret. *Journal of the Royal Society of New Zealand* 31: 185–203.

Clarke, C. M. H. 1976. Eruption, deterioration and decline of the Nelson red deer herd. *New Zealand Journal of Forestry Science* 5: 235–49.

Clout, M. N. and J. L. Craig. 1994. The conservation of critically endangered flightless birds in New Zealand. *Ibis* 137 (supplement): 181–90.

Clout, M. N. and J. R. Hay. 1981. South Island kokako (*Callaeas cinerea cinerea*) in *Nothofagus* forest. *Notornis* 28: 256–59.

Clout, M. N. and J. R. Hay. 1989. The importance of birds as browsers, pollinators and seed dispersers in New Zealand forests. *New Zealand Journal of Ecology* 12 (supplement): 27–33.

Clout, M. N., K. Denyer, R. E. James and I. G. McFadden. 1995. Breeding success of New Zealand pigeons (*Hemiphaga novaeseelandiae*) in relation to control of introduced mammals. *New Zealand Journal of Ecology* 19: 209–12.

Cockayne, L. 1926. *Monograph on the New Zealand beech forests*. New Zealand State Forest Service Bulletin 4.

Collar, N. J., M. J. Crosby and A. J. Stattersfield. 1994. *Birds to watch 2: The world list of threatened birds*. Birdlife Conservation Series 4. Birdlife International, Cambridge.

Cooper, A. and R. A. Cooper. 1995. The Oligocene bottleneck and New Zealand biota: Genetic record of a past environmental crisis. *Proceedings of the Royal Society of London B* 261: 293–302.

Cooper, A. and D. Penny. 1997. Mass survival of birds across the Cretaceous–Tertiary boundary: Molecular evidence. *Science* 275: 1109–113.

Cooper, A., C. Mourer-Chauvire, G. K. Chambers, A. von Haeseler, A. C. Wilson and S. Paabo. 1992. Independent origins of New Zealand moas and kiwis. *Proceedings of the National Academy of Sciences of the USA* 89: 8741–44.

Cooper, A., I. A. E. Atkinson, W. G. Lee and T. H. Worthy. 1993. Evolution of moa and their affect on the New Zealand flora. *Trends in Ecology and Evolution* 8: 433–37.

Cooper, A., C. Lalueza-Fox, S. Anderson, A. Rambaut, J. Austin and R. Ward. 2001. Complete mitochondrial genome sequences of two extinct moas clarify ratite evolution. *Nature* 409: 704–07.

Cooper, R. A. and P. R. Millener. 1993. The New Zealand biota: Historical background and new research. *Trends in Ecology and Evolution* 8: 429–33.

Cooper, W. J., C. M. Miskelly, K. Morrison and R. J. Peacock. 1986. Birds of the Solander Islands. *Notornis* 33: 77–89.

Coulbourne, R. M. and H. A. Robertson. 1997. Successful translocations of little spotted kiwi (*Apteryx owenii*) between offshore islands of New Zealand. *Notornis* 44: 253–58.

Cowan, P. E. 1991. The ecological effects of possums on the New Zealand environment, pp. 73–88 in R. Jackson (ed.), *Proceedings from a symposium on tuberculosis*. Veternary Continuing Education Publication 182, Massey University, Palmerston North.

Cowan, P. E. 2001. Advances in New Zealand mammalogy 1900–2000: Brushtail possum. *Journal of the Royal Society of New Zealand* 31: 15–29.

Cox, G. J. 1994. *Mountains of fire: The volcanic past of Banks Peninsula*. Canterbury University Press, Christchurch.

Cox, J. E., R. H. Taylor and R. Mason. 1967. *Motunau Island Canterbury, New Zealand: An ecological survey*. DSIR Bulletin 178.

Craig, J. L. 1991a. Communal breeding along the changing face of theory. *Acta XX Congressus Internationalis Ornithologici*: 233–46.
Craig, J. L. 1991b. Are small populations viable? *Acta XX Congressus Internationalis Ornithologici*: 2546–52.
Craig, J. L. and I. G. Jamieson. 1990. Pukeko: Different approaches and some different answers, pp. 385–412 in P. B. Stacey and W. D. Koenig (eds), *Cooperative breeding in birds*. Cambridge University Press, Cambridge.
Craig, J. L. and A. M. Stewart. 1994. Conservation: a starfish without a central disk? Pacific *Conservation Biology* 1: 163–68.
Craig, J., S Anderson, M. Clout, B. Creese, N. Mitchell, J. Ogden, M. Roberts and G. Ussher. 2000. Conservation issues in New Zealand. *Annual Review of Ecology and Systematics* 31: 61–78
Crawley, M. C. and G. J. Wilson. 1976. The natural history and behaviour of the New Zealand fur seal (*Arctocephalus forsteri*). *Tuatara* 22: 1–29.
Cree, A. 1994. Low annual reproductive output in female reptiles from New Zealand. *New Zealand Journal of Zoology* 21: 351–372.
Cree, A. and D. Butler. 1993. *Tuatara recovery plan* (Sphenodon *spp.*). Department of Conservation, Wellington.
Cree, A. and L. J. Guillette. 1995. Biennial reproduction with a fourteen-month pregnancy in the gecko *Hoplodactylus maculatus* from southern New Zealand. *Journal of Herpetology* 29: 163–73.
Cree, A., J. F. Cockrem and L. J. Guillette. 1992. Reproductive cycles of male and female tuatara (*Sphenodon punctatus*) on Stephens Island, New Zealand. *Journal of Zoology (London)* 226: 199–217.
Cree, A., C. H. Daugherty, D. R. Towns and B. Blanchard. 1994. The contribution of captive management to the conservation of tuatara (*Sphenodon*) in New Zealand, pp. 377–85 in J. B. Murphy, K. Adler and J. T. Collins (eds), *Captive management and conservation of amphibians and reptiles*. *Contributions to Herpetology*, vol. 11. Society for the Study of Amphibians and Reptiles, Ithaca.
Cree, A., C. H. Daugherty and J. M. Hay. 1995. Reproduction of a rare New Zealand reptile the tuatara *Sphenodon punctatus*, on rat-free and rat-inhabited islands. *Conservation Biology* 9: 373–83.
Cresswell, M. 1996. *Kakapo recovery plan 1996–2005*. Department of Conservation, Wellington.
Crook, I. G. 1973. The tuatara, *Sphenodon punctatus* Gray, on islands with and without populations of the Polynesian rat, *Rattus exulans* (Peale). *Proceedings of the New Zealand Ecological Society* 20: 115–20.
Crook, I. G. 1975. The tuatara, pp. 331–52 in G. Kuschel (ed.), *Biogeography and ecology in New Zealand*. Junk, The Hague.
Crosby, A. W. 1986. *Ecological imperialism: The biological expansion of Europe, 900–1900*. Cambridge University Press, Cambridge.
Crossland, A. 1996. *Port Hills birdlife: Inventory, analysis and restoration potential*. Christchurch City Council, Christchurch.
Croxall, J. P. 1997. Research and conservation: a future for albatrosses, pp. 269–90 in G. Robertson and R. Gales (eds), *Albatross biology and conservation*. Surrey Beatty, Chipping Norton.
Croxall, J. P. and R. Gales. 1997. An assessment of the conservation status of albatrosses, pp. 46–65 in G. Robertson and R. Gales (eds), *Albatross biology and conservation*. Surrey Beatty, Chipping Norton.
Croxall, J. P., P. Rothery, S. P. C. Pickering and P. A. Prince. 1990. Reproductive performance, recruitment and survival of wandering albatrosses *Diomedia exulans* at Bird Island, South Georgia. *Journal of Animal Ecology* 59: 775–96.

Cuthbert, R., E. Sommer and L. S. Davis. 2000. Seasonal variation in the diet of stoats in a breeding colony of Hutton's shearwaters. *New Zealand Journal of Zoology* 27: 367–73.

Daniel, M. J. 1973. Seasonal diet of the ship rat (*Rattus r. rattus*) in lowland forest in New Zealand. *Proceedings of the New Zealand Ecological Society* 20: 21–30.

Daniel, M. J. 1976. Feeding by the short-tailed bat (*Mystacina tuberculata*) on fruit and possibly nectar. *New Zealand Journal of Zoology* 3: 391–98.

Daniel, M.J. 1979. The New Zealand short-tailed bat *Mystacina tuberculata*: A review of present knowledge. *New Zealand Journal of Zoology* 6: 357–70.

Darwin, C. 1845. *The voyage of the Beagle . . .* , John Murray, London

Darwin, C. 1871. *The descent of man and selection in relation to sex*. John Murray, London.

Daugherty, C. H., A. Cree, J. M. Hay and M. B. Thompson. 1990a. Neglected taxonomy and continuing extinctions of tuatara (*Sphenodon*). *Nature* 347: 177–79.

Daugherty, C. H., D. R. Towns, I. A. E. Atkinson and G. W. Gibbs. 1990b. The significance of the biological resources of New Zealand islands for ecological restoration, pp. 9–21 in D. R. Towns, C. H. Daugherty and I. A. E. Atkinson (eds), *Ecological restoration of New Zealand islands*. Conservation Sciences Publication 2, Department of Conservation, Wellington.

Daugherty, C. H., G. W. Gibbs and R. A. Hitchmough. 1993. Mega-island or micro-continent? New Zealand and its fauna. *Tree* 8: 437–42.

Daugherty, C. H., G. B. Patterson and R. A. Hitchmough. 1994. Taxonomic and conservation review of the New Zealand herpetofauna. *New Zealand Journal of Zoology* 21: 317–23.

Davidson, J. 1984. *The prehistory of New Zealand*. Longman Paul, Auckland.

Dawbin, W. H. 1982. The tuatara *Sphenodon punctatus* (Reptilia: Rhynchocephalia): A review, pp. 149–81 in D. G. Newman (ed.), *New Zealand herpetology*. Wildlife Service Occasional Publication 2.

Dawson, D. G. and P. C. Bull. 1970. A questionnaire survey of bird damage to fruit. *New Zealand Journal of Agricultural Research* 13: 362–71.

Dawson, E. and I. D. R. Creswell. 1949. Bird life at Governors Bay, Banks Peninsula. *NZ Bird Notes* 3:141–46.

Dawson, S. M. 1985. *The New Zealand whale and dolphin digest*. Brick Row, Auckland.

Dawson, S. and E. Slooten. 1996. *Down-under dolphins: The story of Hector's dolphin*. Canterbury University Press, Christchurch.

Day, T. D. 1995. Bird species composition and abundance in relation to native plants in urban gardens, Hamilton, New Zealand. *Notornis* 42: 175–86.

De Lange, P. J., P. B. Heenan, D. R. Given, D. A. Norton and C. C. Ogle. 1999. Threatened and uncommon plants of New Zealand. *New Zealand Journal of Botany* 37: 603–28.

Dennis, A. 1986 *Banks Peninsula reserves*. Department of Lands and Survey, Christchurch.

Department of Conservation 1995. *Kaimanawa wild horses plan*. Department of Conservation, Wanganui.

Department of Conservation 1997. *Issues and options for managing the impacts of deer on native forests and other ecosystems: A public discussion document*. Department of Conservation, Wellington.

Dewar, R. E. 1984. Extinctions in Madagascar: the loss of the subfossil fauna, pp. 574–93 in P. S. Martin and R. G. Klein (eds), *Quaternary extinctions: A prehistoric revolution*. University of Arizona Press, Tucson.

Diamond, Jared M. 1981. Flightlessness and fear of flying in island species. *Nature* 293: 507–08.

Diamond, J. M. 1984. Distributions of New Zealand birds on real and virtual islands. *New Zealand Journal of Ecology* 7: 37–55.

Diamond, J. M. 1990. New Zealand as an archipelago: An international perspective, pp. 3–8 in D. R. Towns, C. H. Daugherty and I. A. E. Atkinson, *Ecological restoration of New*

Zealand islands. Conservation Sciences Publication 2, Department of Conservation, Wellington.

Diamond, J. M. 1991. A new species of rail from the Solomon Islands and convergent evolution of insular flightlessness. *Auk* 108: 461–70.

Diamond, J. 1992. *The rise and fall of the third chimpanzee*. Vintage, London.

Diamond, J. M. and C. R. Veitch. 1981. Extinctions and introductions in the New Zealand avifauna: cause and effect? *Science* 211: 499–501.

Diamond, Judy. and A. B. Bond. 1999. *Kea, bird of paradox: The evolution and behavior of a New Zealand parrot*. University of California Press, Berkeley.

Doughty, C., N. Day and A. Plant. 1999. *Birds of the Solomons, Vanuatu and New Caledonia*. Christopher Helm, London.

Druett, J. 1983. *Exotic intruders: The introduction of plants and animals into New Zealand*. Heinemann, Auckland.

Drummond, J. 1907. *Our feathered immigrants: Svidence for and against introduced birds in New Zealand together with note on the native avifauna*. Department of Agriculture Bulletin 16.

Duncan, R. P. 1997. The role of competition and introduction effort in the success of Passeriforme birds introduced to New Zealand. *The American Naturalist* 149: 903–15.

Duncan, R. P., T. M. Blackburn and C. J. Veltman. 1999. Determinants of geographic range sizes: A test using introduced New Zealand birds. *Journal of Animal Ecology* 68: 963–75.

Eason, D. K. and M. Willans. 2001. Captive rearing: A management tool for the recovery of the endangered takahe, pp. 80–95 in W. G. Lee and I. G. Jamieson (eds), *Fifty years of conservation management and research*. University of Otago Press, Dunedin.

East, R. and G. R. Williams. 1984. Island biogeography and the conservation of New Zealand's indigenous forest-dwelling avifauna. *New Zealand Journal of Ecology* 7: 27–35.

Ecroyd, C. 1993. In search of the wood rose. *Forest & Bird* 267: 24–28.

Ecroyd, C. E. 1995. The wood rose and bats: The link between two unique endangered New Zealand species. *Horticulture in New Zealand* 6: 39–41.

Edwards, S. V. and W. E. Boles. 2002. Out of Gondwana: The origin of passerine birds. *Trends in Ecology and Evolution* 17: 347–49.

Ehrlich, P. R. 1995. The scale of the human enterprise and biodiversity loss, pp. 214–26 in J. H. Lawton and R. M. May (eds), *Extinction Rates*. Oxford University Press, Oxford.

Elliott, G. P. 1996a. Productivity and mortality of mohua (*Mohoua ochrocephala*). *New Zealand Journal of Zoology* 23: 229–37.

Elliott, G. P. 1996b. Mohua and stoats: A population viability analysis. *New Zealand Journal of Zoology* 23: 239–47.

Elliott, G. P., P. J. Dilks and C. J. F. O'Donnell. 1996. The ecology of yellow-crowned parakeets (*Cyanoramphus auriceps*) in *Nothofagus* forests in Fiordland, New Zealand. *New Zealand Journal of Zoology* 23: 249–65.

Fenwick, G. D. and W. M. M. Browne. 1975. Breeding of the spotted shag at Whitewash Head, Banks Peninsula. *Journal of the Royal Society of New Zealand* 5: 31–45.

Fineran, B. A. 1973. A botanical survey of seven mutton-bird islands south-west Stewart Island. *Journal of the Royal Society of New Zealand* 3: 475–526.

Fitzgerald, B. M. and B. J. Karl. 1979. Foods of feral house cats (*Felis catus* L) in forest of the Orongorongo Valley, Wellington. *New Zealand Journal of Zoology* 6: 107–26.

Fitzgerald, B. M. and C. R. Veitch. 1985. The cats of Herekopare Island, New Zealand: Their history, ecology and affects on birdlife. *New Zealand Journal of Zoology* 12: 319–30.

Fitzgerald, B. M, M. J. Daniel, A. E. Fitzgerald, B. J. Karl, M. J. Meads and P. R. Notman. 1996. Factors affecting the numbers of house mice (*Mus musculus*) in hard beech (*Nothofagus truncata*) forest. *Journal of the Royal Society of New Zealand* 26: 237–49.

Flack, J. A. D. 1976. New Zealand robins. *Wildlife: A Review* 7: 15–19.

Flack, J. A. D. and B. D. Lloyd. 1978. The effect of rodents on the breeding success of the South Island robin, pp. 59–66 in P. R. Dingall, I. A. E.Atkinson and C. H. Hay (eds), *The ecology and control of rodents in New Zealand nature reserves*. Department of Lands and Survey Information Series 4.

Flannery, T. F. 1994. *The future eaters*. Reed, Port Melbourne

Fleming, C. A. 1976. New Zealand as a minor source of terrestrial plants and animals in the Pacific. *Tuatara* 22: 30–37.

Fleming, C. A. 1979. *The geological history of New Zealand and its life*. Auckland University Press, Auckland.

Flux, J. E. C. 1994. World distribution, pp. 8–21 in H. V. Thompson and C. M. King (eds), *The European rabbit: The history and biology of a successful colonizer*. Oxford University Press, Oxford.

Foggo, M. N., R. A. Hitchmough and C. H. Daugherty. 1997. Systematic and conservation implications of geographic variation in pipits (*Anthus*: Motacillidae) in New Zealand and some offshore islands. *Ibis* 139: 366–73.

Ford, H. A. 1989. *Ecology of birds: An Australian perspective*. Surrey Beatty, Chipping Norton.

Forsyth, D. M. and G. J. Hickling 1998. Increasing Himalayan thar and decreasing chamois densities in the eastern Southern Alps, New Zealand: Evidence for interspecific competition. *Oecologica* 113: 377–82.

Forsyth, D. M. and K. G. Tustin. 2001. Advances in New Zealand mammalogy 1900–2000: Himalayan tahr. *Journal of the Royal Society of New Zealand* 31: 251–61.

Forsyth, D. M., J. P. Parkes and G. J. Hickling. 2000. A case for integrated management of sympatric herbivore pest impacts in the central Southern Alps, New Zealand. *New Zealand Journal of Ecology* 24: 97–103.

Freeman, Alastair B. 1993. Size difference between populations of *Hoplodactylus maculatus* in Canterbury, New Zealand. *Herpetofauna* 23: 9–15.

Freeman, A. B. 1997a. Comparative ecology of two *Oligosoma* skinks in coastal Canterbury: A contrast with Central Otago. *New Zealand Journal of Ecology* 21: 153–60.

Freeman, A. B. 1997b. The conservation status of a coastal duneland lizard fauna at Kaitorete Spit, Canterbury, New Zealand. *Herpetofauna* 27: 25–30.

Freeman, A. B. 1997c. *The distribution of lizards in Christchurch and its environs*. Lincoln University Wildlife Management Report No. 11.

Freeman, Amanda N. D. 1997. The influence of hoki fishing vessels on Westland petrel (*Procellaria westlandica*) distribution at sea. *Notornis* 44: 159–64.

Freeman, A. N. D. 1998. Diet of Westland petrels *Procellaria westlandica*: The importance of fisheries waste during chick rearing. *Emu* 98: 36–43.

Freeman, A. N. D. 1999. Bird counts in Kennedys Bush Scenic Reserve, Port Hills, Christchurch. *Notornis* 46: 388–404.

Freeman, A. N. D., D. G. Nicholls, K–J. Wilson and J. A. Bartle. 1997. Radio and satellite tracking Westland petrels *Procellaria westlandica*. *Marine Ornithology* 25: 31–36.

Freeman, A. N. D., K-J. Wilson and D. G. Nicholls. 2001. Westland petrels and the hoki fishery: determining co-occurrence using satellite telemetry. *Emu* 101: 47–56.

Frith C. B. 1997. Huia (*Heteralocha acutirostris*: Callaeidae)-like sexual bill dimorphism in some birds of paradise (Paradisaeidae) and its significance. *Notornis* 44: 177–84.

Fuller, E. (ed.) 1990. *Kiwis: A monograph of the family Apterygidae*. Seto Publishing, Auckland.

Gage, M. 1980. *Legends in the rocks: An outline of New Zealand geology*. Whitcoulls, Christchurch.

Galbraith, M. P. and C. R. Hayson. 1995. Tiritiri Matangi Island, New Zealand: public participation in species translocation to an open sanctuary, pp. 149–154 in M. Serena (ed.), *Reintroduction biology of Australian and New Zealand fauna*. Surrey Beatty, Chipping Norton.

Galbreath, R. 1989. *Walter Buller: The reluctant conservationist.* Government Printing Office, Wellington.
Galbreath, R. 1993. *Working for Wildlife: A history of the New Zealand Wildlife Service.* Bridget Williams Books, Wellington.
Gales, R. 1997. Albatross populations: status and threats, pp. 20–45 in G. Robertson and R. Gales (eds), *Albatross biology and conservation.* Surrey Beatty, Chipping Norton.
Gardner, P. and K-J Wilson. 1999. *Chatham petrel* (Pterodroma axillaris) *studies: Breeding biology and burrow blockading.* Science for Conservation 131, Department of Conservation, Wellington.
Gaskin, D. E. 1968. Distribution of Delphinidae (Cetacea) in relation to sea surface temperatures off eastern and southern New Zealand. *New Zealand Journal of Marine and Freshwater Research* 2: 527–34.
Gaze, P. 2000. The response of a colony of sooty shearwaters (*Puffinus griseus*) and flesh-footed shearwaters (*P. carneipes*) to the cessation of harvesting and the eradication of Norway rats (*Rattus norvegicus*). *New Zealand Journal of Zoology* 27: 375–79.
Gaze, P. 2001. *Tuatara recovery plan.* Department of Conservation, Wellington.
Gaze, P. D. and M. N. Clout. 1983. Honeydew and its importance to birds in beech forests of South Island, New Zealand. *New Zealand Journal of Ecology.* 6: 33–37.
Gibb, J. A. 2000. Activity of birds in the western Hutt hills, New Zealand. *Notornis* 47: 13–35.
Gibb, J. A. and J. M. Williams. 1994. The rabbit in New Zealand, pp. 158–204 in H. V. Thompson and C. M. King (eds), *The European rabbit: The history and biology of a successful colonizer.* Oxford University Press, Oxford.
Gibbs, G. 1999. Insects at risk. *Forest & Bird* 294: 32–35.
Gilham, M. E. 1957. Ecology of some New Zealand seabird colonies. *Proceedings of the New Zealand Ecological Society* 5: 9–10.
Gilham, M. E. 1960. Plant communities of the Mokohinau Islands, northern NZ. *Transactions of the Royal Society of New Zealand* 88: 79–98.
Gill, B. J. 1976. Aspects of the ecology, morphology and taxonomy of two skinks (Reptilia: Lacertillia) in the coastal Manawatu area of New Zealand. *New Zealand Journal of Zoology* 3: 141–57.
Gill, B. J. 1980. Abundance, feeding, and morphology of passerine birds at Kowhai Bush, Kaikoura, New Zealand. *New Zealand Journal of Zoology* 7: 235–46.
Gill, B. J. 1989. Bird counts in regenerated urban forest at Auckland Domain. *Notornis* 36: 81–87.
Gill, B. J. 1998. Prehistoric breeding sites of New Zealand sea lions (*Phocarctos hookeri*, Carnivora: Otatiidae) at North Cape. *Records of the Auckland Museum* 35: 55–64.
Gill, B. and P. Martinson. 1991. *New Zealand's extinct birds.* Random Century, Auckland.
Gill, B. J. and I. G. McLean. 1992. Population dynamics of the New Zealand whitehead (Pachycephalidae) – a communal breeder. *Condor* 94: 628–35.
Gill, B. and G. Moon. 1999. *New Zealand's unique birds.* Reed, Auckland.
Gill, B. and T. Whitaker 1996. *New Zealand frogs and reptiles.* David Bateman, Auckland.
Gill, B. J., R. G. Powlesland and M. H. Powlesland. 1983. Laying seasons of three insectivorous song-birds at Kowhai Bush, Kaikoura. *Notornis* 30:81–85.
Gillies, C. 2001. Advances in New Zealand mammalogy 1900–2000: House cat. *Journal of the Royal Society of New Zealand* 31: 205–18.
Gleeson, D. M., R. L. J. Howitt and N. Ling. 1999. Genetic variation, population structure and cryptic species within black mudfish, *Neochanna diversus*, an endemic galaxiid from New Zealand. *Molecular Ecology* 8: 47–57.
Gosler, A. 1993. *The great tit.* Hamlyn, London.
Gravatt, D. J. 1971. Aspects of habitat use by New Zealand honeyeaters, with reference to other forest species. *Emu* 71: 65–72.

Gray R. D. and J. L. Craig. 1991. Theory really matters: Hidden assumptions in the concept of 'habitat requirements'. *Acta XX Congressus Internationalis Ornithologici*: 2553–60.

Green, D. M. 1994 Genetic and cytogenetic diversity in Hochstetter's frog *Leiopelma hochstetteri*, and its importance for conservation management. *New Zealand Journal of Zoology* 21: 417–24.

Green, D. M. and D. C. Cannatella. 1993. Phylogenetic significance of the amphicoelous frogs, Ascaphidae and Leiopelmatidae. *Ethology Ecology and Evolution* 5: 233–45.

Greene B. 1999. Genetic variation and hybridisation of black stilts (*Himantopus novaezelandiae*) and pied stilts (*H. h. leucocephalus*), Order Charadriiformes. *New Zealand Journal of Zoology* 26: 271–77.

Greene, T. C. and J. R. Fraser 1998. Sex ratio of North Island kaka (*Nestor meridionalis septentrionalis*), Waihaha Ecological Area, Pureora Forest Park. *New Zealand Journal of Ecology* 22: 11–16.

Greenwood, R. M. and I. A. E. Atkinson. 1977. Evolution of divaricating plants in New Zealand in relation to moa browsing. *Proceedings of the New Zealand Ecological Society* 24: 21–29.

Gummer, H. and M. Williams. 1999. Campbell Island teal: Conservation update. *Wildfowl* 50: 133–38.

Guthrie-Smith. H. 1910. *Birds of the water, wood and waste*. Whitcombe and Tombs, Wellington.

Guthrie-Smith. H. 1914. *Mutton birds and other birds*. Whitcombe and Tombs, Wellington.

Guthrie-Smith, H. 1925. *Bird life on island and shore*. William Blackwood, Edinburgh.

Guthrie-Smith, H. 1936. *Sorrows and joys of a New Zealand naturalist*. A. H. and A. W. Reed, Dunedin.

Guthrie-Smith, H. 1969. *Tutira: The story of a New Zealand sheep station*. 4th edn. A. H. and A. W. Reed, Wellington.

Hand, S. J., P. Murray, D. Megirian, M. Archer and H. Godthelp. 1998. Mystacinid bats (Microchiroptera) from the Australian Tertiary. *Journal of Paleontology* 72: 538–45.

HANZAB. *Handbook of Australian, New Zealand and Antarctic birds*, vols 1 (1990), 2 (1993), 3 (1996), 4 (1999), 5 (2001) and 6 (2002). Oxford University Press, Melbourne.

Harcourt, R. G. 2001. Advances in New Zealand mammalogy 1990–2000: Pinnipeds. *Journal of the Royal Society of New Zealand* 31: 135–60.

Hardy, G. S. 1977. The New Zealand Scincidae (Reptilia: Lacertilia): A taxonomic and zoogeographic study. *New Zealand Journal of Zoology* 4: 221–325.

Harper, P. C. 1983. Biology of the Buller's shearwater (*Puffinus bulleri*) at the Poor Knights Islands, New Zealand. *Notornis* 30: 299–318.

Harper, P. C., J. P. Croxall and J. Cooper. 1985. *A guide to foraging methods used by marine birds in Antarctic and subantarctic Seas*. BIOMASS Handbook 24.

Haw, J. M. and M. N. Clout. 1999. Diet of morepork (*Ninox novaeseelandiae*) throughout New Zealand by analysis of stomach contents. *Notornis* 46: 333–45.

Hawke, D. J., R. N. Holdaway, J. E. Causer and S. Ogden. 1999. Soil indicators of pre-European seabird breeding in New Zealand at sites identified by predator deposits. *Australian Journal of Soil Research* 37: 103–13.

Hay, J. 1915. *Reminisces of earliest Canterbury (principally Bank's Peninsula) and its settlers*. The Press, Christchurch.

Hay, R. 1984. The kokako: Perspective and prospect. *Forest & Bird* 15(1): 6–11.

Heath, S. 1986. Rock wrens in the Southern Alps of New Zealand, pp. 277–88 in B. A. Barlow (ed.), *Flora and fauna of alpine Australasia*. CSIRO, Melbourne.

Heather, B. D. and H. A. Robertson. 1996. *The field guide to the birds of New Zealand*. Viking, Auckland.

Higham, T., A. Anderson and C. Jacomb. 1999. Dating the first New Zealanders: The

chronology of the Wairau Bar. *Antiquity* 73: 420–27.
Hill, J. E. and M. J. Daniel. 1985. Systematics of the New Zealand short-tailed bat *Mysticina* Gray, 1843 (Chiroptera: Mystacinidae). *Bulletin of British Museum of Natural History (Zoology)* 48: 279–300.
Hill, S. and J. Hill 1987. *Richard Henry of Resolution Island*. John McIndoe, Dunedin.
Hitchmough, R. A. 1997. A systematic review of the New Zealand Gekkonidae. PhD thesis, Victoria University of Wellington.
Holdaway, R. N. 1988. The New Zealand passerine list: What if Sibley and Ahlquist are right? *Notornis* 35: 63–70.
Holdaway, R. 1989. New Zealand's pre-human avifauna and its vulnerability. *New Zealand Journal of Ecology* 12 (supplement): 11–25.
Holdaway, R. N. 1991. Sibley et al.'s (1988) classification of living birds applied to the New Zealand list. *Notornis* 38: 152–64.
Holdaway, R. N. 1996 Arrival of rats in New Zealand. *Nature* 384: 225–26
Holdaway, R. N. 1999a. A spatio-temporal model for the invasion of the New Zealand archipelago by the Pacific rat *Rattus exulans*. *Journal of the Royal Society of New Zealand* 29: 91–105.
Holdaway, R. N. 1999b. Introduced predators and avifaunal extinctions in New Zealand, pp. 189–238 in R. D. E. MacPhee (ed.), *Extinctions in near time: Causes, contexts and consequences*. Kluwer Academic/Plenum, New York.
Holdaway, R. N. and N. R. Beavan. 1999. Reliable 14C AMS dates on bird and Pacific rat *Rattus exulans* bone gelatin from a $CaCO_2$-rich deposit. *Journal of the Royal Society of New Zealand* 29: 185–211.
Holdaway, R. N. and C. Jacomb. 2000. Rapid extinction of the moas (Aves: Dinornithiformes): Model, test and implications. *Science* 287: 2250–54.
Holdaway, R. and T. Worthy. 1991. Lost in time. *New Zealand Geographic* 12: 51–68.
Holdaway, R. N. and T.H. Worthy. 1996. Diet and biology of the laughing owl, *Sceloglaux albifacies* (Aves: Strigidae) on Takaka Hill, Nelson, New Zealand. *Journal of Zoology (London)* 239: 545–72.
Holdaway, R. N. and T. H. Worthy. 1997. A reappraisal of the late Quaternary fossil vertebrates of Pyramid Valley Swamp, North Canterbury, New Zealand. *New Zealand Journal of Zoology* 24: 69–121.
Holdaway, R. N., T. H. Worthy and Tennyson. 2001. A working list of breeding bird species of the New Zealand region at first human contact. *New Zealand Journal of Zoology* 28: 119–87.
Holmes, B. 1998. The day of the sparrow. *New Scientist*, 27 June 1998: 32–35.
Holyoak, A., B. Waldman and N. Gemmell. 1999. A re-examination of the specific status of *Leiopelma hamiltoni* and *L. pakeka*. *New Zealand Journal of Zoology* 26: 258.
Howard, B. 1940. *Rakiura: A history of Stewart Island, New Zealand*. A. H. and A. W. Reed, Dunedin.
Hunt, D. M. and B. J. Gill (eds). 1979. *Ecology of Kowhai Bush, Kaikoura*. Mauri Ora Special Publication 2.
Hutton, F. W. and J. Drummond. 1905. *The animals of New Zealand: An account of the colony's air-breathing vertebrates*. Whitcombe and Tombs, Christshurch.
Imber, M. J. 1973. The food of grey-faced petrels (*Pterodroma macroptera gouldi* (Hutton)), with special reference to diurnal vertical migration of their prey. *Journal of Animal Ecology* 42: 645–62.
Imber, M. J. 1975. Petrels and predators. *Bulletin of the International Council for Bird Preservation* 12: 260–63.
Imber, M. J. 1999. Diet and feeding ecology of the royal albatross *Diomedia epomophora* – king of the shelf break and inner slope. *Emu* 99: 200–11.
Imboden Ch. 1982. Opening address, pp 21–22 in D. G. Newman (ed.), *New Zealand*

herpetology. Wildlife Service Occasional Publication 2.

Innes, J. 1995. The impacts of possums on native fauna, pp. 11–15 in C. F. J. O'Donnell (ed.), *Possums as conservation pests*. Department of Conservation, Wellington.

Innes, J. 2001. Advances in New Zealand mammalogy 1900–2000: European rats. *Journal of the Royal Society of New Zealand* 31: 111–25.

Innes, J. and G. Barker. 1999. Ecological consequences of toxin use for mammalian pest control in New Zealand: An overview. *New Zealand Journal of Ecology* 23: 111–27.

Innes, J. and I. Flux. 1999. *North Island kokako recovery plan 1999–2009*. Department of Conservation, Wellington.

Innes, J., R. Hay, I. Flux, P. Bradfield, H. Speed and P. Jansen. 1999. Successful recovery of North Island kokako *Callaeas cinerea wilsoni* populations by adaptive management. *Biological Conservation* 87: 201–14.

James, H. F. 1991. The contribution of fossils to knowledge of Hawaiian birds. *Acta XX Congressus Internationalis Ornithologici*: 420–24.

James, H. F. 1995. Prehistoric extinctions and ecological changes on oceanic islands, pp. 87–102 in P. M. Vitousek, L. L. Loope and H. Adsersen (eds), *Islands biological diversity and ecosystem function*. Springer-Verlag, Berlin.

Jamieson, I. G. and C. J. Ryan. 2001. Island takahe: Closure of the debate over the merits of introducing Fiordland takahe to predator-free islands, pp 96–113 in W. G. Lee and I. G. Jamieson (eds), *Fifty years of conservation management and research*. University of Otago Press. Dunedin.

Jamieson, I. G. and H. G. Spencer. 1996. The bill and foraging behaviour of the huia (*Heteralocha acutirostris*): Were they unique? *Notornis* 43: 14–18.

Johnson, K. P. 2001. Taxon sampling and the phylogenetic position of Passeriformes: Evidence from 916 avian Cytochrome *b* sequences. *Systematic Biology* 50: 128–36.

Jolly, J. N. 1989. A field study of the breeding biology of the little spotted kiwi (*Apteryx owenii*) with emphasis on the causes of nest failures. *Journal of the Royal Society of New Zealand* 19: 433–48.

Jolly, J. N. and R. M. Coulbourne. 1991. Translocations of little spotted kiwi (*Apteryx owenii*) between offshore islands of New Zealand. *Journal of the Royal Society of New Zealand* 21: 143–49.

Jones, C. 2000. Sooty shearwater (*Puffinus griseus*) breeding colonies on mainland South Island, New Zealand: Evidence of decline and predictors of persistence. *New Zealand Journal of Zoology* 27: 327–34.

Kear, J. and R. J. Scarlett. 1970. The Auckland Islands merganser. *Wildfowl* 21: 78–86.

Kennedy, M., A. M. Paterson, J. C. Morales, S. Parsons, A. P. Winnington and H. G. Spencer. 1999. The long and the short of it: Branch lengths and the problem of placing the New Zealand short-tailed bat, Mystacina. *Molecular Phylogenetics and Evolution* 13: 405–16.

King, C. M. (Kim). 1983. The relationship between beech (*Nothofagus* spp.) seedfall and populations of mice (*Mus musculus*), and the demographic and dietary responses of stoats (*Mustela ermina*), in three New Zealand forests. *Journal of Animal Ecology* 52: 141–66

King, C. 1984. *Immigrant killers: Introduced predators and the conservation of birds in New Zealand.* Oxford University Press, Auckland.

King, C. M. (ed.) 1990. *The handbook of New Zealand mammals.* Oxford University Press, Auckland.

King, C. M. and J. E Moody. 1982a. The biology of the stoat (*Mustela erminea*) in the national parks of New Zealand. I: General introduction. *New Zealand Journal of Zoology* 9: 49–55.

King, C. M. and J. E Moody. 1982b. The biology of the stoat (*Mustela erminea*) in the national parks of New Zealand. II: Food habits. *New Zealand Journal of Zoology* 9: 57–80.

King, C. M., M. Flux, J. G. Innes and B. M. Fitzgerald. 1996a. Population biology of small mammals in Pureora Forest Park: 1: Carnivores (*Mustela erminea, M. furo, M. nivalis,* and *Felis catus*). *New Zealand Journal of Ecology* 20: 241–51.
King, C. M., J. G. Innes, M. Flux, M. O. Kimberley, J. R. Leathwick and D. S. Williams. 1996b. Distribution and abundance of small mammals in relation to habitat in Pureora Forest Park. *New Zealand Journal of Ecology* 20: 215–40.
King, C. M., K. Griffiths and E. C. Murphy. 2001. Advances in New Zealand mammalogy 1900–2000: Stoat and weasel. *Journal of the Royal Society of New Zealand* 31: 165–83.
King, J. E. 1983. *Seals of the world*. British Museum (Natural History), London.
King, M. 1981. *The collector: A biography of Andreas Reischek*. Hodder & Stoughton, Auckland.
Kirikiri, R. and G. Nugent. 1995. Harvesting of New Zealand native birds by Maori, pp. 54–59 in Grigg, G. C., P. T. Hale and D. Lunney (eds), *Conservation through sustainable use of wildlife*. Centre for Conservation Biology, University of Queensland, Brisbane.
Kirk, H. B. 1920. Opossums in New Zealand. *Appendices to the Journal of the New Zealand House of Representatives*, Session 1, H-28: 1-12.
Kirsch, J. A. W., J. M. Hutcheon, D. G. P. Byrnes and B. D. Lloyd. 1998. Affinities and historical zoogeography of the New Zealand short-tailed bat, *Mystacina tuberculata* Gray 1843, inferred from DNA-hybridization comparisons. *Journal of Mammalian Evolution* 4: 33–64.
Kitson, J. C., J. B. Cruz, C. Lalas, J. B. Jillett, J. Newman and P. O'B. Lyver. 2000. Interannual variations in the diet of breeding sooty shearwaters (*Puffinus griseus*). *New Zealand Journal of Zoology* 27: 347–55.
Knegtmans, J. W. and R. G. Powlesland. 1999. Breeding biology of the North Island tomtit (*Petroica macrocephala toitoi*) at Pureora Forest Park. *Notornis* 46: 446–56.
Knox, G. A. 1975. The marine benthic ecology and biogeography, pp. 353–403 in G. Kuschel (ed.), *Biogeography and ecology in New Zealand*. Junk, The Hague.
Ladley, J. J. and D. Kelly. 1995. Explosive New Zealand mistletoe. *Nature* 378: 766.
Ladley, J. J. and D. Kelly. 1996. Dispersal, germination and survival of New Zealand mistletoes (Loranthaceae): Dependence on birds. *New Zealand Journal of Ecology* 20: 69–79.
Ladley, J. J., D. Kelly and A. W. Robertson. 1997. Explosive flowering, nectar production, breeding systems and pollinators of New Zealand mistletoes (Loranthaceae): *New Zealand Journal of Botany* 35: 345–60.
Lamb, R. C. 1964. *Birds, beasts and fishes: The first hundred years of the North Canterbury Acclimatisation Society*. North Canterbury Acclimatisation Society, Christchurch.
Lambert, D. M. and C. D. Millar. 1995. DNA science and conservation. *Pacific Conservation Biology* 2: 21–38.
Lawton, J. H. and R. M. May. 1995. *Extinction rates*. Oxford University Press, Oxford.
Leathwick, J. R., J. R. Hay and A. E. Fitzgerald. 1983. The influence of browsing by introduced mammals on the decline of North Island kokako. *New Zealand Journal of Ecology* 6: 55–70.
Lee, W. G. 2001. Fifty years of takahe conservation, research and management: What have we learnt? pp. 49–60 in W. G. Lee and I. G. Jamieson (eds), *Fifty years of conservation management and research*. University of Otago Press, Dunedin.
Lee, W. G. and I. G. Jamieson (eds). 2001. *The takahe: Fifty years of conservation management and research*. University of Otago Press. Dunedin.
Lee, W. G., J. B. Wilson and P. N. Johnson. 1988. Fruit colour in relation to the ecology and habit of *Coprosma* (Rubiacae) species in New Zealand. *Oikos* 53: 325–31.
Lee, W. G., M. N. Clout, H. A. Robertson and J. B. Wilson. 1991. Avian dispersers and fleshy fruits in New Zealand. *Acta XX Congressus Internationalis Ornithologici*: 1617–23.
Lever, C. 1992. *They dined on Eland: The story of the acclimatisation societies*. Quiller Press, London.

Lever, C. 1994. *Naturalised animals: The ecology of successfully introduced species*. T. and A. D. Poyser, London.
Linklater, W. L., E. Z. Cameron, K. J. Stafford and C. J. Veltman. 2000. Social and spatial structure and range use by Kaimanawa wild horses (*Equus caballus*: Equidae). *New Zealand Journal of Ecology* 24: 139–52.
Livezey, B. C. 1989. Pylogenetic relationships and incipient flightlessness of the extinct Auckland Islands merganser. *Wilson Bulletin* 101: 410–35.
Livezey, B. C. 1990. Evolutionary morphology of flightlessness in the Auckland Islands teal. *Condor* 92: 639–73.
Livezey, B. C. 1992. Morphological corollaries and ecological implications of flightlessness in the kakapo (Psittaciformes: *Strigops habroptilus*). *Journal of Morphology* 213: 105–45.
Lloyd, B. D. 2001. Advances in New Zealand mammalogy 1990–2000: Short-tailed bats. *Journal of the Royal Society of New Zealand* 311: 59–81.
Lloyd, B. D. and R. G. Powlesland. 1994. The decline of kakapo *Strigops habroptilus* and attempts at conservation by translocation. *Biological Conservation* 69: 75–85.
Lloyd, D. G. 1985. Progress in understanding the natural history of New Zealand plants. *New Zealand Journal of Botany* 23: 707–22.
Long, J. A. 1998. *Dinosaurs of Australia and New Zealand and other animals of the Mesozoic era*. University of New South Wales Press, Sydney.
Lord, J. M. 1991. Pollination and seed dispersal in *Freycinetia baueriana*, a dioecious liane that has lost its bat pollinator. *New Zealand Journal of Botany* 29: 83–86.
Lovegrove, T. G. 1996. Island releases of saddlebacks *Philesternus carunculatus* in New Zealand. *Biological Conservation* 77: 151–57.
Lyver, P. O'B. 2000. Sooty shearwater (*Puffinus griseus*) harvest intensity and selectivity on Poutama Island, New Zealand. *New Zealand Journal of Ecology* 24: 169–80.
Lyver, P. O'B. and H. Moller. 1999. Modern technology and customary use of wildlife: The harvest of sooty shearwaters by Rakiura Maori as a case study. *Environmental Conservation* 26: 280–88.
Lyver, P. O'B., H. Moller and C. Thompson. 1999. Changes in sooty shearwater *Puffinus griseus* chick production and harvest precede ENSO events. *Marine Ecology Progress Series* 188: 237–48.
Lyver, P. O'B., H. Moller and C. J. R. Robertson. 2000. Predation at sooty shearwater *Puffinus griseus* colonies on the New Zealand mainland: Is there safety in numbers? *Pacific Conservation Biology* 5: 347–357.
McCann, C. 1963. External features of the tongues of New Zealand psittaciformes. *Notornis* 10: 326–28, 341–45.
McCulloch, B. 1987. The Polynesian impact, pp. 29–39 in *The natural and human history of Akaroa and Wairewa Counties*. Queen Elizabeth II National Trust, Wellington.
McDowall, R. M. 1994. *Gamekeepers for the nation: The story of New Zealand's acclimatisation societies 1861–1990*. Canterbury University Press, Christchurch.
McDowall, R. M. 2000. *The Reed field guide to New Zealand freshwater fishes*. Reed, Auckland.
McGlone, M. S. 1989. The Polynesian settlement of New Zealand in relation to environmental and biotic changes. *New Zealand Journal of Ecology* 12 (supplement): 115–29.
McGlone, M. S. and C. J. Webb. 1981. Selective forces influencing the evolution of divaricating plants. *New Zealand Journal of Ecology* 4: 20–28.
McGlone, M. S. and J. M. Wilmshurst. 1999. Dating initial Maori environmental impact in New Zealand. *Quaternary International* 59: 5–16.
McGlone, M. S., A. J. Anderson and R. N. Holdaway. 1994. An ecological approach to the Polynesian settlement of New Zealand, pp. 136–63 in D. G. Sutton (ed.), *The origins of the first New Zealanders*. Auckland University Press, Auckland.
McIntyre, M. 1997. *Conservation of the tuatara*. Victoria University Press, Wellington.

McKee, J. W. A. and J. Wiffen. 1998. *Mangahouanga Stream – New Zealand's Cretaceous dinosaur and marine reptile site*. Geological Society of New Zealand Miscellaneous Publication 96.

MacKinnon, J. and K. Phillipps. 1993. *A field guide to the birds of Borneo, Sumatra, Java and Bali*. Oxford University Press, Oxford.

McLean, I. G. and B. J. Gill. 1988. Breeding of an island-endemic bird: the New Zealand whitehead *Mohoua albicilla*: Pachycephalinae. *Emu* 88:177–82.

McLennan, J. A. 1988. Breeding of North Island brown kiwi, *Apteryx australis mantelli*, in Hawke's Bay, New Zealand. *New Zealand Journal of Ecology* 11: 89–97.

McLennan, J. A. and A. J. McCann. 1991. Incubation temperatures of great spotted kiwi, *Apteryx haastii*. *New Zealand Journal of Ecology* 15: 163–66.

McLennan, J. A., and M. A. Potter. 1992. Distribution, population changes and management of brown kiwi in Hawke's Bay. *New Zealand Journal of Ecology* 16: 91–102.

McLennan, J. A., M. A. Potter, H. A. Robertson, G. C. Wake, R. Colbourne, L. Dew, L. Joyce, A. J. McCann, J. Miles, P. J. Miller and J. Reid. 1996. Role of predation in the decline of kiwi, *Apteryx* spp., in New Zealand. *New Zealand Journal of Ecology* 20: 27–35.

McSaveney, M. J. and I. E. Whitehouse. 1989. Anthropic erosion of mountain land in Canterbury. *New Zealand Journal of Ecology* 12 (supplement): 151–63.

Maloney, R. F. 1999. Bird populations in nine braided rivers of the Upper Waitaki Basin, South Island, New Zealand: Changes after 30 years. *Notornis* 46: 243–56.

Maloney, R. F., A. L. Rebergen, R. J. Nilsson and N. J. Wells. 1997. Bird density and diversity in braided river beds in the Upper Waitaki Basin, South Island, New Zealand. *Notornis* 44: 219–32.

Maloney, R. F., R. J. Keedwell, N. J. Wells, A. L. Rebergen and R. J. Nilsson 1999. Effect of willow removal on habitat use by five birds of braided rivers, Mackenzie Basin, New Zealand. *New Zealand Journal of Ecology* 23: 53–60.

Mansfield, B. 1996. *Moving from successful restoration of islands to ecosystem restoration on mainland New Zealand*. Department of Conservation, Wellington.

Mark, A. F. 1989. Responses of indigenous vegetation to contrasting trends in utilization by red deer in two southwestern New Zealand national parks. *New Zealand Journal of Ecology* 12 (supplement): 103–14.

Martin, P. S. and R. G. Klein (eds). 1984. *Quaternary extinctions: A prehistoric revolution*. University of Arizona Press, Tucson, Arizona.

Maxwell, J. M. 2001. Fiordland takahe: population trends, dynamics and problems, pp. 60–79 in W. G. Lee and I. G. Jamieson (eds), *Fifty years of conservation management and research*. University of Otago Press, Dunedin.

May, R. M. 1990. Taxonomy as destiny. *Nature* 347: 129–30.

Maynard-Smith, J. 1978. *The evolution of sex*. Cambridge University Press, Cambridge.

Meads, M. J., K. J. Walker and G. P. Elliott. 1984. Status, conservation and management of the land snails of the genus *Powelliphanta* (Mollusca, Pulmonata). *New Zealand Journal of Zoology* 11: 277–306.

Merton, D. V. 1965. Transfer of saddlebacks from Hen Island to Middle Chicken Island January 1964. *Notornis* 12: 213–22.

Merton, D. V. 1975. Success in re-establishing a threatened species: the saddleback – its status and conservation. XII *Bulletin of the International Council for Bird Preservation*: 150–58.

Merton, D. 1992. The legacy of 'Old Blue'. *New Zealand Journal of Ecology* 16: 65–68.

Merton, D. and M. Clout. 1999. Kakapo, back from the brink. *Wingspan* 9 (2): 14–17.

Merton, D. and R. Empson. 1989. But it doesn't look like a parrot. *Birds International* 1: 60–72.

Merton, D. V., R. B. Morris and I. A. E. Atkinson. 1984. Lek behaviour in a parrot: The kakapo *Strigops habroptilus* of New Zealand. *Ibis* 126: 277–83.

Millener, P. R. 1989. The only flightless passerine: The Stephens Island wren (*Traversia lyalli*: Acanthisittidae). *Notornis* 36: 280–84.
Millener, P. 1996. Extinct birds, pp. 113–20 in *The Chatham Islands: Heritage and conservation*. Canterbury University Press, Christchurch.
Millener, P. R. and T. H. Worthy. 1991. Contributions to New Zealand's late Quaternary avifauna II: *Dendroscansor decurvirostris*, a new genus and species of wren (Aves: Acanthisittidae). *Journal of the Royal Society of New Zealand* 21: 179–200.
Miller, C. J., J. L. Craig and N. D. Mitchell. 1994. Ark 2020: A conservation vision for Rangitoto and Motutapu Islands. *Journal of the Royal Society of New Zealand* 24: 65–90.
Mills, J. A. and A. F. Mark 1977. Food preferences of takahe in Fiordland National Park, New Zealand, and the effect of competition from introduced red deer. *Journal of Animal Ecology* 46: 939–58.
Mills, J. A., R. B. Lavers and M.C. Crawley. 1982a. Takahe and the wapiti issue. *Forest & Bird* 14 (3): 2–5.
Mills, J. A., R. B. Lavers, W. G Lee and A. S. Garrick. 1982b. Management recommendations for the conservation of takahe. Unpublished report for Wildlife Service.
Mills, J. A., R. B. Lavers and W. G. Lee. 1984. The takahe – a relict of the Pleistocene grassland avifauna of New Zealand. *New Zealand Journal of Ecology* 7: 57–70.
Mills, J. A., R. B. Lavers, W. G. Lee and M. K. Mara. 1991. Food selection by takahe *Notornis mantelli* in relation to chemical composition. *Ornis Scandinavica* 22: 111–28.
Ministry for the Environment. 1997. *The state of New Zealand's environment 1997*. Ministry for the Environment, Wellington.
Molloy, B. 1995. *Riccarton Bush: Putaringamotu: Natural history and management*. Riccarton Bush Trust, Christchurch.
Molloy, J. 1995. *Bat (peka peka) recovery plan* (Mystacina, Chalinolobus). Department of Conservation, Wellington.
Molloy, J and A. Davis 1994. *Setting priorities for the conservation of New Zealand's threatened plants and animals*. 2nd edn. Department of Conservation, Wellington.
Molloy, J., B. Bell, M. Clout, P. de Lange, G. Gibbs, D. Given, D. Norton, N. Smith and T. Stephens. 2001 draft. *Classifying species according threat of extinction: A system for New Zealand*. Department of Conservation, Wellington.
Moore, P. J. 2001. Historical records of yellow-eyed penguin (*Megadyptes antipodes*) in southern New Zealand. *Notornis* 48: 145–56.
Moorhouse, R. J. 1996. The extraordinary bill dimorphism of the huia (*Heteralocha acutirostris*): Sexual selection or intersexual competition? *Notornis* 43: 19–34.
Moorhouse, R., L. Moran, G. Taylor and D. Butler. Unpublished manuscript. Significant improvement in kaka (*Nestor meridionalis*) breeding success following control on introduced mammalian predators in Nelson Lakes National Park, New Zealand: Can predator control reverse the decline of a threatened parrot?
Moors, P. J. 1985. Eradication campaigns against *Rattus norvegicus* on the Noises Islands, New Zealand, using brodifacoum and 1080. *ICBP Technical Publications* 3: 143–55.
Mundy, P. Date unknown. *The travels of Peter Mundy in Europe and Asia 1608–1667*.
Murphy, E.C. 1996 Ferrets as threats to conservation values, pp. 48–51 in *Ferrets as vectors of tuberculosis and threats to conservation*. Royal Society of New Zealand Miscellaneous Series 36.
Murphy, E and P. Bradfield. 1992. Change in diet of stoats following poisoning of rats in a New Zealand forest. *New Zealand Journal of Ecology* 16: 137–40.
Murphy, E. C. and J. E. Dowding 1994. Range and diet of stoats (*Mustela erminea*) in a New Zealand beech forest. *New Zealand Journal of Ecology* 18: 11–18.
Murphy, E. C. and J. E. Dowding 1995. Ecology of the stoat in *Nothofagus* forest: Home range, habitat use and diet at different stages of the beech mast cycle. *New Zealand Journal of Ecology* 19: 97–109.

Murphy, E. C., B. K. Clapperton, P. M. F. Bradfield and H. J. Speed. 1998. Effects of rat-poisoning operations on abundance and diet of mustelids in New Zealand podocarp forests. *New Zealand Journal of Zoology* 25: 315–28.

Murray, T. E., J. A. Bartle, S. R. Kalish and P. R. Taylor. 1993. Incidental capture of seabirds by Japanese southern bluefin tuna longline vessels in New Zealand waters, 1988–1992. *Bird Conservation International* 3: 181–210.

Nelson, N. J. and C. H. Daugherty. 1997. The first experimental translocation of a tuatara population: conservation of *Sphenodon guntheri*. *New Zealand Journal of Zoology* 24: 328.

Newman, D. G. 1977. Some evidence of the predation of Hamilton's frog (*Leiopelma hamiltoni* (McCulloch)) by tuatara (*Sphenodon punctatus* (Grey)) on Stephens Island. *Proceedings of the New Zealand Ecological Society* 24: 43–47.

Newman, D. 1987a. *Tuatara*. John McIndoe, Dunedin.

Newman, D. G. 1987b. Burrow use and population densities of tuatara (*Sphenodon punctatus*) and how they are influenced by fairy prions (*Pachyptila turtur*) on Stephens Island, New Zealand. *Herpetologica* 43: 336–44.

Newman, D. G. and I. McFadden 1990. Seasonal fluctuations of numbers, breeding and food of kiore (*Rattus exulans*) on Lady Alice Island (Hen and Chickens group), with a consideration of kiore:tuatara (*Sphenodon punctatus*) relationships in New Zealand. *New Zealand Journal of Zoology* 17: 55–63.

Newman, D. G., P. R. Watson and I. McFadden. 1994. Egg production by tuatara on Lady Alice and Stephens Island, New Zealand. *New Zealand Journal of Zoology* 21: 387–98.

Nicholls, D. G., M. D. Murray and C. J. R. Robertson. 1994. Oceanic flights of the northern royal albatross *Diomedea epomophora sanfordi* using satellite telemetry. *Corella* 18: 50–52.

Niethammer, G. 1970. Clutch sizes of introduced European passeriformes in New Zealand. *Notornis* 17: 214–22

Norbury, G. 2001. Advances in New Zealand mammalogy 1900–2000.: Lagomorphs. *Journal of the Royal Society of New Zealand* 31: 83–97.

Norton, D. A. 1995. Vegetation on goat-free islands in a low-alpine lake, Paparoa Range, and implications for monitoring goat control operations. *New Zealand Journal of Ecology* 19: 67–72.

Nugent, G. 1992. Big-game, small-game, and game bird hunting in New Zealand: Hunting effort, harvest and expenditure in 1988. *New Zealand Journal of Zoology* 19: 75–90.

Nugent, G. 1994. Effects of possums on the native flora, pp. 5–10 in C. J. F. O'Donnell (ed.), *Possums as conservation pests*. Department of Conservation, Wellington.

Nugent, G., K. W. Fraser and P. J. Sweetapple. 1997. *Comparison of red deer and possum diets and impacts in podocarp-hardwood forest, Waihaha catchment, Pureora Conservation Park*. Science for Conservation 50. Department of Conservation, Wellington.

Nugent, G., K. W. Fraser, G W. Asher and K. G. Tustin. 2001. Advances in New Zealand mammalogy 1900–2000: Deer. *Journal of the Royal Society of New Zealand* 31: 263–98.

Numata, M., L. S. Davis and M. Renner. 2000. Prolonged foraging trips and egg desertion in little penguins (*Eudyptula minor*). *New Zealand Journal of Zoology* 27: 277–89.

O'Donnell, C. F. J. (ed.) 1994. *Possums as conservation pests*. Department of Conservation, Wellington.

O'Donnell, C. 1995. Birdlife of Riccarton Bush, pp. 247–259 in B. Molloy (ed.), *Riccarton Bush: Putaringamotu*. Riccarton Bush Trust, Christchurch.

O'Donnell, C. F. J. 1996a. Predators and the decline of New Zealand forest birds: An introduction to the hole-nesting bird and predator programme. *New Zealand Journal of Zoology* 23: 213–19.

O'Donnell, C. F. J. 1996b. Monitoring mohua (yellowhead) populations in the South Island, New Zealand, 1983–93. *New Zealand Journal of Zoology* 23: 221–28.

O'Donnell, C. F. J. 2000a. Conservation status and causes of decline of the threatened

New Zealand long-tailed bat *Chalinolobus tuberculatus* (Chiroptera: Vespertillionidae). *Mammal Review* 30: 89–106.
O'Donnell, C. F. J. 2000b. Influence of season, habitat, temperature and invertebrate availability on nocturnal activity of the New Zealand long-tailed bat (*Chalinolobus tuberculatus*). *New Zealand Journal of Zoology* 27: 207–21.
O'Donnell, C. F. J. 2001. Advances in New Zealand mammalogy 1990–2000: Long-tailed bat. *Journal of the Royal Society of New Zealand* 31: 43–57.
O'Donnell, C. F. J. and P. R. Dilks. 1994. Foods and foraging of forest birds in temperate rainforest, South Westland, New Zealand. *New Zealand Journal of Ecology* 18: 87–107.
O'Donnell, C. F. J. and G. Rasch. 1991. *Conservation of kaka in New Zealand*. Science and Research Internal Report 101. Department of Conservation, Wellington.
O'Donnell, C. F. J. and J. A. Sedgeley. 1999. Use of roosts by the long-tailed bat *Chalinolobus tuberculatus*, in temperate rainforest in New Zealand. *Journal of Mammalogy* 880: 913–23.
O'Donnell, C. F. J., P. J. Dilks and G. P. Elliott. 1996. Control of a stoat (*Mustela erminea*) population irruption to enhance mohua (yellowhead) (*Mohoua ochrocephala*) breeding success in New Zealand. *New Zealand Journal of Zoology* 23: 279–86.
O'Donnnell, C. F. J., J. Christie, C. Corben, J. A. Sedgeley and W. Simpson. 1999. Rediscovery of short-tailed bats (*Mystacina* sp.) in Fiordland, New Zealand: Preliminary observations of taxonomy, echolocation calls, population size, home range and habitat use. *New Zealand Journal of Ecology* 23: 21–30.
Ogle, C. C. and J. Cheyne. 1981. *The wildlife and wildlife values of the Whangamarino wetlands*. Wildlife Service Fauna Survey Report 28.
Onley, D. J. 1980. Bird counts in lowland forests in the western Paparoas. *Notornis* 27: 335–62.
Onley, D. J. 1983. The effects of logging on winter bird populations near Karamea. *Notornis* 30: 187–197.
OSNZ 1970. *Annotated checklist of the birds of New Zealand*. Ornithological Society of New Zealand, Wellington.
OSNZ 1990. *Checklist of the birds of New Zealand*. 3rd edn. Ornithological Society of New Zealand, Wellington.
Parkes, J. 1995. Other animals, pp. 292–97 in B. Molloy (ed.), *Riccarton Bush: Putaringamotu*. Riccarton Bush Trust, Christchurch.
Parkes, J. 2001. Advances in New Zealand mammalogy 1900–2000: Feral livestock. *Journal of the Royal Society of New Zealand* 31: 233–41.
Parkes J. P. and C. Thomson. 1995. *Management of tahr*. Science for Conservation 7. Department of Conservation Wellington.
Parliamentary Commissioner for the Environment 1998. *The rabbit calicivirus disease (RCD) saga: A biosecurity/bio-control fiasco*. Parliamentary Commissioner for the Environment, Wellington.
Parrish, R. and M. Williams. 2001. Decline of brown teal (*Anas chlorotis*) in Northland, New Zealand, 1988–99. *Notornis* 48: 131–36.
Patterson, G. B. 1992. The ecology of a New Zealand grassland lizard guild. *Journal of the Royal Society of New Zealand* 22: 91–106.
Patterson, G. B. 1997. South Island skinks of the genus *Oligosoma*: Description of *O. longipes* n.sp. with redescription of *O. otagense* (McCann) and *O. waimatense* (McCann). *Journal of the Royal Society of New Zealand* 27: 439–450.
Patterson, G. B. and C. H. Daugherty. 1990. Four new species and one new subspecies of skinks, genus *Leiolopisma* (Reptilia: Lacertilia: Scincidae) from New Zealand. *Journal of the Royal Society of New Zealand* 20: 65–84.
Patterson, G. B. and C. H. Daugherty. 1995. Reinstatement of the genus *Oligosoma* (Reptilia: Lacertilia: Scincidae). *Journal of the Royal Society of New Zealand* 25: 327–31.

Pekelkaring, C. J. and R. N. Reynolds. 1983. Distribution and abundance of browsing mammals in Westland National Park in 1978, and some observations on their impact on the vegetation. *New Zealand Journal of Forestry Science* 13: 247–65.

Phillipps, W. J. 1963. *The book of the huia*. Whitcomb and Tombs, Auckland.

Pierce, R. J. 1979. Foods and feeding of the wrybill (*Anarhychus frontalis*) on its riverbed breeding grounds. *Notornis* 26: 1–21.

Pierce, R. J. 1983. The Charadriiforms of a high-country river valley. *Notornis* 30: 169–85.

Pierce, R. J. 1984a. The changed distribution of stilts in New Zealand. *Notornis* 31: 7–18.

Pierce, R. J. 1984b. Plumage, morphology and hybridisation of New Zealand stilts *Himantopus* spp. *Notornis* 3: 106–30.

Pierce, R. J. 1985. Feeding methods of stilts (*Himantopus* spp.). *New Zealand Journal of Zoology* 12: 467–72.

Pierce, R. J. 1986. Differences in susceptibility to predation during nesting between pied and black stilts (*Himantopus* spp). *Auk* 103: 273–80.

Pierre, J. P. 2000. Foraging behaviour and diet of a reintroduced population of the South Island saddleback (*Philesturnus carunculatus carunculatus*). *Notornis* 47: 7–12.

Porter, R. 1987. An ecological comparison of two *Cyclodina* skinks (Reptilia: Lacertilia) in Auckland, New Zealand. *New Zealand Journal of Zoology* 14: 493–507.

Porter, R. E. R., M. R. Rudge and J. A. McLennan. 1994. *Birds and small mammals: A pest control manual*. Manaaki Whenua Press, Lincoln.

Potter, M. A. 1990. Movement of North Island brown kiwi (*Apteryx australis mantelli*) between forest remnants. *New Zealand Journal of Ecology* 14: 17–24.

Potts, T. H. 1873. On the birds of New Zealand (Part 3). *Transactions of the New Zealand Institute* 5:171–205.

Potts, T. H. 1874. On the birds of New Zealand (Part 4). *Transactions of the New Zealand Institute* 6: 139–53.

Potts, T. H. 1882. *Out in the open: A budget of scraps of natural history gathered in New Zealand*. Lyttelton Times, Christchurch.

Powlesland, R. G. 1981. The foraging behaviour of the South Island robin. *Notornis* 28: 89–102.

Powlesland, R. G. 1987. The foods, foraging behaviour and habitat use of the North Island kokako in Puketi State Forest, Northland. *New Zealand Journal of Ecology* 10: 117–28.

Powlesland, R. G. and B. D. Lloyd. 1994. Use of supplimentary feeding to induce breeding in free-living kakapo *Strigops habroptilus*, in New Zealand. *Biological Conservation* 69: 97–106.

Powlesland, R. G., B. D. Lloyd, H. A. Best and D. V. Merton. 1992. Breeding biology of the kakapo, *Strigops habroptilus* on Stewart Island, New Zealand. Ibis 134: 361–373.

Powlesland, R. G., A. Roberts, B. D. Lloyd and D. V. Merton. 1995. Number, fate and distribution of kakapo (*Strigops habroptilus*) on Stewart Island, New Zealand, 1979–92. *New Zealand Journal of Zoology* 22: 239–48.

Powlesland, R. G., J. W. Knegtmans and S. J. Marshall. 1999. Costs and benefits of aerial 1080 possum control operations using carrot baits to North Island robins (*Petroica australis longipes*), Pureora Forest Park. *New Zealand Journal of Ecology* 23: 149–159.

Pratt, H. D., P. L. Bruner and D. G. Berrett. 1987. *A field guide to the birds of Hawaii and the tropical Pacific*. Princeton University Press. Princeton.

Pyke, G. H. and A. W. White. 1997. Conservation and management of bell frogs in Australia and New Zealand. *New Zealand Journal of Zoology* 24: 328–29.

Raikow, R. J. and A. H. Bledsoe. 2000. Phylogeny and evolution of the passerine birds. *BioScience* 50: 487–499.

Ramsay, G. W. 1978. A review of the effect of rodents on the New Zealand invertebrate fauna, pp. 89–95 in P. R. Dingall, I. A. E.Átkinson and C. H. Hay (eds), *The ecology and control of rodents in New Zealand nature reserves*. Department of Lands and Survey Information Series 4.

Rando, J. C., M. Lopez and B. Segui. 1999. A new species of extinct flightless passerine (Emberizidae: *Emberiza*) from the Canary Islands. *Condor* 101: 1–13.
Recher, H. F. 1996. The role of conservation biology in the new millenium. *Pacific Conservation Biology* 2: 311.
Reed, C. and D. Merton. 1991. Behavioural manipulation of endangered New Zealand birds as an aid toward species recovery. *Acta XX Congressus Internationalis Ornithologici*: 2514–22.
Reed, C. E. M., D. P. Murray and D. J. Butler. 1993a. *Black stilt recovery plan* (Himantopus novaezealandiae). Department of Conservation, Wellington.
Reed, C. E. M., R. J. Nilsson and D. P. Murray. 1993b. Cross-fostering New Zealand's black stilt. *Journal of Wildlife Management* 57: 608–11.
Reid, B. 1977. The energy value of the yolk reserve in a North Island brown kiwi chick (*Apteryx australis mantelli*). *Notornis* 24: 194–95.
Reid, B. and G.R. Williams. 1975. The kiwi, pp. 301–330 in G. Kuschel (ed.), *Biogeography and ecology in New Zealand*. Junk, The Hague.
Reilly, P. 1994. *Penguins of the world*. Oxford University Press, Melbourne.
Richdale, L. E. 1946. Maori and mutton-bird. Otago Daily Times Wildlife Series 7, Dunedin.
Richdale, L. E. 1957. *A population study of penguins*. Oxford University Press, Oxford.
Richdale, L. 1965. Biology of the birds of Whero Island, New Zealand with special reference to the diving petrel and the white-faced storm petrel. *Transactions of the Zoological Society of London* 25: 4–86.
Riney, T. 1956. Comparison of occurrence of introduced animals with critical conservation areas to determine priorities for control. *New Zealand Journal of Science and Technology* 38B: 1–18.
Roberts, M., W. Norman, N. Minhinnick, D. Wihongi and C. Kirkwood. 1995. Kaitiakitanga: Maori perspectives on conservation. *Pacific Conservation Biology* 2: 7–20.
Robertson, C. J. R. 1991. *Questions on the harvesting of toroa in the Chatham Islands*. Science and Research Series 35, Department of Conservation, Wellington.
Robertson, C. J. R. 1993. Survival and longevity of the northern royal albatross *Diomedea epomophora sandfordi* at Taiaroa Head 1937–1992. *Emu* 93: 269–76.
Robertson, C. J. R. 1997. Factors influencing the breeding performance of the northern royal albatross. Pages 99–104 in G. Robertson and R. Gales (eds), *Albatross biology and conservation*. Surrey Beatty, Chipping Norton.
Robertson, C. J. R. and D. G. Nicholls. 2000. Round the world with the northern royal albatross (*Diomedia sanfordi*). *Notornis* 47: 176.
Robertson, C. J. R. and G. B. Nunn. 1997. Towards a new taxonomy for albatrosses, pp. 13–19 in G. Robertson and R. Gales (eds), *Albatross biology and conservation*. Surrey Beatty, Chipping Norton.
Robertson, C. J. R. and G. F. van Tets. 1982. The status of birds at the Bounty Islands. *Notornis* 29: 311–36.
Robertson, C. J. R., D. Bell and D. G. Nicholls. 2000. The Chatham albatross (*Thalassarche eremita*): At home and abroad. *Notornis* 47: 174.
Rogers, G. M. 1991. Kaimanawa feral horses and their environmental impacts. *New Zealand Journal of Ecology* 15: 49–64.
Rose, A. B., C. J. Pekelharing and K. H. Platt. 1992. Magnitude of canopy dieback and implications for conservation of southern rata-kamahi (*Metrosideros umbellata-Weimannia racemosa*) forests, Central Westland, New Zealand. *New Zealand Journal of Ecology* 16: 23–32.
Rudge, M. R. 1982. Conserving feral farm mammals in New Zealand. *New Zealand Agricultural Science* 166: 157–60.
Rudge, M. R. 1983. A reserve for feral sheep on Pitt Island, Chatham group, New Zealand. *New Zealand Journal of Zoology* 10: 349–64.

Sadleir, R. M. F. and B. Warburton. 2001. Advances in New Zealand mammalogy 1900–2000: Wallabies. *Journal of the Royal Society of New Zealand* 31: 7–14.
Sagar, P. M. and C. J. F. O'Donnell. 1982. Seasonal movements and population of the southern crested grebe in Canterbury. *Notornis* 29: 143–49.
Sagar, P. M. and J. Warham. 1993. A long-lived southern Buller's mollymawk (*Diomedea bulleri*) with a small egg. *Notornis* 40: 303–04.
Sagar, P. M. and H. Weimerskirch. 1996. Satellite tracking of southern Buller's albatrosses from the Snares, New Zealand. *Condor* 98: 649–52.
Sagar, P. M., J. Molloy, A. J. D. Tennyson and D. Butler. 1994. Numbers of Buller's mollymawks breeding at the Snares Islands. *Notornis* 41: 85–92.
Sagar, P. M., J. C. Stahl, J. Molloy, G. A. Taylor and A. J. D. Tennyson. 1999. Population size and trends within two populations of southern Buller's albatross *Diomedia bulleri bulleri*. *Biological Conservation* 89: 11–19.
Sagar, P. M., J. Molloy, H. Weimerskirch and J. Warham. 2000. Temporal and age-related changes in survival rates of southern Buller's albatrosses (*Thalassarche bulleri bulleri*) at the Snares, New Zealand, 1948 to 1997. *Auk* 117: 699–708.
Sanders, M. D. 1999. Effect of changes in water level on numbers of black stilts (*Himantopus novaezelandiae*) using delta of Lake Benmore. *New Zealand Journal of Zoology* 26: 155–63.
Saul, E. K., H. A. Robertson and A. Tiraa. 1998. Breeding biology of the kakerori (*Pomarea dimidiata*) on Rarotonga, Cook Islands. *Notornis* 45: 255–69.
Saunders, A. 1995. Translocations in New Zealand: An overview, pp 43–46 in M. Serena (ed.), *Reintroduction biology of Australian and New Zealand fauna*. Surrey Beatty, Chipping Norton.
Saunders, A. and D. A. Norton. 2001. Ecological restoration at mainland islands in New Zealand. *Biological Conservation* 99: 109–19.
Saunders, D. A. and R. J. Hobbs (eds). 1991. *Nature Conservation 2: The role of bush corridors*. Surrey Beatty, Chipping Norton.
Schodde, R. and D. P. Faith. 1991. The development of modern avifaunulas. *Acta XX Congressus Internationalis Ornithologici*: 404–42.
Sedgeley, J. A. 2001. Winter activity in the tree-roosting lesser short-tailed bat, *Mystacina tuberculata*, in a cold-temperate climate in New Zealand. *Acta Chiropterologica* 3: 179–95.
Sedgeley, J. A. and C. J. F. O'Donnell, 1999. Roost selection by the long–tailed bat, *Chalinolobus tuberculatus*, in temperate New Zealand rainforest and its implications for the conservation of bats in managed forests. *Biological Conservation* 88: 261–76.
Serena, M. (ed.) 1995. *Reintroduction biology of Australian and New Zealand fauna*. Surrey Beatty, Chipping Norton.
Sherley, G. H. 1990. Co–operative breeding in riflemen (*Acanthissitta chloris*) benefits to parents, offspring and helpers. *Behaviour* 112: 1–22.
Sibley, C. G. 1991. Phylogeny and classification of birds from DNA comparisons. *Acta XX Congressus Internationalis Ornithologici*: 111–26.
Sibley, C. G. and J. E. Ahlquist. 1990. Phylogeny and classification of birds: A study in molecular evolution. Yale University Press, New Haven.
Sibley, C. G., G. R. Williams and J. E. Ahlquist. 1982. The relationships of the New Zealand wrens (Acanthisittidae) as indicated by DNA-DNA hybridization. *Notornis* 29: 113–30.
Simpson, K. and N. Day. 1993. *Field guide to the birds of Australia*. 4th edn. Viking O'Neil, Ringwood.
Siriwardena, G. M., S. R. Baillie, S. T. Buckland, R. M. Fewster, J. H. Merchant and J. D. Wilson. 1998. Trends in the abundance of farmland birds: a quantitative comparison of smoothed common birds census indices. *Journal of Applied Ecology* 35: 24–43.
Slooten, E. and S. M. Dawson. 1995. Conservation of marine mammals in New Zealand. *Pacific Conservation Biology* 2: 64–76.

Smith, G. P., J. R. Ragg, H. Moller and K. A. Waldrup. 1995. Diet of feral ferrets (*Mustela furo*) from pastoral habitats in Otago and Southland, New Zealand. *New Zealand Journal of Zoology* 22: 363–69.

Smith, I. W. G. 1989. Maori impact on the marine megafauna: Pre-European distributions of New Zealand sea mammals, pp 76–108 in D. G. Sutton (ed.), *Saying so doesn't make it so*. New Zealand Archaeological Association Monograph 17.

Sorensen, J. H. 1969. *New Zealand fur seals with special reference to the 1946 open season*. Fisheries Technical Report 42. New Zealand Marine Department, Wellington.

Sowman, W. C. R. 1981. *Meadow, mountain, forest and stream: The provincial history of the Nelson Acclimatisation Society 1863–1968*. Nelson Acclimatisation Society, Nelson.

Stahl, J. C. and P. M. Sagar. 2000a. Foraging strategies of southern Buller's albatrosses *Diomedea b. bulleri* breeding on the Snares, New Zealand. *Journal of the Royal Society of New Zealand* 30: 299–318.

Stahl, J. C. and P. M. Sagar. 2000b. Foraging strategies and migration of southern Buller's albatrosses *Diomedea b. bulleri* breeding on the Solander Is, New Zealand. *Journal of the Royal Society of New Zealand* 30: 319–34.

Stattersfield, A. J., M. J. Crosby, A. J. Long and D. C. Wege. 1998. *Endemic bird areas of the world: Priorities for biodiversity conservation*. Birdlife Conservation Series 7. Birdlife International, Cambridge.

St. Clair, C. C. and R. C. St. Clair. 1992. Weka predation on eggs and chicks of Fiordland crested penguins. *Notornis* 39: 60–63.

Stead, E. F. 1927. The native and introduced birds of Canterbury, pp. 204–25 in R. Speight, A. Wall and R. M. Lang (eds), *Natural history of Canterbury*. Philosophical Institute of Canterbury, Christchurch.

Stead, E. F. 1932. *The life histories of New Zealand birds*. Search, London.

Steadman, D. W. 1991a. Ecological impact of the human depletion of frugivorous birds in Polynesia. *Acta XX Congressus Internationalis Ornithologici*: 424.

Steadman, D. W. 1991b. Extinction of species: past, present and future, pp. 156–69 in R. L. Wyman (ed.), *Global climate change and life on earth*. Chapman and Hall, New York.

Steadman, D. W. 1997. Human-caused extinction of birds, pp. 139–61 in M. L. Reaka-Kudla, D. E. Wilson and E. O. Wilson (eds), *Biodiversity II*. Joseph Henry Press, Washington, DC.

Stephenson, B. M., E. O. Minot and D. P. Armstrong. 1999. Fate of moreporks (*Ninox novaeseelandiae*) during a pest control operation on Mokoia Island, Lake Rotorua, North Island, New Zealand. 23: 233–40.

Stephenson, G. K. 1986. *Wetlands: Discovering New Zealand's shy places*. Government Printing Office, Wellington.

Stevens, G. 1985. *Lands in collision: Discovering New Zealand's past geography*. DSIR, Wellington.

Stevens, G. R. 1991. Geological evolution and biotic links in the Mesozoic and Cenozoic of the southwest Pacific. *Acta XX Congressus Internationalis Ornithologici*: 361–82.

Stevens, G., M. McGlone and B. McCulloch. 1988. *Prehistoric New Zealand*. Heinemann Reed, Auckland.

Stewart, G. H. and L. E. Burrows. 1989. The impact of white-tailed deer *Odocoileus virginianus* on regeneration in the coastal forests of Stewart Island, New Zealand. *Biological Conservation* 49: 275–93.

Stewart, G. H. and A. B. Rose 1988. Factors predisposing rata-kamihi (*Metrosideros umbellata-Weinmannia racemosa*) forests to canopy dieback, Westland, New Zealand. *GeoJournal* 17: 217–23.

Stork, N. E. 1997. Measuring global biodiversity and its decline, pp. 41–68 in M. L. Reaka-Kudla, D. E. Wilson and E. O. Wilson (eds), *Biodiversity II*. Joseph Henry Press, Washington, DC.

Strahan, R. (ed.) 1995. *The mammals of Australia*. Reed, Chatswoood.
Sullivan, W. and K-J. Wilson. 2001. Use of burrow entrance flaps to minimise interference to Chatham petrel (*Pterodroma axillaris*) chicks by broad-billed prions (*Pachyptila vittata*). *New Zealand Journal of Ecology* 25 (2): 71–75.
Taborsky, M. 1988. Kiwis and dog predation: Observations in Waitangi State Forest. *Notornis* 35:197–202.
Taiepa, T., P. Lyver, P. Horseley, J. Davis, M. Bragg and H. Moller. 1997. Co-management of New Zealand's conservation estate by Maori and Pakeha: A review. *Environmental Conservation* 24: 236–50.
Taylor, G. A. 2000. *Action plan for seabird conservation in New Zealand*. Threatened Species Occasional Publication 16. Department of Conservation, Wellington.
Taylor, R. H. 1975. What limits kiore (*Rattus exulans*) distribution in New Zealand. *New Zealand Journal of Zoology* 2: 473–77.
Taylor, R. H. 1979. Predation on sooty terns at Raoul Island by cats and rats. *Notornis* 26: 199–202.
Taylor, R. H. 1982. New Zealand fur seals at the Bounty Islands. *New Zealand Journal of Marine and Freshwater Research* 16: 1–9.
Taylor, R. H. 1984. Distribution and interactions of introduced rodents and carnivores in New Zealand. *Acta Zool. Fennica* 172: 103–05.
Taylor, R. H. 1985. Status, habits and conservation of *Cyanoramphus* parakeets of the New Zealand region. *ICBP Technical Publication* 3: 195–211.
Taylor, R. H. 1992. New Zealand fur seals at the Antipodes Islands. *Journal of the Royal Society of New Zealand* 22: 107–22.
Taylor, R. H. 1998. A reappraisal of the orange-fronted parakeet (*Cyanoramphus* sp.) – species or colour morph? *Notornis* 45: 49–63.
Taylor, R. H. and B. W. Thomas. 1989. Eradication of Norway rats (*Rattus norvegicus*) from Hawea Island, Fiordland, using brodifacoum. *New Zealand Journal of Ecology* 12: 23–32.
Taylor, R. H. and B. W. Thomas 1993. Rats eradicated from rugged Breaksea Island (170 ha), Fiordland, New Zealand. *Biological Conservation* 65: 191–98.
Taylor, R. H. and J. A. V. Tilley. 1984. Stoats (*Mustela erminea*) on Adele and Fisherman Islands, Abel Tasman National Park, and other offshore islands in New Zealand. *New Zealand Journal of Ecology* 7: 139–45.
Taylor, R. H., K. J. Barton, P. R. Wilson, B. W. Thomas, B. J. Karl. 1995. Population status and breeding of New Zealand fur seals (*Arctocephalus forsteri*) in the Nelson–northern Marlborough region 1991–94. *New Zealand Journal of Marine and Freshwater Research* 29: 223–34.
Tennyson, A. 1992. Marine threats to New Zealand seabirds, pp. 40–46 in *Proceedings of the Marine Conservation and Wildlife Protection Conference 1992*. New Zealand Conservation Authority, Wellington.
Tennyson, A. J. D. and P. R. Millener. 1994. Bird extinctions and fossil bones from Mangere Island, Chatham Islands. *Notornis* 41 (supplement): 165–78.
Thomas, B. W. 1985. Observations on the Fiordland skink (*Leiolopisma acrinasum* Hardy), pp. 17–22 in G. Grigg, R. Shine and H. Ehmann (eds), *The biology of Australasian frogs and reptiles*. Surrey Beatty, Chipping Norton.
Thomas, B .W. and A. H. Whitaker. 1994. Translocation of the Fiordland skink *Leiolopisma acrrinasum* to Hawea Island, Breaksea Sound, Fiordland, New Zealand, pp. 91–95 in M. Serena (ed.), *Reintroduction biology of Australian and New Zealand fauna*. Surrey Beatty, Chipping Norton.
Thompson, M. B., C. H. Daugherty, A. Cree, D. C. French, J. C. Gillingham and R. E. Barwick. 1992. Status and longevity of the tuatara, *Sphenodon guntheri*, and Duvaucel's gecko, *Hoplodactylus duvaucelii*, on North Brother Island, New Zealand. *Journal of the*

Royal Society of New Zealand 22: 123–30.

Thomsen, S., S. D. Wratten and C. M. Frampton. 2001. Skylark (*Alauda arvensis*) winter densities and habitat associations in Canterbury, New Zealand, pp. 139–48 in P. F. Donald and J. A. Vickery (eds), *The ecology and conservation of skylarks* Alauda arvensis. Royal Society for Protection of Birds, Sandy.

Thomson, G. M. 1922. *The naturalisation of animals and plants in New Zealand*. Cambridge University Press, London.

Thurley, T. and B. D. Bell. 1994. Habitat distribution and predation on a western population of terrestrial *Leiopelma* (Anura: Leiopelmatidae) in the northern King Country, New Zealand. *New Zealand Journal of Zoology* 21: 431–36.

Timmins, S. M., I. A. E. Atkinson and C. C. Ogle. 1987. Conservation opportunities on a highly modified island: Mana Island, Wellington, New Zealand. *New Zealand Journal of Ecology* 10: 57–65.

Tocher, M. 1998. Skinks with altitude. Saving Otago's high country lizards. *Forest & Bird* 290, pp. 30–33.

Tocher, M. and D. Newman. 1997. Leaps and bounds: The conservation of New Zealand's native frogs. *Forest & Bird* 285: 14–20.

Towns, D. R. 1991. Response of lizard assemblages in the Mercury Islands, New Zealand, to the removal of an introduced rodent: the kiore (*Rattus exulans*). *Journal of the Royal Society of New Zealand* 21: 119–36.

Towns, D. R. 1994. The role of ecological restoration in the conservation of Whitaker's skink (*Cyclodina whitakeri*), a rare New Zealand lizard (Lacertilia: Scincidae). *New Zealand Journal of Zoology* 21: 457–471.

Towns, D. R. and C. H. Daugherty. 1994. Patterns of range contraction and extinctions in the New Zealand herpetofauna following human colonisation. *New Zealand Journal of Zoology* 21: 325–39.

Towns, D. R. and G. P. Elliott. 1996. Effects of habitat structure on distribution and abundance of lizards at Pukerua Bay, Wellington, New Zealand. *New Zealand Journal of Ecology* 20: 191–206.

Towns, D. R. and M. Williams. 1993. Single species conservation in New Zealand: towards a redefined conceptual approach. *Journal of the Royal Society of New Zealand* 23: 61–78.

Towns, D. R., C. H. Daugherty and D. G. Newman. 1985. An overview of the ecological biogeography of the New Zealand lizards (Gekkonidae, Scincidae), pp. 107–15 in G. Grigg, R. Shine and H. Ehmann (eds), *The biology of Australasian frogs and reptiles*. Surrey Beatty, Chipping Norton.

Towns, D. R., C. H. Daugherty and A. Cree. 2001. Raising the prospects for a forgotten fauna: A review of 10 years of conservation effort for New Zealand reptiles. *Biological Conservation* 99: 3–16.

Trewick, S. A. 1996. Morphology and evolution of two takahe: Flightless rails of New Zealand. *Journal of Zoology (London)* 238: 221–37.

Trewick, S. A. 1997a. Sympatric flightless rails *Gallirallus dieffenbachii* and *G. modestus* on the Chatham Islands, New Zealand: Morphometrics and alternative evolutionary scenarios. *Journal of the Royal Society of New Zealand* 27: 451–64.

Trewick, S. A. 1997b. Flightlessness and phylogeny amongst endemic rails (Aves: Rallidae) of the New Zealand region. *Philosophical Transactions of the Royal Society of London B* 352: 429–446.

Trewick, S. A. and T. H. Worthy 2001. Origins and prehistoric ecology of takahe based on morphometric, molecular, and fossil data, pp. 31–48 in W. G. Lee and I. G. Jamieson (eds), *Fifty years of conservation management and research*. University of Otago Press. Dunedin.

Triggs, S., M. Williams, S. Marshall and G. Chambers. 1991. Genetic relationships within a population of blue duck *Hymenolaimus malacorhynchos*. *Wildfowl* 42: 87–93.

Trotter, M. M. 1975. Archaeological investigations at Redcliffs, Canterbury, New Zealand. *Records of the Canterbury Museum* 9: 189–220.
Turbott, E. G. (ed.) 1967. *Buller's birds of New Zealand*. Whitcombe and Tombs, Christchurch.
Turbott, E. G. 1969. Native birds, pp. 426–34 in G. A. Knox (ed.), *The natural history of Canterbury*. Reed, Wellington.
Ussher, G. T. 1999. Tuatara (*Sphenodon punctatus*) feeding ecology in the presence of kiore (*Rattus exulans*). *New Zealand Journal of Zoology* 26: 117–25.
Veitch, C. R. 1977. Arctic waders wintering in New Zealand. *Proceedings of the New Zealand Ecological Society* 24: 110–12.
Veitch, C. R. 1985. Methods of eradicating feral cats from offshore islands in New Zealand, pp. 125–41 in P. J. Moors (ed.), *Conservation of island birds*. ICBP Technical publication 3.
Veitch, C. R. 1995. Habitat repair: a necessary prerequisite to translocation of threatened birds, pp. 97–104 in M. Serena (ed.), *Reintroduction biology of Australian and New Zealand fauna*. Surrey Beatty, Chipping Norton.
Veitch C. R. and B. D. Bell. 1990. Eradication of introduced animals from the islands of New Zealand, pp. 137–46 in D. R. Towns, C.H. Daugherty, and I. A. E. Atkinson (eds), *Ecological restoration of New Zealand islands*. Conservation Sciences Publication 2, Department of Conservation, Wellington.
Veltman, C. J., S. Nee and M. J. Crawley. 1996. Correlates of introduction success in exotic New Zealand birds. *The American Naturalist* 147: 542–57.
Waldman, B. and M. Tocher. 1998. Behavioural ecology, genetic diversity and declining amphibian populations, pp. 394–443 in T. Caro (ed.), *Behavioural ecology and conservation biology*, Oxford University Press, New York.
Walker, K. and G. Elliott. 1999. Population changes and biology of the wandering albatross *Diomedia exulans gibsoni* at the Auckland Islands. *Emu* 99: 239–47.
Walker, K., G. Elliott, D. Nicholls, D. Murray and P. Dilks. 1995. Satellite tracking of wandering albatross (*Diomedia exulans*) from the Auckland Islands: Preliminary results. *Notornis* 42: 127–37.
Wallace, A. R. 1883. *Australasia: Stanford's compendium of geography and travel*. Edward Stanford, London.
Walls, G. Y. 1981. Feeding ecology of the tuatara (*Sphenodon punctatus*) on Stephens Island, Cook Strait. *New Zealand Journal of Ecology* 4: 89–97.
Wardle, J. A. 1984. *The New Zealand beeches*. New Zealand Forest Service, Wellington.
Warham, J. 1990. *The petrels: Their ecology and breeding systems*. Academic Press, London.
Warham, J. 1996. *The behaviour, population biology and physiology of the petrels*. Academic Press, London.
Warham, J. and G. J. Wilson. 1982. The size of the sooty shearwater population at the Snares Islands, New Zealand. *Notornis* 29: 23–30.
Warham, J., B. R. Keeley and G. J. Wilson. 1977. Breeding of the mottled petrel. *Auk* 94: 1–17.
Warham, J., G. J. Wilson and B. R. Keeley. 1982. The annual cycle of the sooty shearwater *Puffinus griseus* at the Snares Islands, New Zealand. *Notornis* 29: 269–92.
Watt, J. C. 1975 The terrestrial insects, pp. 507–35 in G. Kuschell (ed.), *Biogeography and ecology in New Zealand*. Junk, The Hague.
Watt, J. P. C. 1975. Notes on Whero Island and other roosting and breeding stations of the Stewart Island shag (*Leucocarbo carunculatus chalconotus*). *Notornis* 22: 265–72.
Webb, P. I., J. A. Sedgeley and C. F. J. O'Donnell. 1998. Wing shape in New Zealand lesser short-tailed bats (*Mystacina tuberculata*). *Journal of Zoology (London)* 246: 462–65.
Weimerskirch, H. and P. M. Sagar. 1996. Diving depths of shearwaters *Puffinus griseus*. *Ibis* 138: 786–88.
Wells, S. M., R. M. Pyle and N. M. Collins. 1983. *The IUCN invertebrate red data book*. IUCN, Gland.

West, J. A. and R. J. Nilsson. 1994. Habitat use and burrow densities of burrow-nesting seabirds on South East Island, Chatham Islands, New Zealand. *Notornis* 41 (supplement): 27–37.

Whitaker, A. H. 1968a. *Leiolopisma suteri* (Boulenger), an oviparous skink in New Zealand. *New Zealand Journal of Science* 11: 425–32.

Whitaker, A. H. 1968b. The lizards of the Poor Knights Islands, New Zealand. *New Zealand Journal of Science* 11: 623–51.

Whitaker, A. H. 1973. Lizard populations on islands with and without Polynesian rats, *Rattus exulans* (Peale). *Proceedings of the New Zealand Ecological Society* 20: 121–30.

Whitaker, A. H. 1982. Interim results from a study of *Hoplodactylus maculatus* (Boulenger) at Turakirae Head, Wellington, pp. 363–74 in D. G. Newman (ed.), *New Zealand Herpetology*, Wildlife Service Occasional Publication 2.

Whitaker, A. H. 1984. *Hoplodactylus kahutarae* n.sp. (Reptilia: Gekkonidae) from the Seaward Kaikoura Range, Marlborough, New Zealand. *New Zealand Journal of Zoology* 11: 259–70.

Whitaker, A. H. 1987. The role of lizards in New Zealand plant reproductive strategies. *New Zealand Journal of Botany* 25: 315–28.

Whitaker, T. (A. H.) 1994. New Zealand lizards: Their ecology and conservation. *Ecological Management* 2: 1–7.

Whitaker, A. H. 1996. *Impact of agricultural development on grand skink (*Oligosoma grande*) (Reptilia: Scincidae) populations at Macraes Flat, Otago, New Zealand.* Science for Conservation 33. Department of Conservation, Wellington.

Whitaker, T. (A. H.) 1997. The lizards of the Seaward Kaikoura Range. *New Zealand Journal of Zoology* 24: 330–31.

Whitaker, A. H. and M. R. Rudge. 1976. *The value of feral farm mammals in New Zealand.* Department of Lands and Survey Information Series 1, Wellington

Williams, G. R. 1977. Marooning – a technique for saving threatened species from extinction. *International Zoo Yearbook* 17: 102–06.

Williams, M.J. 1981. *The duckshooters bag: An introduction to New Zealand's wetland game birds.* Wetland Press, Wellington.

Williams, M. 1986. Native bird management. *Forest & Bird*, November 1986: 7–9.

Williams, M. J. 1991a. Introductory remarks: Ecological and behavioural adaptations of southern hemisphere waterfowl. *Acta XX Congressus Internationalis Ornithologici*: 841–42.

Williams, M. 1991b. Social and demographic characteristics of blue duck *Hymenolaimus malacorhynchos*. *Wildfowl* 42: 65–86.

Williams, M. J., F. McKinney and F. I. Norman. 1991. Ecological and behavioural responses of Austral teal to island life. *Acta XX Congressus Internationalis Ornithologici*: 876–84.

Williams, P. A. and B. J. Karl. 1996. Fleshy fruits of indigenous and adventive plants in the diet of birds in forest remnants, Nelson, New Zealand. *New Zealand Journal of Ecology* 20: 127–45.

Williams, P. A. and B. J. Karl, P. Bannister and W. G. Lee. 2000. Small mammals as potential seed dispersers in New Zealand. *Austral Ecology* 25: 523–32.

Williams, W. D. (ed.) 1974. *Biogeography and ecology in Tasmania.* Junk, The Hague.

Wilson, E. O. 1988. The current state of biological diversity, pp. 3–18 in E. O. Wilson (ed.), *Biodiversity*. National Academy Press, Washington, DC.

Wilson, G. J. 1981. *Distribution and abundance of the New Zealand fur seal,* Arctocephalus forsteri. Fisheries Research Division Occasional Publication 20, Ministry of Agriculture and Fisheries, Wellington

Wilson, H. 1998. Living in Raoul country: the changing flora and vegetation of Banks Peninsula, pp. 101–121 in C. J. Burrows (ed.), *Etienne Raoul and Canterbury botany 1840–1996*. Canterbury Botanical Society/Manuka Press, Christchurch.

Wilson, K-J. 1990. Kea, creature of curiosity. *Forest & Bird* 21 (3): 20–26.
Wilson, K-J. 1992. New Zealand fur seals – 1000 years of exploitation, pp. 47–53 in *Proceedings of the Marine Conservation and Wildlife Protection Conference 1992*. New Zealand Conservation Authority, Wellington.
Wilson, K-J. 1997. Extinct and introduced vertebrate species in New Zealand: A loss of biodistinctiveness and gain in biodiversity. *Pacific Conservation Biology* 3: 301–05.
Wilson, K-J. 1999. *Status and conservation of the sooty shearwater colony at Nt Oneone, Wanganui River, Westland*. Conservation Advisory Science Notes 250. Department of Conservation, Wellington.
Wilson, K-J. 2000. Mainland muttonbirds. *Southern Bird* 1: 4–5.
Wilson, K-J and R. Brejaart. 1992. The kea: A brief research review, pp. 24–28 in L. Joseph (ed.), *Issues in the conservation of parrots in Australasia and Oceania: Challenges to conservation biology*. RAOU Report 83. Birds Australia, Melbourne.
Wilson, P. R. 1984. The effects of possums on mistletoe on Mt Misery, Nelson Lakes National Park, pp. 53–60 in P. R. Dingwall (comp.), *Protection and parks*, Information Series 12. Department of Lands and Survey, Wellington.
Wilson, P. R., B. J. Karl, R. J. Toft, J. R. Beggs and R. H. Taylor. 1998. The role of introduced predators and competitors in the decline of kaka (*Nestor meridionalis*) populations in New Zealand. *Biological Conservation* 83: 175–85.
Wingham, E. J. 1985. Food and feeding range of the Australasian gannet *Morus serrator* (Gray). *Emu* 85: 231–39.
Wodzicki, K., C. J. R. Robertson, H. R. Thompson and C. J. T. Alderton. 1984. The distribution and numbers of gannets (*Sula serrator*) in New Zealand. *Notornis* 31: 232–61.
Wood, K. A. 1990. Seasonal abundance and marine habits of *Procellaria*, fulmarine and gadfly petrels off central New South Wales. *Notornis* 37: 81–105.
Worthy, T. H. 1987a. Osteology of *Leiopelma* (Amphibia, Leiopelmatidae) and descriptions of three new sub-fossil *Leiopelma* species. *Journal of the Royal Society of New Zealand* 17: 201–51.
Worthy, T. H. 1987b. Palaeoecological information concerning members of the frog genus *Leiopelma*: Leiopelmatidae in New Zealand. *Journal of the Royal Society of New Zealand* 17: 409–20.
Worthy, T. H. 1990. An analysis of the distribution and relative abundance of moa species (Aves: Dinornithiformes). *New Zealand Journal of Zoology* 17: 213–41.
Worthy, T. H. 1991a. Fossil skink bones from Northland, New Zealand, and description of a new species of *Cyclodina*, Scincidae. *Journal of the Royal Society of New Zealand* 21: 329–48.
Worthy, T. H. 1991b. An overview of the taxonomy, fossil history, biology and extinction of moas. *Acta XX Congressus Internationalis Ornithologici*: 555–62.
Worthy T. H. 1993. A review of fossil bird bones from loess deposits in eastern South Island, New Zealand. *Records of the Canterbury Museum* 10: 95–106.
Worthy, T. H.1995. Description of some post-cranial bones of *Malacorhynchus scarletti*, a large extinct pink-eared duck from New Zealand. *Emu* 95: 13–22.
Worthy, T. H. 1997a. Quaternary fossil fauna of South Canterbury, South Island, New Zealand. *Journal of the Royal Society of New Zealand* 27: 67–162.
Worthy, T. H. 1997b. Fossil deposits in the Hodges Creek cave system, on the northern foothills of Mt Arthur, Nelson, South Island, New Zealand. *Notornis* 44: 111–24.
Worthy, T. H. 1998a. Quaternary fossil faunas of Otago, South Island, New Zealand. *Journal of the Royal Society of New Zealand* 28: 421–521.
Worthy, T. H. 1998b. The Quaternary fossil avifauna of Southland, South Island, New Zealand. *Journal of the Royal Society of New Zealand* 28: 537–89.
Worthy, T. H. 1998c. A remarkable fossil and archaeological avifauna from Marfells Beach,

Lake Grassmere, South Island, New Zealand. *Records of the Canterbury Museum* 12: 79–176.

Worthy, T. H. 1998d. Fossil avifaunas from Old Neck and Native Island, Stewart Island: Polynesian middens or natural sites? *Records of the Canterbury Museum* 12: 49–82.

Worthy, T. H. 2001. A fossil vertebrate fauna accumulated by laughing owls (*Sceloglaux albifacies*) on the Gouland Downs, northwest Nelson, South Island. *Notornis* 48: 225–33.

Worthy, T. H. and R. Brassey. 2000 New Zealand pigeon (*Hemiphaga novaeseelandiae*) on Raoul Island, Kermadec group. *Notornis* 47: 36–38.

Worthy, T. H. and R. N. Holdaway. 1993. Quaternary fossil faunas from caves in the Punakaiki area, West Coast, South Island, New Zealand. *Journal of the Royal Society of New Zealand* 23: 147–254.

Worthy, T. H. and R. N. Holdaway. 1994a. Quaternary fossil faunas from caves in Takaka Valley and on Takaka Hill, northwest Nelson, South Island, New Zealand. *Journal of the Royal Society of New Zealand* 24: 297–391.

Worthy, T. H. and R. N. Holdaway. 1994b. Scraps from an owl's table: Predator activity as a significant taphonomic process newly recognised from New Zealand Quaternary deposits. *Alcheringa* 18: 229–45.

Worthy, T. H. and R. N. Holdaway. 1995. Quaternary fossil faunas from caves on Mt Cookson, North Canterbury, South Island, New Zealand. *Journal of the Royal Society of New Zealand* 25: 333–70.

Worthy, T. H. and R. N. Holdaway. 1996. Quaternary fossil faunas, overlapping taphonomies, and palaeofaunal reconstruction in North Canterbury, South Island, New Zealand. *Journal of the Royal Society of New Zealand* 26: 275–361.

Worthy, T. H. and R. N. Holdaway. 2000. Terrestrial fossil vertebrate faunas from inland Hawke's Bay, North Island, New Zealand: Part 1. *Records of the Canterbury Museum* 14: 89–154.

Worthy, T. H. and R. N. Holdaway. 2002. *The lost world of the moa: Prehistoric life in New Zealand*. Canterbury University Press, Christchurch.

Worthy, T. H. and D. C. Mildenhall. 1989. A late Otiran–Holocene paleoenvironment reconstruction based on cave excavations in northwest Nelson, New Zealand. *New Zealand Journal of Geology and Geophysics* 32: 243–53.

Worthy, T. H., A. R. Edwards and P. R. Millener. 1991. The fossil record of moas (Aves: Dinornithiformes) older than the Otira (last) glaciation. *Journal of the Royal Society of New Zealand* 21: 101–18.

Worthy, T. H., M. J. Daniel and J. E. Hill. 1996. An analysis of skeletal size variation in *Mystacina robusta* Dwyer, 1962 (Chiroptera: Mystacinidae). *New Zealand Journal of Zoology* 23: 99–110.

Wragg, G. M. 1985. The comparative biology of fluttering shearwater and Hutton's shearwater and their relationship to other shearwater species. MApplSc thesis, Lincoln University.

Wragg, G. M. and M. I. Weisler. 1994. Extinctions and new records of birds from Henderson Island, Pitcairn Group, South Pacific Ocean. *Notornis* 41: 61–70.

Yerex, D. 2001. *Deer: The New Zealand story*. Canterbury University Press, Christchurch.

Young, E. C. 1978. Behavioural ecology of *lonnbergi* skuas in relation to environment on the Chatham Islands, New Zealand. *New Zealand Journal of Zoology* 5: 401–16.

INDEX